Oxidative Stress and Space Biology: An Organ-Based Approach

Special Issue Editors

Melpo Christofidou-Solomidou
Thomas J. Goodwin

MDPI • Basel • Beijing • Wuhan • Barcelona • Belgrade

MDPI

Special Issue Editors
Melpo Christofidou-Solomidou
University of Pennsylvania Perelman School of Medicine
USA

Thomas J. Goodwin
Infectious Disease Research Center Colorado State University
USA

Editorial Office
MDPI
St. Alban-Anlage 66
Basel, Switzerland

This edition is a reprint of the Special Issue published online in the open access journal *International Journal of Molecular Sciences* (ISSN ISSN 2072-6651) in 2017–2018 (available at: http://www.mdpi.com/journal/ijms/special_issues/space_biology_2016).

For citation purposes, cite each article independently as indicated on the article page online and as indicated below:

Lastname, F.M.; Lastname, F.M. Article title. *Journal Name* **Year**, *Article number*, page range.

First Editon 2018

Cover photo courtesy of Melpo Christofidou-Solomidou.

ISBN 978-3-03842-903-6 (Pbk)
ISBN 978-3-03842-904-3 (PDF)

Table of Contents

About the Special Issue Editors

Melpo Christofidou-Solomidou, PhD, Research Professor of Medicine, received her undergraduate degree in Biological Sciences from the State University of NY (S.U.N.Y.) at Stony Brook and her M.Sc. and Ph.D. in Cell Biology from the University of Bonn, in Germany. After completing post-doctoral training at Albany Medical College in NY, she joined the Pulmonary Division of the Department of Medicine at the University of Pennsylvania, in Philadelphia, where she serves currently. She is an NIH-funded investigator actively involved in preclinical and clinical studies at UPenn. Her research focus involves a) the investigation of novel antioxidant approaches to acute and chronic lung disease with a special interest in pulmonary inflammation and lung fibrosis related to radiation; b) chemoprevention of lung carcinogenesis and evaluation of anticarcinogenic effects of common botanicals and c) ischemia/reperfusion injury related to lung transplantation. Her research group at UPenn has been supported by research awards from the NIH (NCI, NIAID, NIEHS, NCCIH), NASA and other national agencies over the past 25 years and has published extensively in peer-reviewed journals.

Thomas J. Goodwin, PhD— NASA cellular physiologist (Retired), pioneer and authority in 3D biology, tissue engineering, bioelectromagnetic field effects, and synthesized tissue growth. Dr. Goodwin was the Manager of the Disease Modeling and Tissue Analogues Laboratory at the NASA Johnson Space Center and Lead Scientist for the Oxidative Stress and Damage research discipline for more than 15 years. Dr. Goodwin is Honorary President and Chair of the Personalized Medicine panel of the European Society of Translational Medicine (EUSTM). He holds adjunct positions at the University of Texas Medical School, Galveston Texas, Dept. of Surgery, Division of Cardiothoracic Surgery, the University of Houston, Dept. of Health and Human Performance, and is an Adjunct Scientist at the Texas Biomedical Research Institute and the Southwest National Primate Research Center which is sponsored by the National Institutes of Health. He has authored or co-authored over 65 peer review articles, four book chapters on three-dimensional biology and personalized spaceflight medicine, has been awarded 23 U.S. patents, on 3D biology and human physiology, electromagnetic tissue effects, and is the recipient of more than 70 NASA Scientific and Technical awards.

Preface to "Oxidative Stress and Space Biology: An Organ-Based Approach"

Technological advances now allow the planning of deep space exploration missions with the aim to discover new habitats for humankind. The National Aeronautics and Space Administration (NASA) has spearheaded this effort and the research into the identification of risks to crew members associated with such lengthy missions. Exciting work from a multitude of investigators across the US, Europe and Japan have identified oxidative damage as a significant risk to major organs that could pose a threat to the health of the astronauts and the success of the mission. This Special Issue of IJMS is dedicated to providing a comprehensive overview of the identified risks and focus on how oxidative stress specifically could impact major organ systems when exposed to space-relevant conditions such as cosmic/galactic radiation, solar particle events, hypogravity, hyperoxia and hypoxia or a combination of stressors.

Melpo Christofidou-Solomidou, Thomas J. Goodwin
Special Issue Editors

International Journal of
Molecular Sciences

MDPI

Editorial

Oxidative Stress and Space Biology: An Organ-Based Approach

Thomas J. Goodwin [1,*] and Melpo Christofidou-Solomidou [2,*]

[1] The National Aeronautics and Space Administration (NASA, retired) Johnson Space Center, Houston, TX 77058, USA

[2] Division of Pulmonary, Allergy, and Critical Care Medicine and the Department of Medicine, University of Pennsylvania Perelman School of Medicine, 3450 Hamilton Walk, Edward J. Stemmler Hall, 2nd Floor, Office Suite 227, Philadelphia, PA 19104, USA

* Correspondence: tgoodwin3@comcast.net (T.J.G.); melpo@mail.med.upenn.edu (M.C.-S.);
  Tel.: +1-832-524-9651 (T.J.G.); +1-215-573-9917 (M.C.-S.); Fax: +1-215-573-4469 (M.C.-S.)

Received: 2 March 2018; Accepted: 21 March 2018; Published: 23 March 2018

1. Introduction

The environment of space provides many challenges to the human physiology and therefore to extended habitation and exploration. Research to define and countermeasures to address these challenges have been explored for well over 40 years yet success in alleviating these challenges has been inadequate, due in part to the lack of investigative tools (subject number and technology) to monitor and diagnose conditions in real time and by the overall lack of a relevant data base to understand the central causative agent(s) responsible for physiological adaptation and homeostasis in space [1–5]. Space flight adaptation syndromes are a product of the environmental conditions and the synergistic reaction of the systemic human physiology, which together combine to produce a combinatorial syndrome that must be resolved in order to safely inhabit and explore space, especially for extended periods [3–6]. This paper documents an exciting glimpse of current pertinent scientific literature sustaining the involvement of oxidative stress and damage (OSaD) [7,8] as a significant contributing factor in the following areas of Earth-based and space flight-related dysregulation of: (1) bone loss [9–13]; (2) cardiovascular function [14]; (3) immune insufficiency and metabolism [15–17]; (4) neurological impairment [18]; and (5) potential countermeasure implementation [18,19]. The cited literature illuminates the environmental challenges of spaceflight encompassing reduced gravity, radiation, varying atmospheric conditions, such as hyperoxia and hypoxia experienced during extravehicular and intravehicular activity (EVA/IVA), and evidence of synergistic effects that portend substantial consequences for long duration/exploration class missions. Now with 21st century scientific tools and advances in genomics, proteomics, and metabolomics ("omics") we have the resources to define syndromes at the molecular and systemic levels by enlisting a holistic approach to the assessment of the space flight physiology and to countermeasure development. To accomplish this, one must seek a "common denominator" that has systemic effects on the function of the entire physiology and stems from both environmental conditions (reduced gravity, radiation, mental and physical stress, etc.) and the byproducts of the human systems biology (reduction oxidation (REDOX)) equilibrium, free radical balance, and reactive oxygen and nitrogen species (ROS and RNS). One such common denominator is the stabilization of physiological homeostasis and control of OSaD in the body [7,8,20].

REDOX equilibrium is cyclic biofeedback related biochemistry. OSaD control/REDOX homeostasis encompasses many aspects of the human physiological response to the space environment and if mitigated properly may normalize a portion of the adverse phenomena seen in space flight combinatorial events and suggest relevant countermeasures for those situations, thus increasing the level of safety and occupational health for each crew member during long duration flight and facilitate improved performance during exploration class missions.

2. Overview

The generation of ROS/RNS in the human physiology is a normal part of human systems biology [7,8]. Oxidative stress and damage (OSaD) is, however, the result of organic and systemic dysregulation of the free radical normalization and scavenging process. Documented sources of OSaD encompass a range of factors which can be divided into two broad categories, being those comprised of physiological and environmental conditions [7,8]. They include, but are not limited to, physiological factors such as physical and psychological stress, poor or constrained diet and nutrition, exercise and/or lack thereof, immune dysregulation and malfunction (autoimmune and wasting syndromes etc.), cardiovascular insufficiency, endocrine imbalances, genotype, and cancers. In addition, environmental conditions such as radiation exposure (occupational exposure, cancer therapy), cancer chemotherapy (toxicity/poisons), pharmaceutical use, exposure to hypoxia or hyperoxia (industrial divers and pilots, etc.), allergens and environmental pollutants all play significant roles in the assault on the human body's ability to equilibrate the REDOX coefficient systematically, in short to control OSaD. Thus, OSaD is the result of loss of homeostasis in the human physiology regarding the body's REDOX equilibrium [7,8]. These conditions have been documented in the scientific literature and some of that literature is provided herein.

Space travel and persistent habitation of reduced gravity or non-terrestrial environments poses many problems which are in some respects similar to the systems biology of terrestrial living associated with numerous disease states and the progression of aging (Figure 1) [6,7]. However, counter to the avenues and facilities available to humans in 1G, astronauts, although they are generally very healthy, are confined to remarkably smaller volumes, with restricted access to environmental, medical, and recreational facilities. Thus, as we begin to prepare for exploration beyond low Earth orbit (BLEO), it is obligatory to assess the similarities in OSaD loads on Earth to those in space and provide reasonable countermeasures to these imbalances. Furthermore, in the exploration environment, astronauts are and will be faced with a host of simultaneous events, again unlike the average human terrestrial resident.

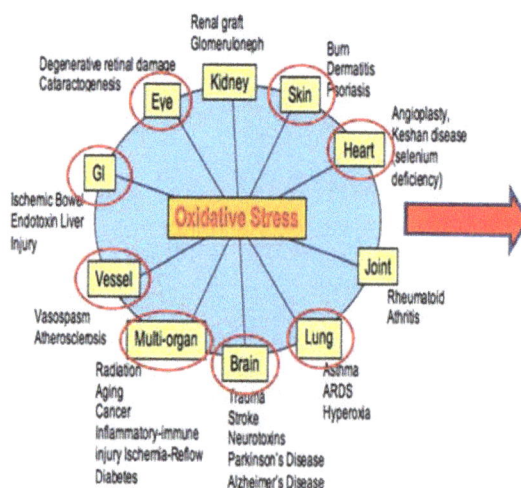

Figure 1. This graphic depiction of human disease states linked to Oxidative Stress and Damage (OSaD) reflects the categories and similarities, although to a lesser degree, experienced by astronauts. Adapted from National Institute of Standards and Technology (NIST)[6].

This Special Issue of IJMS presents some recent supporting data and possible countermeasure approaches. Figures 1 and 2 serves to illustrate the complex myriad of influences on OSaD production and REDOX imbalance which affect the physiology and are central to the theme of this research. The graphic below demonstrates the impact of both environmental and physiological factors that come together to mediate the relationship of OSaD to disease and loss of homeostatic control.

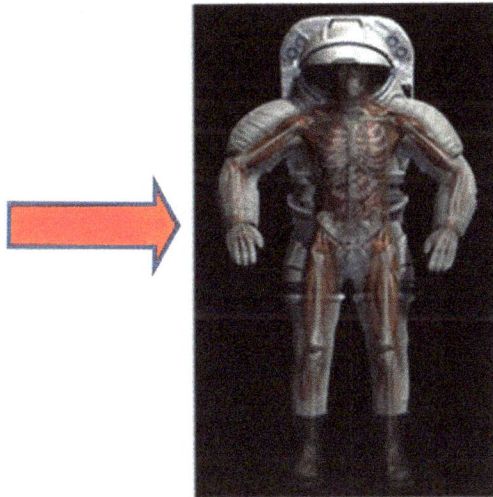

Figure 2. This illustration demonstrates the general construct of the space suit and the lack of substantive shielding leading to increased radiation exposure and generation of OSaD. Courtesy of NASA Johnson Space Center.

Most of the conditions exemplified in Figures 1 and 2 are relevant to space flight, especially those influencing eye, brain, bone, vessel, heart, lung and other multi-organ systems. The figure illustrates the broad reaching influence of OSaD in the human system and thereby the relevance to space habitation. Calculations and accurate assessment of the potential "space hazards" for human survival require the composite expression of multiple factors that present in the space environment, as seen in the figure above, rather than an isolated single upset event approach.

Mammalian, specifically human systems biology depends on a complicated and somewhat fragile series of dynamic checks and balances to constantly provide homeostasis. An essential part of this regulatory process is the production and scavenging of excess ROS and RNS to maintain an appropriate balance of free radicals (FRs) in the physiology. Some of the FR species particularly important for space physiology are nitric oxide (NO) as it has profound cardiovascular and sub-cellar metabolic impacts, hydrogen peroxide (H_2O_2) as it via Haber-Weiss reaction interacts with ferric and other ions reduced by superoxide radicals [7]. This is particularly important as astronauts tend to have high iron content through plasma loss and food content in their blood and Peroxynitrite (–ONOOH) as a downstream molecule capable of depleting sulfhydryl groups. All these represent species responsible for lipid peroxidation, fatty acid loss, protein degradation and DNA damage [2,5,9,20]. Loss of this competency or genetic insufficiency or both normally results in sickness, disease, and eventual death if not resolved.

3. Special Issue Manuscripts

In this Special Issue, a total of 10 excellent papers consisting of six original research studies and four reviews have been published, as detailed in Table 1.

Table 1. Contributions to the Special Issue "Oxidative Stress and Space Biology: An Organ-Based Approach.

Authors	Title	Topic	Type
Alwood et al.	Dose- and Ion-Dependent Effects in the Oxidative Stress Response to Space-Like Radiation Exposure in the Skeletal System	Bone	Original Research
Shanmugarajan et al.	Combined Effects of Simulated Microgravity and Radiation Exposure on Osteoclast Cell Fusion	Bone/Protective	Original Research
Tian et al.	The Impact of Oxidative Stress on the Bone System in Response to the Space Special Environment	Bone	Review
Tahimic et al.	Redox Signaling and Its Impact on Skeletal and Vascular Responses to Spaceflight	Bone/Cardio	Review
Takahashi et al.	Effect of Oxidative Stress on Cardiovascular System in Response to Gravity	Cardiovascular	Review
Blaber et al.	Spaceflight Activates Autophagy Programs and the Proteasome in Mouse Liver	Immune/Metabolism	Original Article
Anslem et al.	Re-adaption on Earth after Spaceflights Affects the Mouse Liver Proteome	Immune/Metabolism	Original Article
Burns and Manda	Metabolic Pathways of the Warburg Effect in Health and Disease: Perspectives of Choice, Chain or Chance	Immune/Metabolism	Review
Endesfelder et al.	Neuroprotection by Caffeine in Hyperoxia-Induced Neonatal Brain Injury	Neuro/Protective	Original Article
Velalopoulou et al.	Synthetic Secoisolariciresinol Diglucoside (LGM2605) Protects Human Lung in an Ex Vivo Model of Proton Radiation Damage	Lung/Protective	Original Article

As initially outlined above, the effects of OSaD are evident in almost all physiological organ subsystems. Here we present the some of the current literature in the field. Published manuscripts are grouped according to the specific organ subsystems affected but also show to a considerable extent the overlapping interactions. This aspect of the research field exemplifies the need for a "common denominator" approach to understanding and treating these physiological effects.

Increased radiation exposure is one of the largest challenges to the prolonged habitation of space, generating substantial amounts of ROS and RNS species. While the increase in radiation exposure in low Earth Orbit is manageable due to the protection of the Van Allen belts, part of the Earth's protective electromagnetic field, it is by no means inconsequential. Each individual astronaut, depending on age, is allowed a defined term in space, thus limiting the amount of time one can fly. Alternatively, prolonged stays in space such as in the International Space Station (ISS) are limited normally to two 6-month tours [3–5]. This restriction is in place in part due to the manner in which radiation affects the bone and muscle.

Alwood et al. observed the effects of radiation on the process of osteoblastogenesis by studying the skeletal structure of C57BL6J mice exposed to either low LET (low energy transfer) protons (50 cGy) or high LET ^{56}Fe ion radiation [9]. They suggest that low energy did not induce significant cellular responses, however the high LET ^{56}Fe ions (200 cGy) were responsible for increases in marrow cells production of mineralized nodules ex vivo regardless of radiation type or dose; and ^{56}Fe (200 cGy) inhibited osteoblastogenesis by more than 90% (5 weeks and 1 year post-IR). After 5 weeks, irradiation (protons or ^{56}Fe) resulted in minimal changes in gene expression levels during osteoblastogenesis, however a high dose ^{56}Fe (200 cGy) resulted in bone loss, increased Catalase and Gadd45 gene expression, and the radiation damage seemed to be mitigated by the additions of superoxide dismutase (SOD) [10].

In a simulated microgravity culture system (RWV) Shanmugarajan et al. investigated the combined effects of gamma radiation and microgravity by monitoring the maturation of a hematopoietic cell line RAW 264.7 into mature osteoclasts. In short, results of the investigation demonstrated radiation alone at 100 cGy may stimulate osteoclast cell fusion determined via giant multinucleated cells GMCs presence and the expression of the signature genes tartrate resistant acid phosphatase (Trap) and dendritic cell-specific transmembrane protein (Dcstamp). Notably, in 1G controls osteoclast cell fusion decreased in doses above 0.5 Gy. By comparison to radiation exposure, simulated microgravity resulted in increased levels of cell fusion, and the effects of radiation and microgravity are additive. Remarkably, simulated microgravity culture effects on osteoclast stimulatory transmembrane protein (Ocstamp) and Dcstamp expression was substantially higher than the radiation effect, inferring that microgravity may increase the synthesis of adhesion molecules more than radiation [11] and thus, demonstrating the interlocking effects of the multiple environmental factors seen in space.

In light of the excellent work by Alwood et al. and Shanmugarajan et al. [10,11] it would seem prudent to consider a holistic approach to ROS and RNS countermeasures. In their review, Tian et al. suggest that the consumption of certain vitamin supplements (e.g., vitamins C and E and carotenoids) may reduce OSaD in bones and, additionally, consuming a diet high in naturally occurring antioxidants, such as carotenoids and flavonoids, may have the ability to mitigate microgravity-induced skeletal involution. Further, natural agents curcumin and turmeric, mayshow promise to attenuate hind-limb unloading (HLU)-induced bone loss by suppressing oxidative stress. Other natural products, many which are under consideration by NASA, have demonstrated skeletal benefits against OSaD. Tanshinol, rescued the decrease of osteoblastic differentiation via down-regulation of FoxO3a signaling and upregulation of Wnt signal under oxidative stress. Further antioxidants, like lipoic acid and *N*-acetyl cysteine, show promise in the arrest of oxidative stress in bone as well. Though these are reports from Earth-based studies, still they suggest important avenues to be pursued in the space flight arena to combat OSaD during prolonged flight [12].

In their review, Tahimic and Globus intertwine another important facet of space flight and the complex environment that consists of changes in the general physiology including bone, muscle and the vascular system; all subject to systemic OSaD effects. Microgravity forces selective environmental pressures, causing a cephalad fluid shift, profound reductions in mechanical loading of bone and muscle, and a reduced immune competency concomitant with inflammation, coupled with galactic cosmic radiation (GCR) and high energy particles (HZE). Here the authors review spaceflight-induced perturbations in calcium homeostasis and specific physiological reductions in bone mass (osteopenia), which pose protracted risks for tissue repair and skeletal health. Space analogue rodent models for these types of studies such as hind-limb unloading (HLU) result in not only bone but also vascular anomalies reminiscent of aging like vascular density (rarefication), and vasodilation responses. These responses tied to excess ROS/RNS, (NO) and inflammation are implicated both in diseases of aging, like osteoporosis and atherosclerosis, and pursuant to insults such as radiation exposure as seen in spaceflight. ROS can directly facilitate bone osteoclastogenesis, leading to resorption and bone loss during aging and estrogen deficiency may be partly attributed to OSaD mechanisms, and thus portends to link the complex phenomena of bone and muscle loss, vascular deconditioning, and immune functions together in an intricate syndrome [13].

To address the cardiovascular system as related to ROS, Takahasi et al. reviewed the effects of OSaD on the cardiovascular system in the presence of reduced gravity environments the central constituents being microgravity and radiation. Prior research has shown that 3–4-week HLU resulted in increased superoxide anion levels in the carotid and adjacent arteries of rats. In this investigation, eNOS expression was upregulated in the carotid artery. The effect of microgravity on the cardiovascular system seems to be different depending on the region. This is not necessarily surprising due to the graded simulation of HLU. Longer HLU studies presented increased superoxide levels, elevation of pro-oxidative enzymes NOX2 and NOX4 and reduction of important anti-oxidative enzymes like Mn-SOD and GPx-1, an effect not seen in mesenteric arteries. Once again, the inflight studies demonstrate synergistic ROS effects by the combination of radiation and microgravity. In other HLU-7-day studies on mouse brains, low dose radiation (LDR) and HLU evidenced lipid oxidation in the brain cortex, but in neither of the independent conditions alone. In 9-month studies of ROS effects, lipid peroxidation and expression of NOX2 were seen in both LDR and HLU yet the combinatorial effect was still greater [14].

Thus, it would appear that the physiology may be resistant to short term single event upset and thereby conceal larger problems in evidence when model studies incorporate multiple stressors as seen in actual microgravity.

We see a myriad of factors impinge on the human physiology in space. Microgravity and increased radiation are environmental catalysts for very complex systemic responses driven to abate a cascade of negative events realized singularly in space. For example, at the metabolic and immune levels the fluid cephalic shift disrupts the entire constituency of the blood and vascular compartment driving loss of body fluids and necessitating the cannibalizing of excess blood cells now too numerous in the reduced plasma volume, consequently launching inflammatory responses and metabolic destabilization [1,2,4,5].

Tahimic and Globus [13] note the increase in pro-inflammatory conditions, which involve immune function and general metabolism. These increases are in part the result of physiological changes enumerated in the sections above [11–14]. Blaber et al. goes further to define the system biology of space flight in a study performed on mice sent to space for only 13.5 days aboard the Space Shuttle Atlantis. This group found unmistakable evidence of increased OSaD and significant changes in the metabolism and production of glutathione signaling impairment in oxidative defense through sophisticated multi-omics enrichment analyses of metabolite and gene sets [15]. These analyses enumerated significant changes in glycerophospholipid and sphingolipid metabolism-related pathways and osmolyte concentrations possibly related to some space related dehydration. Also seen was an enrichment of purine metabolic pathways and aminoacyl-tRNA biosynthesis attendant to enrichment of autophagy associated genes and the ubiquitin-proteasome. These results, in concert with the downregulation in nuclear factor (erythroid-derived 2)-like 2-mediated signaling, suggest a decreased hepatic oxidative defense and could represent aberrant tRNA post-translational processing, induction of degradation programs and the onset of senescence-associated mitochondrial dysfunction as a consequence of the spaceflight environment [15]. Blaber and team have shown short term exposure to the space environment results in elevated ROS, OSaD, and impaired oxidative defense via suppression of NRF2-related pathways in the mouse liver. Over the long term this raises significant concern regarding potential liver damage for astronauts by way of autophagy and systemic immune related inflammation.

Anselm and colleagues followed up on the excellent work of Blaber and team by assessing metabolic pathways and conditions of mouse liver physiology after 30 days of flight on the Russian Bion-M1 spacecraft. After 30 days in space one cohort of the mice were allowed a 7-day re-adaptation while the remainder were not. Analyses of the liver proteomic profiling included shotgun mass spectrometry and label free quantification. The analyses yielded 1086 known proteins and 12,206 unique peptides and statistical testing by ANOVA revealed 218 up-regulated and 224 down-regulated proteins in the post-flight compared to the other groups [16]. Amino acid metabolism related proteins exhibited increased levels after re-adaptation, a possible indicator of elevated gluconeogenesis. In comparison to the non-adaptive flight group, mice allowed to re-adapt demonstrated reduced lipotoxicity marked by normalized levels of the peroxisome proliferator-activated receptor pathway family and bile acid secretion was normalized as a possible consequence of increased levels of transmembrane and CYP superfamily proteins during the recovery. In the non-adapted group however proteomic analyses indicated lipotoxicity, a sign of impaired fatty acid metabolism, along with altered bile secretions and glucose-uptake. These data provide a window into the advent of possible adaptive countermeasures to be used for space flight crews [16].

Investigators Burns and Manda address the vast complexities of the tricarboxylic acid (TCA) cycle and oxidative phosphorylation (OXPHOS) metabolism as inferential of OSaD in their review through an in depth description of the Warburg effect (WE), originally described to yield insights into aberrant cancer metabolism which include features of increased cellular glycolysis and remarkable changes to mitochondrial function and mtDNA repair now known to be associated with other human disease states [17]. Of note, the review illuminates the present understanding of WE regarding global diseases, cancer, diabetes, and, of particular interest, implications for spaceflight and aging. WE may be regulated through a series of events such as chain oncogenic occurrences, chosen responses consequential of restricted or impaired glucose metabolism and finally by the chance manifestation of genetic changes reflective of aging. These chain, choice or chance algorithms may be theoretically extrapolated to interpret the neurodegenerative state, as presents in Alzheimer's and other metabolic diseases with cues to evolving therapeutics. WE pathways investigated in hostile environments with prolonged exposure like space habitation could enlighten medical countermeasures valuable in space and on Earth [17]. For extended missions to Gateway or to Mars, astronauts will require systemic countermeasures to abrogate radiation and microgravity induced OSaD and inflammation mediated by the NFκβ pathway. Representative of this might be to employ cytoprotective agents like Amifostine that initiates a metabolic shift to induction of glycolysis and blockage of mitochondrial pyruvate which can provide radioprotection to normal tissues. Also activators of the Nrf2 pathway like dimethyl fumarate and sulforaphane, capable of enhancing the endogenous antioxidant protection against ROS and RNS, are potential candidates for investigation [17].

As we turn to countermeasures we see that in many of the previous manuscripts and reviews cited potential mitigating agents are either physiologically endemic, naturally occurring, or pharmacologic [12–17]. In this review Endesfelder et al. details the mechanisms of caffeine a potent free radical scavenger and adenosine receptor antagonist. Caffeine has been shown to lower the severity of brain damage in preterm infants and has been evaluated in the research arena regarding effects on apoptosis, redox sensitive transcription factors, OSaD markers, the anti-oxidative response, inflammation and extracellular matrix in neonatal rats subjected to hyperoxia [18]. Six-day-old rats are a good model of the human fetal brain at 28–32 weeks of gestation and thus excellent for the study of OSaD and neuroprotection in the developing human brain. Rats pretreated with a single caffeine treatment demonstrated diminished OSaD markers including (glutamate-cysteine ligase catalytic subunit (GCLC hydrogen peroxide, heme oxygenase-1, and lipid peroxidation)) while promoting anti-OSaD molecules (sulfiredoxin 1, SOD and peroxiredoxin 1). Caffeine also modulated redox-sensitive transcription factor expression (Nrf2/Keap1, and NFκB) suppressed extracellular matrix degeneration (matrix metalloproteinases (MMP) 2, and inhibitor of metalloproteinase (TIMP) 1/2) and down-regulated pro-inflammatory cytokines, reduced pro-apoptotic effectors (poly (ADP-ribose) polymerase-1 (PARP-1), apoptosis inducing factor (AIF), and caspase-3) thus demonstrating caffeine to be a pleomorphic neuro-protective agent [18]. These experiments are particularly relevant to EVA and planetary exploration where astronauts will breathe ~100% O_2 during spacesuit operations and be exposed to mild hyperoxia about 50% of the mission length overall [3,4].

The prospects for robust OSaD countermeasures to combat elevated radiation and microgravity effects are essential for planning long duration space missions. Velalopoulou et al. detail in their publication the prospects for the use of LGM2605, a synthetic derivative of the flaxseed lignan secoisolariciresinol diglucoside (SDG), which has been proven to reduce OSaD-related biomarkers in the presence of radiation [19]. LGM2605 was tested to determine its ability to protect and nullify the harmful effects of proton radiation (which makes up 85% of the constituents of GCR) on human lung slices. In an ex vivo model of human lung, precision-cut lung sections (huPCLS) were subjected to severe tissue toxicity via exposure to 4.0 Gy of proton radiation, having first been treated with LGM2605 then analyzed at 30 min and 24 h. post-exposure. All post exposure samples were surveyed for gene expression changes relevant to inflammation, OSaD, and cell cycle arrest. The researchers determined radiation-induced senescence, oxidative tissue damage associated cell cycle changes, and an associated proinflammatory phenotype in non-treated samples. To summarize, the data provides evidence of the significant protective capability by LGM2605. Additionally, LGM2605-pretreatment of proton-irradiated huPCLS significantly upregulates anti-OSaD genes and protects huPCLS from a senescent-like phenotype, regulated at the gene and protein level by p53, members of the cyclin-dependent kinase (CDK) family, and p21. LGM2506 pretreatment substantially reduced p16 induced by proton radiation, plus downregulates proinflammatory cytokine gene levels [19]. Downstream protective effects LGM2506 may yet be realized and may be an excellent candidate for astronaut protections from radiation, microgravity effects and respiratory exposure to hyperoxia and hypoxia.

In conclusion, the 10 significant and elegant manuscripts published in this Special Issue illustrate the intricacies, relevance, and impacts of unstable ROS generation and persistent OSaD in the space flight environment, which are, in many ways, emblematic of the phenomena of aging. Recent data released by NASA on the human Twins Study further illustrates the changes to and potential damage sustained by the human physiology while inhabiting an unrelentingly hostile environment, especially considering the results of analyses of inflammatory cytokines, RNA and DNA analyses and microbiome research conducted by Drs. Snyder, Mason, and Turek, respectively [21]. We would like to thank each of the authors contributing to this Special Issue, for their insights, talents, and enduring interest in furthering the advancement of humankind's quest to travel to and one day inhabit far-off worlds. This arena of research also serves a remarkably timely application, namely that of potentially understanding and ameliorating the myriad of diseases associated with the inevitable process of aging.

Acknowledgments: This work was funded in part by: 1P42ES023720-01 (MCS), and by pilot project support from 1P30 ES013508-02 awarded to MCS (its contents are solely the responsibility of the authors and do not necessarily represent the official views of the NIEHS, NIH).

Conflicts of Interest: The authors declare no conflict of interest.

Abbreviations

BLEO	Beyond Low Earth Orbit
CDK	Cyclin-Dependent Kinase
DNA	Deoxyribonucleic Acid
EVA/IVA	Extravehicular Activity and Intra-Vehicular Activity
HLU	Hind Limb Unloading
GCR	Galactic Cosmic Radiation
GPx-1	Glutathione Peroxidase
cGy	Centi Gray
Gy	Gray
H_2O_2	Hydrogen Peroxide
LEO	Low Earth Orbit
LDR	Low Dose Radiation
Mn-SOD	Manganese Superoxide Dismutase
NO	Nitric Oxide
NOX_2	Nicotinamide Adenine Dinucleotide Phosphate-oxidase (isoform 2)
NOX_4	Nicotinamide Adenine Dinucleotide Phosphate-oxidase (isoform 4)
eNOS	Endothelial Nitric Oxide Synthase
OSaD	Oxidative Stress and Damage
OXPHOS	Oxidative Phosphorylation
RAW 264.7	Murine Monocyte Cell Line Identification
REDOX	Reduction and Oxidation
RNA	Ribonucleic Acid
ROS/RNS	Reactive Oxygen and Reactive Nitrogen Species
SDG	Flaxseed Lignan Secoisolariciresinol Diglucoside
SOD	Super Oxide Dismutase
TCA	Tricarboxcylic Acid
Wnt	Wingless-Type MMTV (mouse mammary tumor virus)

References

1. Hawkins, W.; Zieglschmid, J. Clinical aspects of crew health. In *Biomedical Results of Apollo*; Johnston, R., Dietlein, L., Berry, C., Eds.; National Aeronautics and Space Administration: Washington, DC, USA, 1975; pp. 43–81.

2. Kimzey, S.L. Hematology and immunology studies. In *Biomedical Results from Skylab*; Johnson, R.S., Dietlein, L.F., Eds.; National Aeronautics and Space Administration: Washington, DC, USA, 1977; pp. 248–282.

3. Thirsk, R.; Kuipers, A.; Mukai, C.; Williams, D. The space-flight environment: The International Space Station and beyond. *CMAJ Can. Med. Assoc. J.* **2009**, *180*, 1216–1220. [CrossRef] [PubMed]

4. Williams, D.; Kuipers, A.; Mukai, C.; Thirsk, R. Acclimation during space flight: Effects on human physiology. *CMAJ Can. Med. Assoc. J.* **2009**, *180*, 1317–1323. [CrossRef] [PubMed]

5. Demontis, G.C.; Germani, M.M.; Caiani, E.G.; Barravecchia, I.; Passino, C.; Angeloni, D. Human Pathophysiological Adaptations to the Space Environment. *Front. Physiol.* **2017**, *8*, 547. [CrossRef] [PubMed]

6. Online Resources for Disorders Caused by Oxidative Stress. Available online: http://www.oxidativestressresource.org/ (accessed on 11 March 2018).

7. Cutler, R.G.; Rodriguez, H. *Critical Reviews in Oxidative Stress and Aging: Advances in Basic Science, Diagnostics and Intervention*; World Scientific: Singapore, 2003; Volume 2, 1523p, ISBN 9814490946, 9789814490948.

8. Kohen, R.; Nyska, A. Oxidation of biological systems: Oxidative stress phenomena, antioxidants, redox reactions, and methods for their quantification. *Toxicol. Pathol.* **2002**, *30*, 620–650. [CrossRef] [PubMed]

9. Smith, S.M.; Zwart, S.R.; Block, G.; Rice, B.L.; Davis-Street, J.E. The Nutritional Status of Astronauts Is Altered after Long-Term Space Flight Aboard the International Space Station. *J. Nutr.* **2005**, *135*, 437–443. [CrossRef] [PubMed]

10. Alwood, J.S.; Tran, L.H.; Schreurs, A.S.; Shirazi-Fard, Y.; Kumar, A.; Hilton, D.; Tahimic, C.G.T.; Globus, R.K. Dose- and Ion-Dependent Effects in the Oxidative Stress Response to Space-Like Radiation Exposure in the Skeletal System. *Int. J. Mol. Sci.* **2017**, *18*, 2117. [CrossRef] [PubMed]

11. Shanmugarajan, S.; Zhang, Y.; Moreno-Villanueva, M.; Clanton, R.; Rohde, L.H.; Ramesh, G.T.; Sibonga, J.D.; Wu, H. Combined Effects of Simulated Microgravity and Radiation Exposure on Osteoclast Cell Fusion. *Int. J. Mol. Sci.* **2017**, *18*, 2443. [CrossRef] [PubMed]

12. Tian, Y.; Ma, X.; Yang, C.; Su, P.; Yin, C.; Qian, A.-R. The Impact of Oxidative Stress on the Bone System in Response to the Space Special Environment. *Int. J. Mol. Sci.* **2017**, *18*, 2132. [CrossRef] [PubMed]

13. Tahimic, C.G.T.; Globus, R.K. Redox Signaling and Its Impact on Skeletal and Vascular Responses to Spaceflight. *Int. J. Mol. Sci.* **2017**, *18*, 2153. [CrossRef] [PubMed]

14. Takahashi, K.; Okumura, H.; Guo, R.; Naruse, K. Effect of Oxidative Stress on Cardiovascular System in Response to Gravity. *Int. J. Mol. Sci.* **2017**, *18*, 1426. [CrossRef] [PubMed]

15. Blaber, E.A.; Pecaut, M.J.; Jonscher, K.R. Spaceflight Activates Autophagy Programs and the Proteasome in Mouse Liver. *Int. J. Mol. Sci.* **2017**, *18*, 2062. [CrossRef] [PubMed]

16. Anselm, V.; Novikova, S.; Zgoda, V. Re-Adaption on Earth after Spaceflights Affects the Mouse Liver Proteome. *Int. J. Mol. Sci.* **2017**, *18*, 1763. [CrossRef] [PubMed]

17. Burns, J.; Manda, G. Metabolic Pathways of the Warburg Effect in Health and Disease: Perspectives of Choice, Chain or Chance. *Int. J. Mol. Sci.* **2017**, *18*, 2755. [CrossRef] [PubMed]

18. Endesfelder, S.; Weichelt, U.; Strauß, E.; Schlör, A.; Sifringer, M.; Scheuer, T.; Bührer, C.; Schmitz, T. Neuroprotection by Caffeine in Hyperoxia-Induced Neonatal Brain Injury. *Int. J Mol. Sci.* **2017**, *18*, 187. [CrossRef] [PubMed]

19. Velalopoulou, A.; Chatterjee, S.; Pietrofesa, R.A. Synthetic Secoisolariciresinol Diglucoside (LGM2605) Protects Human Lung in an Ex Vivo Model of Proton Radiation Damage. *Int. J. Mol. Sci.* **2017**, *18*, 2525. [CrossRef] [PubMed]

20. Höhn, A.; Weber, D.; Jung, T.; Ott, C.; Hugo, M.; Kochlik, B.; Kehm, R.; König, J.; Grune, T.; Castro, J.P. Happily (n)ever after: Aging in the context of oxidative stress, proteostasis loss and cellular senescence. *Redox Biol.* **2017**, *11*, 482–501. [CrossRef] [PubMed]

21. NASA Twins Study. Avialable online: https://www.nasa.gov/feature/nasa-twins-study-investigators-to-release-integrated-paper-in-2018 (accessed on 20 March 2018).

International Journal of
Molecular Sciences

MDPI

Article

Dose- and Ion-Dependent Effects in the Oxidative Stress Response to Space-Like Radiation Exposure in the Skeletal System

Joshua S. Alwood [1,†], Luan H. Tran [1,†], Ann-Sofie Schreurs [1,†], Yasaman Shirazi-Fard [1], Akhilesh Kumar [1], Diane Hilton [1], Candice G. T. Tahimic [1,2] and Ruth K. Globus [1,*]

[1]   Bone and Signaling Laboratory, Space BioSciences Division, NASA Ames Research Center, Mail-Stop 236-7, Moffett Field, CA 94035, USA; joshua.s.alwood@nasa.gov (J.S.A.); luan.h.tran@nasa.gov (L.H.T.); ann-sofie.schreurs@nasa.gov (A.-S.S.); yasaman.shirazi-fard@nasa.gov (Y.S.-F.); akhilesh482@gmail.com (A.K.); Hilton.Diane@att.net (D.H.); candiceginn.t.tahimic@nasa.gov (C.G.T.T.)
[2]   Wyle Laboratories, Mail-Stop 236-7, Moffett Field, CA 94035, USA
*   Correspondence: ruth.k.globus@nasa.gov; Tel.: +1-650-604-5247; Fax: +1-650-604-3159
†   These authors contributed equally to this work.

Received: 2 September 2017; Accepted: 30 September 2017; Published: 10 October 2017

Abstract: Space radiation may pose a risk to skeletal health during subsequent aging. Irradiation acutely stimulates bone remodeling in mice, although the long-term influence of space radiation on bone-forming potential (osteoblastogenesis) and possible adaptive mechanisms are not well understood. We hypothesized that ionizing radiation impairs osteoblastogenesis in an ion-type specific manner, with low doses capable of modulating expression of redox-related genes. 16-weeks old, male, C57BL6/J mice were exposed to low linear-energy-transfer (LET) protons (150 MeV/n) or high-LET ^{56}Fe ions (600 MeV/n) using either low (5 or 10 cGy) or high (50 or 200 cGy) doses at NASA's Space Radiation Lab. Five weeks or one year after irradiation, tissues were harvested and analyzed by microcomputed tomography for cancellous microarchitecture and cortical geometry. Marrow-derived, adherent cells were grown under osteoblastogenic culture conditions. Cell lysates were analyzed by RT-PCR during the proliferative or mineralizing phase of growth, and differentiation was analyzed by imaging mineralized nodules. As expected, a high dose (200 cGy), but not lower doses, of either ^{56}Fe or protons caused a loss of cancellous bone volume/total volume. Marrow cells produced mineralized nodules ex vivo regardless of radiation type or dose; ^{56}Fe (200 cGy) inhibited osteoblastogenesis by more than 90% (5 weeks and 1 year post-IR). After 5 weeks, irradiation (protons or ^{56}Fe) caused few changes in gene expression levels during osteoblastogenesis, although a high dose ^{56}Fe (200 cGy) increased *Catalase* and *Gadd45*. The addition of exogenous superoxide dismutase (SOD) protected marrow-derived osteoprogenitors from the damaging effects of exposure to low-LET (^{137}Cs γ) when irradiated in vitro, but had limited protective effects on high-LET ^{56}Fe-exposed cells. In sum, either protons or ^{56}Fe at a relatively high dose (200 cGy) caused persistent bone loss, whereas only high-LET ^{56}Fe increased redox-related gene expression, albeit to a limited extent, and inhibited osteoblastogenesis. Doses below 50 cGy did not elicit widespread responses in any parameter measured. We conclude that high-LET irradiation at 200 cGy impaired osteoblastogenesis and regulated steady-state gene expression of select redox-related genes during osteoblastogenesis, which may contribute to persistent bone loss.

Keywords: cancellous bone; osteoblast; ionizing radiation; spaceflight; oxidative stress

1. Introduction

Structural degradation and oxidative stress following exposure to space radiation potentially endangers skeletal health of astronauts, both during the mission and later in life. Weightlessness-induced musculoskeletal atrophy, psychological stress, and confinement pose additional risks to astronaut health [1]. For exploration missions into deep space, galactic cosmic-rays (GCR) are of particular concern because of the high-energy and densely ionizing pattern of energy deposition from heavy ions [2–4], which can induce oxidative stress [5,6] in various tissues. Both solar particle-emissions (SPE) and GCR contribute to the mixed composition of low-energy radiation (such as protons and γ rays), high linear-energy-transfer (LET) particles (such as ionized ^{56}Fe- and ^{12}C-nuclei), and secondary constituents, such as Bremmstrahlung, fragmentation products, and neutrons [4,7]. Estimates of the total absorbed dose during a Mars mission (180-day transits with 500-day surface operations) to be between 28–47 cGy, with GCR contributing 27 cGy [8,9]. Currently, risk of exposure-induced death from a fatal cancer drives the NASA astronaut career radiation-exposure limit [10], though NASA is investigating additional risks associated with acute and late tissue degeneration following radiation exposure [11].

Irradiation perturbs the redox balance within the bone and marrow compartment in a complex and transient manner. Total-body irradiation (^{137}Cs, 200 cGy) induces rapid reactive oxidative species formation in the marrow [12] and increases expression of the master transcription factor that binds to the antioxidant response element, nuclear factor (erythroid-derived 2)-like 2 (*Nrf2)* [13]. Excess reactive oxygen species (ROS) can interfere with osteogenesis of bone marrow derived osteoprogenitors [14,15] as well as contribute to bone resorption by osteoclasts [16]. High-LET ^{56}Fe ($\geq$200 cGy) impairs cell proliferation during osteoblastogenesis from marrow progenitors in a dose-dependent manner [17]. Further, irradiation in vitro with ^{56}Fe (100 cGy) arrests the cell cycle and inhibits proliferation of mesenchymal stem cells [18]. At higher doses (400–800 cGy), irradiation with X-rays increases ROS, depletes antioxidant stores, increases Nrf2 protein levels, and reduces differentiation in osteoblast-like cells [19]. However, the role of oxidative stress on the potential for total-body irradiation (TBI) to impair osteoblastogenesis over the long term, in particular low-dose, high-LET species, has not been fully characterized.

We posed three principal questions to address in this study. First, are the severity of acute bone loss [12,13,20,21] and impairment of osteoblastogenesis [17] following TBI dependent on radiation type (LET) or dose? Secondly, which pathways are activated? Finally, do irradiation-induced insults to skeletal structure and osteoblastogenesis persist long after exposure? We hypothesized that: (1) irradiation induces bone loss, shown previously to be dominated by total dose of exposure [22], and leads to temporal perturbations in antioxidant defenses; (2) TBI induces both persistent changes in redox-related genes and cancellous microarchitecture in an ion-dependent manner; and, (3) addition of an antioxidant, here the enzyme superoxide dismutase (SOD), mitigates irradiation-induced damage to osteoprogenitors and stem cells when exposed in vitro, implicating an oxidative-stress dependent mechanism.

To address these questions, we conducted in vivo and in vitro experiments. For in vivo work, we selected protons and ^{56}Fe ions as representative radiation for the abundant low-LET particles in space radiation and the heavy-ion component of galactic cosmic radiation, respectively. For in vitro work, we selected ^{137}Cs γ rays as a low-LET reference radiation to enable high-throughput countermeasure testing. In vivo, there were persistent and profound decrements in mineralized nodules after TBI (at 200 cGy) with ^{56}Fe, while protons had a milder effect. In contrast to our hypothesis, few changes were observed in the steady-state expression of redox-related genes during ex vivo osteoblastogenesis following TBI, although there was some indication of altered expression of select genes during the proliferative phase (at 200 cGy ^{56}Fe, but not lower doses). Furthermore, scavenging ROS with an antioxidant (i.e., SOD) mitigated adverse effects of in vitro irradiation with γ rays, but failed to protect fully from irradiation with an equivalent dose of heavy ions.

2. Results

2.1. TBI Experiment Design

To evaluate the individual effects of radiation type, our study queried a dose response to high-LET ^{56}Fe or low-LET protons (see Methods below for details) at one day, seven days, or five weeks post-exposure. Additionally, we assessed the sequelae of [56]Fe irradiation at a late, 1-year post exposure endpoint (Figure 1A,B). Tibiae and femora were collected for assessment of the cancellous microarchitecture. Bone-marrow cells were cultured ex vivo in osteoblastogenic media for colony counts and mineralization assays. Additionally, the effects of total body irradiation on gene expression were assayed at two points, during the proliferation and mineralization phase, of osteoblastogenesis.

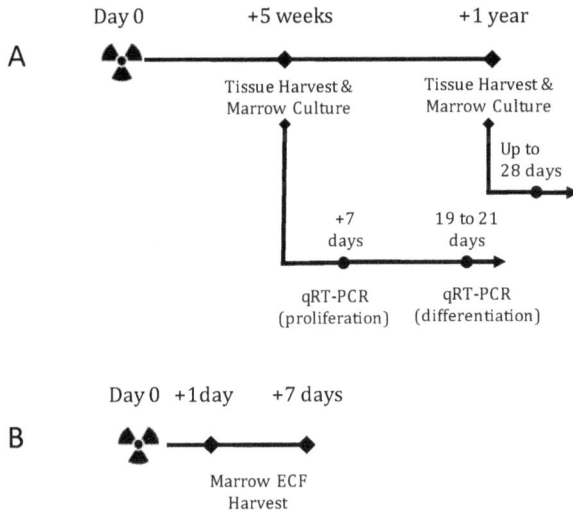

Figure 1. Total Body Irradiation (TBI) experiment designs. (**A**) Experiments to determine whether TBI alters antioxidant gene expression in osteoblasts, osteoblastogenesis, and the structure of cancellous bone in an ion-dependent fashion at five weeks after proton or [56]Fe exposure and to determine the osteoblastogenic and structural sequelae of [56]Fe irradiation one year after exposure. At each endpoint, marrow cells were cultured for osteoblastogenesis. At 5 weeks post-irradiation, osteoblastogenic cultures were halted during the proliferative phase and differentiation phase for gene expression assessment; (**B**) Experiment to determine the acute effect of proton or [56]Fe irradiation on the antioxidant status of the bone marrow extracellular fluid (ECF). Mice were irradiated and euthanized either one or seven days after exposure. Marrow was collected and cells removed through centrifugation. The supernatant, defined as the extracellular fluid (ECF), was assessed for total antioxidant capacity.

2.2. Body Mass Temporal Responses and Coat Color

Body mass was measured (±0.1 g), at least weekly, throughout the study as an index of general health (Figure 2, showing the 1-year post irradiation cohorts). As determined by one-way analysis of variance (ANOVA) and Dunnett's posthoc, within six days of exposure, [56]Fe (200 cGy) reduced the cohort's mean body mass by 4% relative to the starting weights. Other doses and ions had no significant effect relative to starting weights [23]. Shipment of the mice from Brookhaven National Lab (BNL) to NASA Ames between days 6 and 8, caused a 5% decline in body weights of all of the groups.

Following shipment, the body mass of all groups rose over time (+7% for the 5-weeks endpoint and +46% for the 1-year endpoint), with no treatment effect, nor interaction of time × treatment (Figure 2). By the time of euthanasia at 5-weeks or 1-year post-irradiation, body mass did not significantly differ between irradiated and age-matched sham-controls.

Additionally, as an index of general health, the coat color was assessed as grey or black at the end of the experiment (1-year post-irradiation). In sham controls, 1 out of 13 mice had a grey coat (Table 1). In contrast, [56]Fe (200 cGy) caused all of the 11 to have grey coats. The lower doses of [56]Fe irradiation, 10 or 50 cGy, did not modulate coat color compared to sham-controls, respectively.

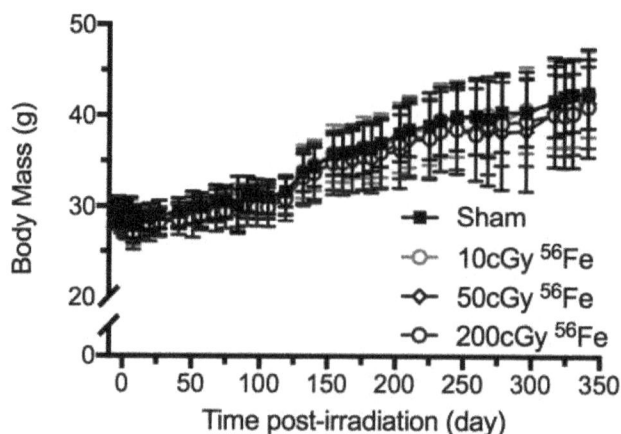

Figure 2. Longitudinal body mass measurements of ^{56}Fe-irradiated mice up to 1 year post-exposure (mice from Figure 1A). Mice were irradiated on day 0 and shipping took place between days 6 and 8. Data are mean ± standard deviation with *n* = 13 per group.

Table 1. Fur coat color and cancellous microarchitecture of the proximal tibial metaphysis 1 year after ^{56}Fe irradiation.

Parameter	Sham	10 cGy	50 cGy	200 cGy
Grey/Not Grey Coat (%)	1/12 (7.7%)	1/12 (7.7%)	1/11 (8.3%)	11/0 (100%)
Percent Bone Volume (%)	9.7 ± 3.1	8.7 ± 2.4	8.2 ± 2.9	7.2 ± 1.5
Trabecular Thickness (μm)	59.3 ± 5.3	58.2 ± 3.2	57.7 ± 2.9	57.1 ± 2.7
Trabecular Separation (μm)	228.8 ± 23.3	245.0 ± 41.8	240.1 ± 30.2	248.5 ± 18.4
Trabecular Number (1/mm)	1.62 ± 0.36	1.49 ± 0.35	1.42 ± 0.48	1.26 ± 0.23

2.3. Bone Structure

Microcomputed tomography was used to assess the effects of aging and irradiation on the cancellous tissue in the proximal tibial metaphysis at the 5 week and 1 year endpoints (Figure 1A) and cortical geometry of the femoral midshaft at the 1-year endpoint. In sham-treated animal cohorts, aging caused a 54% reduction of cancellous bone volume/total volume (BV/TV) between the 5-week (Figure 3A, 21.1 ± 2.2%) and 1-year (Table 1, 9.7 ± 3.1%) post-treatment endpoints.

At 5 weeks post-irradiation, 50 and 200 cGy ^{56}Fe caused a decrement in BV/TV (Figure 3A, −16% and −31%, respectively) and strut number (Tb.N, Figure 3B, −16% and −31%, respectively) in the cancellous region of the proximal tibia, compared to shams. For protons, 200 cGy caused a −22% decrement in BV/TV (Figure 3A) and Tb.N (Figure 3B), while 50 cGy showed a trend towards decreased BV/TV (−11%, *p* = 0.12) and Tb.N (−13%, *p* = 0.06). Additionally, irradiation at 200 cGy increased trabecular separation, but did not affect trabecular thickness [23]. At 5 or 10 cGy, neither ^{56}Fe, nor protons, showed any detectable bone loss, by any structural measure.

At 1 year post-irradiation, ^{56}Fe (200 cGy) showed a trend of decreased BV/TV and Tb.N (−25%, *p* = 0.12, Table 1) in the cancellous region of the proximal tibia, as compared to age-matched controls. No changes were observed at 10 or 50 cGy doses when compared to controls. Additionally, no significant changes in cortical geometry of the femur at midshaft (bone volume and cortical thickness, [23]) were detected, as compared to age-matched controls.

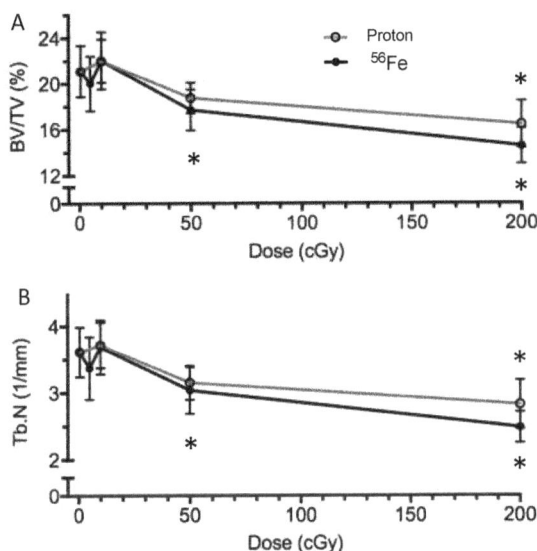

Figure 3. Dose response of cancellous bone microarchitecture following proton or [56]Fe irradiation at 5 weeks after exposure (mice from Figure 1A). Microarchitecture was assessed using microcomputed tomography and the parameters of (**A**) bone volume / total volume (BV/TV, %) and (**B**) trabecular number (Tb.N, 1/mm). Data are mean ± standard deviation, with $n = 8$/group. * denotes $p \leq 0.05$ vs. age-matched sham control.

2.4. Ex Vivo Osteoblastogenesis

To assess the cellular responses to irradiation during ex vivo osteoblast growth and maturation, we cultured marrow progenitor cells at the 5 week and 1 year post-irradiation endpoints. For cultures plated 5 weeks post-irradiation, gene expression, alkaline phosphatase activity, colony counts, and DNA content analyses were performed during the proliferative phase (seven days after plating) and mineralization phase (19 to 21 days after plating). At the end of the latter phase, the mineralized area was quantified to assess the functional outcome of differentiation.

During the mineralization phase, we observed well-formed nodules in the sham and proton-irradiated (200 cGy) groups (Figure 4A,B), with sham values displaying wide variance. In contrast, we observed very few, yet fully mineralized, nodules in the [56]Fe-irradiated (200 cGy) group (Figure 4A,B). After quantifying the nodule area, irradiation with [56]Fe (200 cGy) tended to cause a −91% decrement ($p = 0.06$, Figure 4A,B) in the mineralization area (median) as compared to sham-irradiated controls. This inhibitory effect is consistent with findings from our other experiments with this dose of [56]Fe at various times after exposure [23]. Although, mineralization area of cultures from animals exposed to protons (200 cGy) tended to be lower than sham-controls (−70%, $p = 0.10$, Figure 4A,B), nodules from proton-irradiated mice appeared larger and more widespread than in cultures from [56]Fe-irradiated mice (Figure 4B). Lower doses of either species did not affect mineralized area or any other measured parameters [23].

Radiation exposure did not modify expression levels for most of the genes analyzed during the mineralization phase (Table 2), although a high dose of [56]Fe (200 cGy) tended to decrease expression of the osteoblast differentiation gene *Alpl* by −60% ($p = 0.12$), Figure 4C. Additionally, a low dose of [56]Fe (10 cGy) increased the expression of the antioxidant gene *CuZnSOD* (+90%) as compared to sham-controls, whereas proton exposure did not modify expression levels for any of the genes assayed [23]. Taken together, these data suggest mild effects late in culture, but only for [56]Fe, not protons.

During the proliferative phase, most doses and types of radiation exposure did not modify the expression levels for the genes analyzed when compared to sham-controls (Table 2). A high-dose of ^{56}Fe (200 cGy) did however increase the expression levels of genes for the cell cycle arrest marker *Gadd45* (+223%) and the antioxidant *Catalase* (+81%), and also tended to increase the late-osteoblastic marker *Bglap* (+196%, $p = 0.08$) (Figure 4D). A lower dose of ^{56}Fe (5 cGy) decreased the gene expression of the nitric oxide generator *iNOS* by −53% and increased Gadd45 by +91%, respectively, when compared to sham controls.

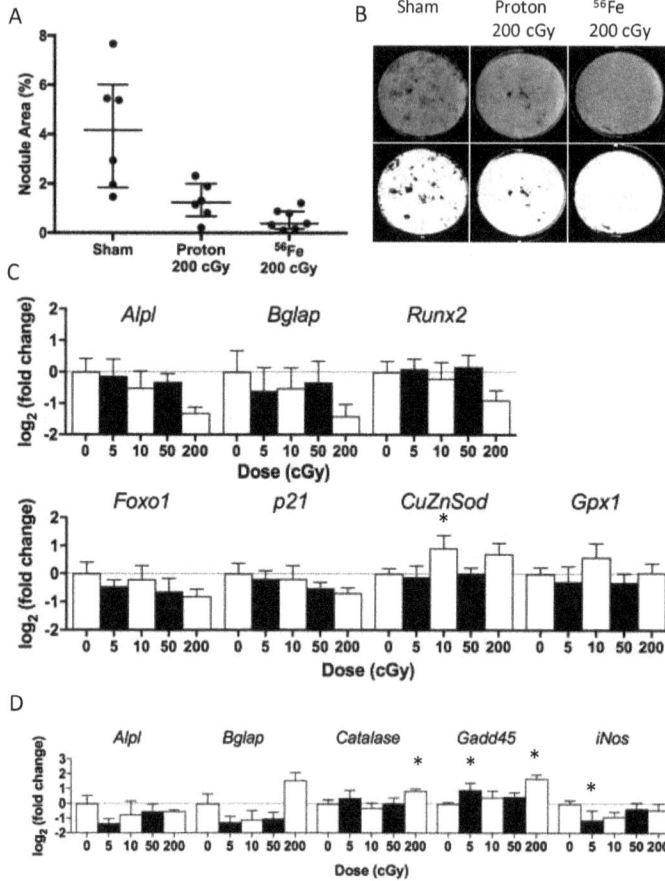

Figure 4. Differentiation and proliferation responses of osteoprogenitors cultured from bone marrow of mice that received proton or ^{56}Fe irradiation (5 weeks post-irradiation from mice in Figure 1A). (**A**) Nodule area of cultures from animals irradiated with 200 cGy proton or ^{56}Fe compared to sham control. Data are summarized by median and interquartile range with $n = 6$–7/group. Each dot represents an individual mouse. (**B**) Grayscale (top row) and binarized (bottom row) images of representative wells (i.e., near the group median), showing nodules during the mineralization phase for 200 cGy proton or iron or sham. Gene expression assessed during (**C**) terminal differentiation or (**D**) proliferation for markers of differentiation, damage response, cell cycle, and redox response in cells cultured from the marrow of previously irradiated mice. Data are mean ± standard deviation of Log$_2$ (Fold Change) transformed values relative to the endogenous gene *HPRT* and normalized to the sham control (i.e., $\Delta\Delta C_t$). The dashed line at 1.0 indicates the sham control level. $n = 4$/group, * denotes $p \leq 0.05$ vs. sham-control.

For the ex vivo marrow culture performed 1-year post irradiation, the numbers of colonies were counted eight days after plating and nodule areas were assessed 28 days after plating. ^{56}Fe irradiation (200 cGy) tended to reduce colony counts at day eight (−45%, Figure 5A) when compared to controls, whereas lower doses of ^{56}Fe (10 and 50 cGy) had no effect. At 28 days after plating, cells from ^{56}Fe (200 cGy)-irradiated mice tended to show reduced mineralized area (−96% reduction of median, $p < 0.06$, Figure 5B), whereas lower doses had no effect as compared to controls. Taken together, these data suggest that only the highest dose of ^{56}Fe had a persistent effect to inhibit osteoblastogenesis, as assessed by colony formation.

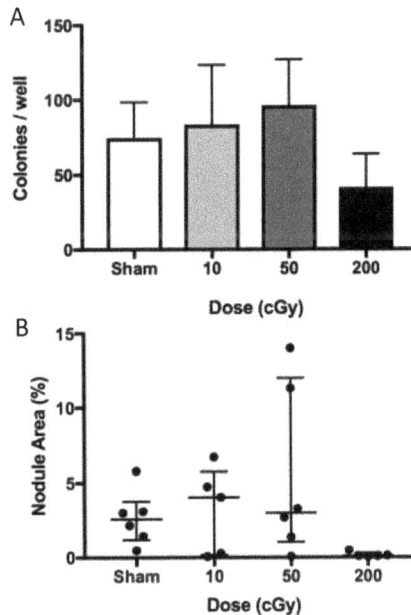

Figure 5. Persistent effects of ^{56}Fe (200 cGy) TBI on osteoblastogenesis at 1 year post-exposure (mice from Figure 1A). Marrow was cultured in osteoblastogenic conditions. We quantified (**A**) colony number on day 8 in vitro and (**B**) nodule area on day 28 in vitro (with dots representing individual mice). Data are summarized by mean and standard deviation (**A**) and median and interquartile range (**B**), with $n = 5$–6/group.

Table 2. List of genes quantified from ex vivo culture, which did not show significant changes in expression levels. Cells were isolated and cultured five weeks post-irradiation and analyzed at proliferative and terminally differentiated stages in growth.

Proliferative Stage	Terminal Differentiation
Caspase 3	*Caspase 3*
Cdk2	*Catalase*
CuZnSOD	*Cdk2*
Foxo1	*Gadd45*
GPX	*iNos*
MnSOD	*MnSOD*
p21	*p53*
p53	*PCNA*
PCNA	-
Runx2	-

2.5. Total Antioxidant Capacity of the Marrow Extracellular Fluid

To assess the changes in the redox microenvironment in the marrow cavity following irradiation, we measured the antioxidant capacity of the extracellular fluid from the marrow, (normalized by protein concentration) at one and seven days after exposure to ^{56}Fe or protons (Figure 1B). Protons (200 cGy) or ^{56}Fe (200 cGy) decreased the total antioxidant capacity of the marrow extracellular fluid by over 35% within one day of exposure (Figure 6). In contrast, seven days after exposure, protons caused a 62% elevation of antioxidant capacity (Figure 6), whereas ^{56}Fe did not differ from sham controls. Lower doses of radiation (5 cGy ^{56}Fe or 10 cGy proton), did not modify antioxidant capacity per proton concentration with sham-controls at either timepoint.

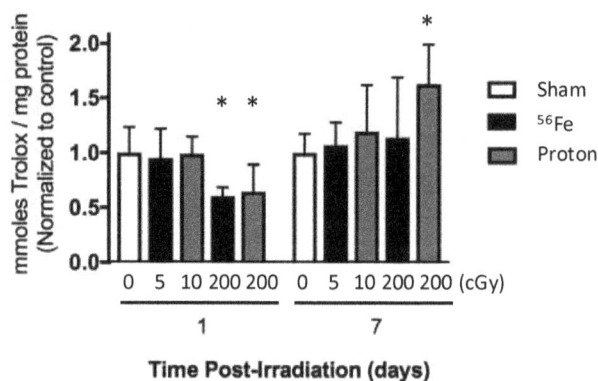

Figure 6. Irradiation perturbation of the marrow oxidative milieu, assessed by the antioxidant capacity of the extracellular fluid (ECF), in mice at one and seven days post exposure (following timeline in Figure 1B). Mice were irradiated with ^{56}Fe at 5 or 200 cGy or protons at 10 or 200 cGy. Data are mean ± standard deviation. * $p \leq 0.05$ versus sham control.

2.6. In Vitro Radiation Effects on Osteoblastogenesis

We hypothesized that a radiation-induced rise in ROS damages osteoprogenitors, leading to a decrease in number and activity of differentiated progeny and therefore exogenous antioxidant is expected to mitigate adverse effects of radiation on the proliferation and subsequent differentiation of osteoprogenitors derived from the marrow. The influence of a low-LET species (^{137}Cs) was compared to that of high-LET ^{56}Fe. Adherent marrow cells grown under osteoblastogenic culture conditions were irradiated at day 3 in culture and the percentage change in DNA content between day 3 (day of irradiation) and day 10 calculated as a surrogate of culture growth (proliferation). Analysis of the dose-response to ^{137}Cs and ^{56}Fe revealed significant decrements in growth after exposure to ^{137}Cs (200 cGy and 500 cGy) or ^{56}Fe (100 cGy, 200 cGy), but not after exposure to lower doses (Figure 7A). Colony counts also were affected at the higher doses of radiation [23]. In summary, as shown in Figure 7A, osteoprogenitors appeared more sensitive to high-LET ^{56}Fe than low-LET ^{137}Cs γ.

To begin to assess the contribution of oxidative stress after in vitro radiation exposure of osteoprogenitors, several antioxidants (including SOD with and without polyethylene glycol (PEG); SOD with nanoparticles or liposomes; Catalase with and without PEG) were screened by addition to the culture media, with SOD (without carrier) showing the most promising results [23]. Addition of exogenous SOD (200 U/mL) provided twice a day (on day 3 and day 4) effectively protected cell growth from irradiation with ^{137}Cs, whereas higher doses of SOD or longer periods of treatment were less effective [23]. As shown in Figure 7B, addition of SOD prevented the inhibition of growth by ^{137}Cs (200 cGy). In contrast, addition of SOD was not as sufficient to rescue the cell death incurred after exposure to an equivalent dose of ^{56}Fe (200 cGy).

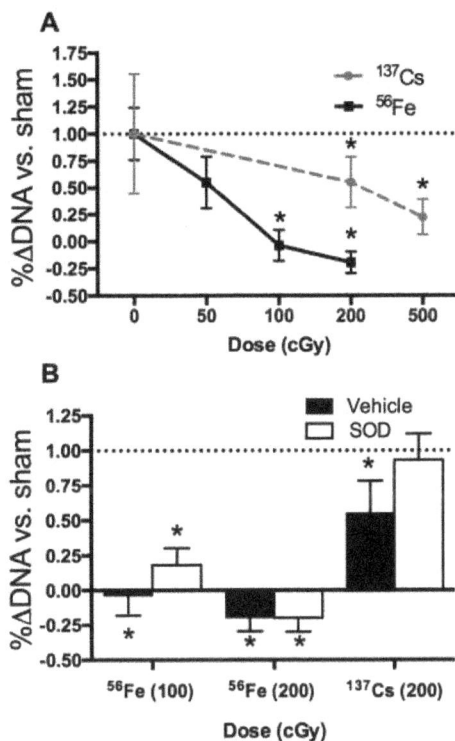

Figure 7. In vitro irradiation effects on osteoblastogenesis. (**A**) Dose response curve of osteoprogenitor growth after exposure to radiation using low-LET ^{137}Cs or high-LET ^{56}Fe; (**B**) Addition of the antioxidant superoxide dismutase (SOD) and its prevention of the irradiation-induced decrement in cell growth. Data are shown normalized to the sham control (which is depicted by the dashed line at 1.0), expressed as mean $\pm$ standard deviation. * denotes $p \leq 0.05$ versus the sham-irradiated control.

3. Discussion

Heavy-ion irradiation during space missions is a risk to the skeletal health of astronauts, causing skeletal degeneration and cellular damage in simulations with rodents [12,20,24]. In this study, we sought to elucidate the role of radiation type and changes in the oxidative milieu of bone marrow related to marrow-derived osteoprogenitor differentiation. We found that low doses (below 50 cGy) of either low-LET protons or high-LET ^{56}Fe did not cause observable adverse effects on structure and osteoblastogenesis at any endpoint, and, further, did not strongly perturb antioxidant capacity at seven days of exposure (Figure 6) nor expression levels of redox-related genes (Figure 4, Table 2) at 35 days of exposure. In contrast, only a high dose of ^{56}Fe radiation (200 cGy) was sufficient to induce detrimental effects on total antioxidant capacity, bone structure, and osteoprogenitor populations of the marrow (Figures 3–6). These changes were associated with transient elevations in mRNA levels (Figure 4C,D) of an antioxidant gene (*Catalase*) and a DNA-damage/cell cycle arrest marker (*Gadd45*) during ex vivo osteoblastogenesis, which we speculate indicates persistent oxidative stress or enhanced antioxidant defenses during the proliferative phase. Interestingly, cultures showed a trend towards elevation of osteocalcin (*Bglap*, Figure 4D), a late marker of differentiation, during the proliferative phase, which may indicate that ^{56}Fe exposure accelerated differentiation.

Body mass and coat color were assessed as general indices of animal health. Body mass indicated that animals showed a modest and transient decline in response to ^{56}Fe irradiation with 200 cGy

(Figure 2), consistent with our previous studies [17]. Additionally, ^{56}Fe irradiation at 200 cGy uniquely caused a shift in coat color from black to grey (Table 1). We interpret this as a hallmark of oxidative stress in the hair follicle [25], and, when coupled with the osteoprogenitor effects, this may signal premature aging in a second tissue, in line with the free-radical theory of cellular aging [26,27].

In rodents, radiation injury to bone between with ^{56}Fe (10–200 cGy, 1 GeV/n) [17] manifests as acute cell death of marrow cells and concomitant cancellous bone loss driven by osteoclasts [12,22,28]. In this work, we showed that high-dose ^{56}Fe or proton (200 cGy) caused a transient reduction in antioxidant capacity in the marrow microenvironment (Figure 6), potentially indicative of a depletion of extracellular antioxidant stores with subsequent recovery of the extracellular antioxidant defense system [29]. These findings are consistent with our data showing the marrow cell response to increase mRNA expression levels of the master antioxidant transcription factor, *Nrf2*, shown previously [13], and with effects also evident in a cell line at even higher doses [19]. Our group showed that a dietary supplement rich in antioxidants and polyphenols (dried plum) prevents bone loss and *Nrf2* induction from γ irradiation in mice [30], supportive of the role of oxidative stress in driving acute bone loss. Additionally, in vitro, 200 cGy ^{56}Fe impairs osteoblastogenesis [17]. Our work extends this high-dose impairment in vivo and to five weeks and one year post-exposure, suggesting an irreversible insult to osteoblastogenesis.

Gene expression analyses during the proliferative phase (Figure 4D) revealed that a high dose of ^{56}Fe (200 cGy) increased the expression of genes for the cell cycle arrest marker *Gadd45* and antioxidant *Catalase* relative to sham. These data are consistent with a G2/M arrest from high-LET irradiation of mesenchymal stem cells [18], a pre-osteoblast cell line [31], and many other cell types [32]. Interestingly, these two genes are targets of *Foxo* transcription factors [33–35]. Nrf2, Foxos, and p21 are critical to the antioxidant response element [36]. Oxidative stress activates Foxo transcription factors [37] and interferes with Wnt signaling [14,38–40]. Hence, our data suggest Foxo activation may play a role in ^{56}Fe irradiation, decreasing osteoprogenitor cell proliferation, and, ultimately, mineralization levels.

During the mineralization phase (Figure 4A,B), ^{56}Fe irradiation (200 cGy) tended to reduce nodule formation and alkaline phosphatase gene expression, suggesting a reduction in early markers of early-stage osteoblast differentiation and fewer cells contributing to nodule formation, consistent with in vitro evidence for high-dose effects [19]. Taken together with gene expression data, these findings suggest that irradiation induced persistent oxidative stress that impaired colony formation via reduced proliferation (e.g., delayed or halted cell division) leading to lower extent of mineralization. Higher doses of γ irradiation (>500 cGy) are required to recapitulate the impairment of osteoblastogenesis [41,42], where changes are also associated with oxidative damage [43]. Taking these findings into consideration, the sequelae of a high-dose of heavy-ion irradiation are likely impaired osteoblastogenesis, which may be attributable to persistent oxidative stress.

For evaluating structural responses, we analyzed cancellous regions in the tibia (Figure 3) and the cortical midshaft of the femur (Table 1) at two time points for parameters indicative of structural integrity. At 5-weeks after irradiation, exposure to high-LET ^{56}Fe irradiation at or above 50 cGy caused decrements in fractional bone volume and trabecular number, and increases in trabecular spacing (Figure 3), indicating a significant deterioration of cancellous microarchitecture, consistent with previous findings [20,22,44]. The magnitude of the changes caused by protons or ^{56}Fe were similar, although the effect of protons at 50 cGy on BV/TV was not statistically significant ($p = 0.12$). Thus, these findings indicate that the threshold for a persistent, ^{56}Fe radiation-induced response lies between 10 and 50 cGy—doses relevant to exploration missions [11]. Overall, we show that a high dose of radiation was able to induce persistent decrements in cancellous bone, regardless of radiation type (Figure 3). Others have shown that these changes persist to four months post-exposure [22]. In this study, the structural decrement affected by ^{56}Fe (200 cGy) was resolved one year after exposure due to the expected age-related decline in sham-control animals changes over time (Table 1), likely the product of age-induced elevations of osteoclasts overtaking waning bone formation by osteoblasts [45]. These data are consistent with our previous findings from growing female mice after exposure to low-LET γ radiation [46]. Although others have shown cortical changes following irradiation with either low-LET γ at high doses (500 cGy) [47] or high-LET ^{56}Fe at 50 cGy [48], no irradiation-induced cortical responses at 1-year post exposure were evident in this study [23], or in other experiments we have performed in the past. Thus, the influence of ionizing radiation at 200 cGy or below appears to be restricted to the more metabolically-active cancellous tissue.

In vitro irradiation studies were conducted to further assess the role of oxidative stress in radiation damage to osteoprogenitors. Our approach was to test antioxidants and their capacity to rescue radiation effects to provide insight into the different mechanisms between low-LET radiation and high-LET radiation [31]. Although similar doses of low-LET radiation are generally less deleterious than high-LET radiation, the mechanisms for radiation damage to primary, marrow-derived osteoprogenitors in the context of oxidative stress are not fully elucidated [49–51]. When comparing the effects of low-LET radiation (^{137}Cs) and high-LET radiation (^{56}Fe), we found high-LET irradiation was more damaging with respect to growth of the cell population, as assessed by a surrogate assay (changes in DNA content over time). Furthermore, exogenous SOD effectively protected osteoprogenitors from low-LET radiation damage, but only minimally from ^{56}Fe at 100 cGy, and not at all from ^{56}Fe at 200 cGy. Taken together, these findings indicate excess extracellular reactive oxygen species (ROS) inhibited growth, and that oxidative stress is possibly the main mechanism involved in low-LET radiodamage to osteoprogenitors; in contrast, high-LET species likely affect the cells via additional detrimental mechanisms, such as double-strand DNA breaks.

Although others [52,53] have shown the beneficial effects of various antioxidants to confer radioprotection for various osteoblast cell lines, we show here a direct comparison of antioxidant treatment with different radiation species when added to primary, marrow-derived osteoprogenitors. Interestingly, adding SOD at higher doses or for longer periods was not advantageous, emphasizing the importance of carefully titrating radical scavengers such as SOD, possibly to maintain endogenous ROS-dependent signalling.

In sum, we showed the ability of the antioxidant superoxide dismutase (SOD) to mitigate the adverse effects of in vitro irradiation with γ rays (used for the low-LET radiation), but not to an equivalent dose of high-LET ^{56}Fe. Our results strengthen the potential application of SOD as a countermeasure for low-LET irradiation-induced damage to marrow stem cells and osteoprogenitors.

Inherent limitations of these experiments with respect to radiation exposures include: (1) exposures were delivered at high dose-rate and fluence in a matter of minutes, whereas in space, the same dose is delivered at a lower rate, and (2) space exposures are mixed doses, rather than a single species (e.g., SPE can deliver 100–200 cGy over days and GCR over years). In addition, with respect to relevance to spaceflight, these experiments were performed in mice that were normally ambulatory, thus not subjected to musculoskeletal disuse and fluid shifts, which are induced by microgravity; groundbased models for spaceflight have shown that unloading can influence radiosensitivity [17,24,54,55].

In summary, our findings indicate that although low doses of radiation do not show lasting effects on skeletal health, a high dose of high-LET ^{56}Fe impairs osteoblast growth and maturation by adversely affecting growth of osteoprogenitors. There were few, select changes in steady-state expression of redox-related genes during osteoblastogenesis ex vivo after total body irradiation with ^{56}Fe (but not protons), with some indication of altered expression of select genes during the proliferative phase. Furthermore, scavenging ROS with antioxidants, such as SOD, mitigated the adverse effects of in vitro irradiation with γ rays (200 cGy), but failed to protect from irradiation with an equivalent dose of heavy ions. Together, our findings indicate that high-dose radiation has a persistent effect to impair bone formation by adversely affecting growth, but not differentiation, of osteoprogenitors.

4. Materials and Methods

4.1. Animals

Male C57BL/6J mice (Jackson Laboratories, Bar Harbor, ME, USA) were individually housed under standard conditions and provided food (LabDiet 5001, Purina, St. Louis, MO, USA) and water *ad libitum* as described by [13]. As a general measure of animal health, we measured body mass weekly for the duration of the study. Additionally, we blindly scored coat color as black or grey at the one-year post-irradiation endpoint. Animals were euthanized by cardiac puncture while under isoflurane overdose followed by cervical dislocation. The Institutional Animal Care and Use Committees for NASA Ames Research Center (Protocol NAS-10-001-Y2, 17 February 2011) and Brookhaven National Laboratory (Protocol #431, 8 February 2011) approved all of the procedures.

4.2. In Vivo Experiment Design and Radiation Exposure

Two experiments using mice were conducted to assess the effect of radiation type on skeletal antioxidant response, osteoblastogenesis, and bone structure. In the first experiment (Figure 1A), conscious 16-weeks old, male, C57BL6/J mice were exposed to low-LET protons (150 MeV/n, LET ~0.52 keV/μm) or high-LET ^{56}Fe ions (600 MeV/n, LET ~175 keV/μm) at 5, 10, 50, or 200 cGy at NASA's Space Radiation Lab at Brookhaven National Lab (Upton, NY, USA). Between 7–9 days after irradiation, mice were shipped overnight from Brookhaven to NASA Ames Research Center, Moffett Field, CA, USA. Select groups of mice were euthanized and tissues harvested after 35–38 days for ^{56}Fe, 36–39 days for proton, or 36–38 days for the sham (these three groups are deemed 5 weeks post-irradiation and had $n = 8$ per group) and late (358 or 360 days, here deemed 1 year post-irradiation and had $n = 13$ per group) for ex vivo osteoprogenitor culture of marrow cells. Tibiae or femora were collected for cancellous microarchitecture assessment with microcomputed tomography. For the 5-week endpoint, osteoblast cultures ($n = 6$–7 mice/group) were analyzed by real-time qPCR ($n = 4$ mice/group, see below for details), total DNA quantification, and alkaline phosphatase activity during the proliferative phase (seven days after plating) or the mineralization phase (19–21 days after plating) stages of osteoblast differentiation. At the termination of cultures, mineralized nodules were imaged and analyzed for percent surface area at the end of the culture period. For the 1-year endpoint, osteoblast cultures ($n = 6$/group) were analyzed for colony number during the proliferative phase (eight days after plating) or for mineralized nodule area during the mineralization phase (28 days after plating) of osteoblast differentiation.

In the second experiment (Figure 1B), we aimed to determine the effects of total-body irradiation on the antioxidant capacity of the extracellular fluid surrounding the marrow cells [56]. To evaluate ion effects, conscious 16-wk old, male, C57BL6/J mice mice were exposed to 10 or 200 cGy of low-LET protons ($n = 6$–8 per group) or 5 or 200 cGy of high-LET ^{56}Fe ions ($n = 8$ per group) as described above, or were sham irradiated ($n = 13$ per group). Mice were euthanized and tissues harvested at one or seven days after exposure, at which point tibial marrow was processed for extracellular fluid collection and quantification of total antioxidant capacity, described in more detail below.

4.3. In Vitro Experiment Design and Radiation Exposure

Osteoprogenitors were grown in vitro from the bone marrow of 16-week old, male C57Bl/6 mice, as previously described [17]. Cells were irradiated three days after plating (day 3) with either γ (^{137}Cs, sham or 10–500 cGy, JL Shepherd Mark I) or ^{56}Fe (sham or 50–200 cGy, NSRL), and then grown until day 10. Cell behavior was assessed by measuring changes in DNA content between day three and day 10 (as a surrogate for cell growth) with the Cyquant assay as previously described [17]. For the SOD experiment, multiple doses were tested in preliminary experiments (100–600 U/mL of SOD or vehicle, for 2 or 3 days of treatment at different times). The protocol selected for more extensive dose-response experiments entailed the addition of SOD (200 U/mL), or vehicle (α-MEM) 2 h before irradiation (day 3), twice a day, through the 4th day after exposure to radiation. Experiments were repeated 2–3 times. Data were normalized to the sham-control.

4.4. Total Antioxidant Capacity

We measured the antioxidant capacity of the extracellular fluid of the marrow (i.e., the marrow plasma) with a colorimetric assay based on the reduction potential of Cu^{2+} to Cu^{1+} as compared to Trolox equivalent standards (Oxford Biomedical, Oxford, MI, USA). In brief, we isolated marrow ex vivo, using 150 µL of sterile saline to flush the marrow from the tibial diaphysis, as described previously [56]. The solution was centrifuged and supernatant preserved as the extracellular fluid at $-80\,^{\circ}C$ for further assessment. The BCA assay assessed the protein concentration of the extracellular fluid against albumin standards, which enabled the normalization of the antioxidant capacity by the protein concentration to account for changes in protein levels with treatment [56].

4.5. Bone Microarchitecture and Geometry

We assessed the bone volume and microarchitecture of the cancellous tissue in the proximal tibial metaphysis using microcomputed tomography (6.7 µm pixel size, 3500 ms integration time, 50 kV, Skyscan 1174; Bruker microCT, Kontich, Belgium) as described in [13]. For cancellous quantification, a 1.0-mm thick region located 0.24 mm distal to the proximal growth plate was semi-autonomously contoured to include cancellous tissue. We assessed changes in cancellous bone using the three-dimensional bone volume to total volume fraction (BV/TV, %), trabecular thickness (Tb.Th, mm), trabecular number (Tb.N, 1/mm), trabecular thickness (Tb.Th, mm), and trabecular separation (Tb.Sp, mm). We assessed the changes in cortical bone at the midshaft of the femur using bone volume and cortical thickness. All analyses follow conventional guidelines [57].

4.6. Ex Vivo Osteoblastogenic Assays and qRT-PCR Analyses

For osteoblastogenic culture, femur marrow was flushed using saline and treated with lysis buffer specific for red blood cells (Sigma-Aldrich, St. Louis, MO, USA), centrifuged for ten minutes at $1000 \times g$, and supernatant removed. Cells were then re-suspended and plated at 3.0×10^5 cells/cm^2 in 6-well plates. Osteoblastogenic growth medium (alpha minimum essential medium with 15% FBS (Gibco, Gaithersburg, MD, USA), $1 \times$ anti-biotic/mycotic, 50 µg/mL ascorbic acid, 10 mM β-glycerophosphate) was replenished every 2–3 days.

For qPCR analyses ($n = 4$/group), cells were collected in a solution of 1% β-mercapthoethanol in RLT Buffer (Qiagen, Valencia, CA, USA) and stored at $-80\,^{\circ}C$ until RNA isolation via RNeasy minikit (Qiagen). RNA quantity and purity was measured using a nanodrop spectrophotometer (Nanodrop 2000, Thermoscientific, Waltham, MA, USA) while RNA quality was assessed using a 0.8% non-denaturing gel. A RT^{2}PCR custom array (PAMM-999A-1, SYBR probes, Qiagen, Hilden, Germany) was processed with an Applied Biosystems 7500 cycler to generate gene expression data. The $\Delta\Delta C_t$ method was used to analyze the raw data. Cycle values above 35 were defined as non-detectable. Statistical analyses were performed on log$_2$-transformed data. Fifteen representative genes were analyzed to assess osteoblast differentiation, proliferation, oxidative metabolism, or apoptosis, see Table 3. All genes were normalized to housekeeping gene *Hprt*. Technical issues forced the loss of one specimen from the ^{56}Fe, 200 cGy group during the proliferative phase and the loss of all four specimens in the proton, 10 cGy group from mineralization phase.

Table 3. SYBR-based custom array to query changes in expression levels of key genes.

Gene Name	Accession No.	Official Name	Process
Runx2	NM_009820	Runt Related Transcription Factor 2	Differentiation
Bglap	NM_007541	Bone γ carboxyglutamate protein (osteocalcin)	Differentiation
Alpl	NM_007431	Alkaline Phosphatase Ligand	Differentiation
Pcna	NM_011045	Proliferating Cell Nuclear Antigen	Proliferation
Cdk2	NM_016756	Cyclin-dependent Kinase 2	Proliferation
p21	NM_007669	Cyclin-dependent Kinase Inhibitor 1	Proliferation
p53	NM_011640	Transformation-related protein 53	Proliferation
Cat	NM_009804	Catalase	Oxidative Metabolism
Gpx1	NM_008160	Glutathione Peroxidase	Oxidative Metabolism
MnSod	NM_013671	Superoxide Dismutase 2, Mitochondrial	Oxidative Metabolism
CuZnSod	NM_011434	Superoxide Dismutase	Oxidative Metabolism
iNos [Nos2]	NM_010927	Nitric Oxide Synthase 2, Inducible	Oxidative Metabolism
Foxo1	NM_019739	Forkhead Box	Oxidative Metabolism
Gadd45a	NM_007836	Growth Arrest and DNA Damage	Oxidative Metabolism
Caspase 3	NM_009810	Caspase 3	Apoptosis
Hprt	NM_013556	Hypoxanthine-guanine phosphoribosyltransferase	Housekeeping

For ALP and DNA analyses, cells were collected in a solution of Tris buffer containing Triton-X and MgCl$_2$ and stored at -80 °C. Quantification of ALP enzymatic activity and total DNA content was performed using Alkaline Phosphatase Activity Colorimetric Assay (Biovision, Mountain View, CA, USA) and CyQuant Cell Proliferation Assay Kit (Thermo Fisher Scientific, Waltham, MA, USA), respectively.

Colony counts early in culture assessed the successful adherence and proliferation of colony forming units. Parameters defining a colony included having distinct region of radial colony growth and having more than 30 cells with a majority of cells having of morphological homogeneity.

To quantify densely mineralizing area (i.e., the nodule area), we harvested cultures in the differentiation phase. First, we washed cultures with saline and scanned the plates at 300 dpi resolution. We quantified the percentage of well area having mineralizing nodules with Image J, using a custom algorithm to threshold the grayscale image.

4.7. Statistics

We report parametric data with mean and standard deviation (SD). For these data, to determine significant differences, a 1-way analysis of variance was used followed by Dunnett's post-hoc test, which allowed for the comparison of respective radiation types and doses to the age-matched sham control (0 cGy). For non-parametric data (i.e., nodule area data, as values near zero display skew), we report the median with interquartile range. For these data, to determine significant differences, we used a Kruskal-Wallis test with a Steel post-hoc test, which allowed for the comparison of respective radiation types and doses to the age-matched, sham control (0 cGy). We assessed temporal (within subjects) treatment (between subjects) effects in body mass, and their interaction (i.e., time × treatment), using Repeated Measures ANOVA and Wilks' Lamda test. Throughout, a p-value ≤ 0.05 was accepted as significant. We performed all analyses using JMP 13 (SAS).

Acknowledgments: Research was supported by National Space Biomedical Research Institute grant # MA02501 (RKG, JSA) under NASA cooperative agreement NCC 9-58, a DOE-NASA Interagency Award #DE-SC0001507, supported by the Office of Science (BER), U.S. Department of Energy (RKG), and NASA Postdoctoral Program fellowships (AS, AK, JSA). The authors thank Charles Limoli (UC Irvine) for his valuable scientific input and consultation. We thank the Brookhaven National Laboratory Animal Care Facility and NASA Space Radiation Laboratory staff Peter Guida, Adam Rusek, MaryAnn Petry, Kerry Bonti, Angela Kim, and Laura Loudenslager for their assistance with this experiment. We thank Nicholas Thomas for careful reading and editing of the manuscript and Samantha Torres for supporting the animal experiments.

Author Contributions: Conceived and Designed Experiments: Joshua S. Alwood, Ruth K. Globus, Ann-Sofie Schreurs, Akhilesh Kumar, Luan H. Tran. Supported/Conducted Experiments: Joshua S. Alwood, Luan H. Tran, Ann-Sofie Schreurs, Yasaman Shirazi-Fard, Akhilesh Kumar, Candice G. T. Tahimic, Ruth K. Globus. Collected Data: Joshua S. Alwood, Luan H. Tran, Ann-Sofie Schreurs, Akhilesh Kumar, Diane Hilton. Interpreted Data: Joshua S. Alwood, Luan H. Tran, Ann-Sofie Schreurs, Yasaman Shirazi-Fard, Akhilesh Kumar, Diane Hilton, Candice G. T. Tahimic, Ruth K. Globus. Wrote and edited the Manuscript: Joshua S. Alwood, Ann-Sofie Schreurs, Luan H. Tran, Ruth K. Globus.

Conflicts of Interest: The authors declare no conflict of interest.

References

1. McPhee, J.C.; Charles, J.B. *Human Health and Peformance Risks of Space Exploration Missions*; Government Printing Office: Washington, DC, USA, 2009.
2. Guo, J.; Zeitlin, C.; Wimmer-Schweingruber, R.F.; Hassler, D.M.; Ehresmann, B.; Kohler, J.; Bohm, E.; Bottcher, S.; Brinza, D.; Burmeister, S.; et al. Msl-rad radiation environment measurements. *Radiat. Prot. Dosimetry* **2015**, *166*, 290–294. [CrossRef] [PubMed]
3. Dietze, G.; Bartlett, D.T.; Cool, D.A.; Cucinotta, F.A.; Jia, X.; McAulay, I.R.; Pelliccioni, M.; Petrov, V.; Reitz, G.; Sato, T. Icrp publication 123: Assessment of radiation exposure of astronauts in space. *Ann. ICRP* **2013**. [CrossRef] [PubMed]
4. Durante, M.; Cucinotta, F.A. Physical basis of radiation protection in space travel. *Rev. Mod. Phys.* **2011**, *83*, 1245–1281. [CrossRef]

5. Tseng, B.P.; Giedzinski, E.; Izadi, A.; Suarez, T.; Lan, M.L.; Tran, K.K.; Acharya, M.M.; Nelson, G.A.; Raber, J.; Parihar, V.K.; et al. Functional consequences of radiation-induced oxidative stress in cultured neural stem cells and the brain exposed to charged particle irradiation. *Antioxid. Redox Signal.* **2014**, *20*, 1410–1422. [CrossRef] [PubMed]

6. Datta, K.; Suman, S.; Fornace, A.J., Jr. Radiation persistently promoted oxidative stress, activated mtor via pi3k/akt, and downregulated autophagy pathway in mouse intestine. *Int. J. Biochem. Cell Biol.* **2014**, *57*, 167–176. [CrossRef] [PubMed]

7. Norbury, J.W.; Slaba, T.C. Space radiation accelerator experiments – the role of neutrons and light ions. *Life Sci. Space Res.* **2014**, *3*, 90–94. [CrossRef]

8. Hassler, D.M.; Zeitlin, C.; Wimmer-Schweingruber, R.F.; Ehresmann, B.; Rafkin, S.; Eigenbrode, J.L.; Brinza, D.E.; Weigle, G.; Bottcher, S.; Bohm, E.; et al. Mars' surface radiation environment measured with the mars science laboratory's curiosity rover. *Science* **2014**, *343*, 1244797. [CrossRef] [PubMed]

9. Zeitlin, C.; Hassler, D.M.; Cucinotta, F.A.; Ehresmann, B.; Wimmer-Schweingruber, R.F.; Brinza, D.E.; Kang, S.; Weigle, G.; Bottcher, S.; Bohm, E.; et al. Measurements of energetic particle radiation in transit to mars on the mars science laboratory. *Science* **2013**, *340*, 1080–1084. [CrossRef] [PubMed]

10. NASA. *Nasa std-3001, Space Flight Human-System Standard Volume 1, Revision a, Change 1: Crew Health*; NASA: Washington, DC, USA, 2015.

11. Townsend, L.W.; Badhwar, G.D.; Braby, L.A.; Blakely, E.A.; Cucinotta, F.A.; Curtis, S.B.; Fry, R.J.M.; Land, C.E.; Smart, D.F. *Report No. 153-Information Needed to Make Radiation Protection Recommendations for Space Missions Beyond Low-Earth Orbit*; US National Council for Radiation Protection and Measurements: Bethesda, MD, USA, 2006.

12. Kondo, H.; Yumoto, K.; Alwood, J.S.; Mojarrab, R.; Wang, A.; Almeida, E.A.; Searby, N.D.; Limoli, C.L.; Globus, R.K. Oxidative stress and gamma radiation-induced cancellous bone loss with musculoskeletal disuse. *J. Appl. Physiol.* **2010**, *108*, 152–161. [CrossRef] [PubMed]

13. Alwood, J.S.; Shahnazari, M.; Chicana, B.; Schreurs, A.S.; Kumar, A.; Bartolini, A.; Shirazi-Fard, Y.; Globus, R.K. Ionizing radiation stimulates expression of pro-osteoclastogenic genes in marrow and skeletal tissue. *J. Interferon Cytokine Res.* **2015**, *35*, 480–487. [CrossRef] [PubMed]

14. Almeida, M.; Ambrogini, E.; Han, L.; Manolagas, S.C.; Jilka, R.L. Increased lipid oxidation causes oxidative stress, increased peroxisome proliferator-activated receptor-gamma expression, and diminished pro-osteogenic wnt signaling in the skeleton. *J. Biol. Chem.* **2009**, *284*, 27438–27448. [CrossRef] [PubMed]

15. Manolagas, S.C.; Almeida, M. Gone with the wnts: Beta-catenin, t-cell factor, forkhead box o, and oxidative stress in age-dependent diseases of bone, lipid, and glucose metabolism. *Mol. Endocrinol.* **2007**, *21*, 2605–2614. [CrossRef] [PubMed]

16. Lean, J.M.; Jagger, C.J.; Kirstein, B.; Fuller, K.; Chambers, T.J. Hydrogen peroxide is essential for estrogen-deficiency bone loss and osteoclast formation. *Endocrinology* **2005**, *146*, 728–735. [CrossRef] [PubMed]

17. Yumoto, K.; Globus, R.K.; Mojarrab, R.; Arakaki, J.; Wang, A.; Searby, N.D.; Almeida, E.A.; Limoli, C.L. Short-term effects of whole-body exposure to (56)fe ions in combination with musculoskeletal disuse on bone cells. *Radiat. Res.* **2010**, *173*, 494–504. [CrossRef] [PubMed]

18. Kurpinski, K.; Jang, D.J.; Bhattacharya, S.; Rydberg, B.; Chu, J.; So, J.; Wyrobek, A.; Li, S.; Wang, D. Differential effects of x-rays and high-energy 56fe ions on human mesenchymal stem cells. *Int.J.Radiat.Oncol.Biol.Phys.* **2009**, *73*, 869–877. [CrossRef] [PubMed]

19. Kook, S.H.; Kim, K.A.; Ji, H.; Lee, D.; Lee, J.C. Irradiation inhibits the maturation and mineralization of osteoblasts via the activation of nrf2/ho-1 pathway. *Mol. Cell. Biochem.* **2015**, *410*, 255–266. [CrossRef] [PubMed]

20. Willey, J.S.; Livingston, E.W.; Robbins, M.E.; Bourland, J.D.; Tirado-Lee, L.; Smith-Sielicki, H.; Bateman, T.A. Risedronate prevents early radiation-induced osteoporosis in mice at multiple skeletal locations. *Bone* **2010**, *46*, 101–111. [CrossRef] [PubMed]

21. Lima, F.; Swift, J.M.; Greene, E.S.; Allen, M.R.; Cunningham, D.A.; Braby, L.A.; Bloomfield, S.A. Exposure to low-dose x-ray radiation alters bone progenitor cells and bone microarchitecture. *Radiat. Res.* **2017**, *188*, 433–442. [CrossRef] [PubMed]

22. Hamilton, S.A.; Pecaut, M.J.; Gridley, D.S.; Travis, N.D.; Bandstra, E.R.; Willey, J.S.; Nelson, G.A.; Bateman, T.A. A murine model for bone loss from therapeutic and space-relevant sources of radiation. *J. Appl. Physiol.* **2006**, *101*, 789–793. [CrossRef] [PubMed]

23. Alwood, J.S.; Tran, L.H.; Schreurs, A.-S.; Shirazi-Fard, Y.; Kumar, A.; Hilton, D.; Tahimic, C.G.T.; Globus, R.K. *Data not Shown*; NASA Ames Research Center: Moffett Field, CA, USA, 2017.

24. Macias, B.R.; Lima, F.; Swift, J.M.; Shirazi-Fard, Y.; Greene, E.S.; Allen, M.R.; Fluckey, J.; Hogan, H.A.; Braby, L.; Wang, S.; et al. Simulating the lunar environment: Partial weightbearing and high-let radiation-induce bone loss and increase sclerostin-positive osteocytes. *Radiat. Res.* **2016**, *186*, 254–263. [CrossRef] [PubMed]

25. Wood, J.M.; Decker, H.; Hartmann, H.; Chavan, B.; Rokos, H.; Spencer, J.D.; Hasse, S.; Thornton, M.J.; Shalbaf, M.; Paus, R.; et al. Senile hair graying: H2o2-mediated oxidative stress affects human hair color by blunting methionine sulfoxide repair. *FASEB J.* **2009**, *23*, 2065–2075. [CrossRef] [PubMed]

26. Harman, D. Free radical theory of aging: An update: Increasing the functional life span. *Ann. N. Y. Acad. Sci.* **2006**, *1067*, 10–21. [CrossRef] [PubMed]

27. Harman, D. Aging: A theory based on free radical and radiation chemistry. *J. Gerontol.* **1956**, *11*, 298–300. [CrossRef] [PubMed]

28. Kondo, H.; Searby, N.D.; Mojarrab, R.; Phillips, J.; Alwood, J.; Yumoto, K.; Almeida, E.A.C.; Limoli, C.L.; Globus, R.K. Total-body irradiation of postpubertal mice with 137cs acutely compromises the microarchitecture of cancellous bone and increases osteoclasts. *Radiat. Res.* **2009**, *171*, 283–289. [CrossRef] [PubMed]

29. Greenberger, J.S.; Epperly, M. Bone marrow-derived stem cells and radiation response. *Semin. Radiat. Oncol.* **2009**, *19*, 133–139. [CrossRef] [PubMed]

30. Schreurs, A.S.; Shirazi-Fard, Y.; Shahnazari, M.; Alwood, J.S.; Truong, T.A.; Tahimic, C.G.; Limoli, C.L.; Turner, N.D.; Halloran, B.; Globus, R.K. Dried plum diet protects from bone loss caused by ionizing radiation. *Sci. Rep.* **2016**, *6*, 21343. [CrossRef] [PubMed]

31. Hu, Y.; Hellweg, C.E.; Baumstark-Khan, C.; Reitz, G.; Lau, P. Cell cycle delay in murine pre-osteoblasts is more pronounced after exposure to high-let compared to low-let radiation. *Radiat. Environ. Biophys.* **2014**, *53*, 73–81. [CrossRef] [PubMed]

32. Snyder, A.R.; Morgan, W.F. Gene expression profiling after irradiation: Clues to understanding acute and persistent responses? *Cancer Metastasis Rev.* **2004**, *23*, 259–268. [CrossRef] [PubMed]

33. Tran, H.; Brunet, A.; Grenier, J.M.; Datta, S.R.; Fornace, A.J.; DiStefano, P.S.; Chiang, L.W.; Greenberg, M.E. DNA repair pathway stimulated by the forkhead transcription factor foxo3a through the gadd45 protein. *Science* **2002**, *296*, 530–534. [CrossRef] [PubMed]

34. Furukawa-Hibi, Y.; Yoshida-Araki, K.; Ohta, T.; Ikeda, K.; Motoyama, N. Foxo forkhead transcription factors induce g(2)-m checkpoint in response to oxidative stress. *J. Biol. Chem.* **2002**, *277*, 26729–26732. [CrossRef] [PubMed]

35. Bartell, S.M.; Kim, H.N.; Ambrogini, E.; Han, L.; Iyer, S.; Serra Ucer, S.; Rabinovitch, P.; Jilka, R.L.; Weinstein, R.S.; Zhao, H.; et al. Foxo proteins restrain osteoclastogenesis and bone resorption by attenuating h2o2 accumulation. *Nat. Commun.* **2014**, *5*, 3773. [CrossRef] [PubMed]

36. Gorrini, C.; Harris, I.S.; Mak, T.W. Modulation of oxidative stress as an anticancer strategy. *Nat. Rev. Drug Discov.* **2013**, *12*, 931–947. [CrossRef] [PubMed]

37. Klotz, L.O.; Sanchez-Ramos, C.; Prieto-Arroyo, I.; Urbanek, P.; Steinbrenner, H.; Monsalve, M. Redox regulation of foxo transcription factors. *Redox. Biol.* **2015**, *6*, 51–72. [CrossRef] [PubMed]

38. Hoogeboom, D.; Essers, M.A.; Polderman, P.E.; Voets, E.; Smits, L.M.; Burgering, B.M. Interaction of foxo with beta-catenin inhibits beta-catenin/t cell factor activity. *J. Biol. Chem.* **2008**, *283*, 9224–9230. [CrossRef] [PubMed]

39. Essers, M.A.; de Vries-Smits, L.M.; Barker, N.; Polderman, P.E.; Burgering, B.M.; Korswagen, H.C. Functional interaction between beta-catenin and foxo in oxidative stress signaling. *Science* **2005**, *308*, 1181–1184. [CrossRef] [PubMed]

40. Maiese, K.; Chong, Z.Z.; Shang, Y.C.; Hou, J. Rogue proliferation versus restorative protection: Where do we draw the line for wnt and forkhead signaling? *Expert Opin. Ther. Targets* **2008**, *12*, 905–916. [CrossRef] [PubMed]

41. Turner, R.T.; Iwaniec, U.T.; Wong, C.P.; Lindenmaier, L.B.; Wagner, L.A.; Branscum, A.J.; Menn, S.A.; Taylor, J.; Zhang, Y.; Wu, H.; et al. Acute exposure to high dose gamma-radiation results in transient activation of bone lining cells. *Bone* **2013**, *57*, 164–173. [CrossRef] [PubMed]

42. Green, D.E.; Adler, B.J.; Chan, M.E.; Rubin, C.T. Devastation of adult stem cell pools by irradiation precedes collapse of trabecular bone quality and quantity. *J. Bone Miner. Res.* **2012**, *27*, 749. [CrossRef] [PubMed]

43. Cao, X.; Wu, X.; Frassica, D.; Yu, B.; Pang, L.; Xian, L.; Wan, M.; Lei, W.; Armour, M.; Tryggestad, E.; et al. Irradiation induces bone injury by damaging bone marrow microenvironment for stem cells. *Proc. Natl. Acad. Sci. USA* **2011**, *108*, 1609–1614. [CrossRef] [PubMed]

44. Alwood, J.S.; Yumoto, K.; Mojarrab, R.; Limoli, C.L.; Almeida, E.A.; Searby, N.D.; Globus, R.K. Heavy ion irradiation and unloading effects on mouse lumbar vertebral microarchitecture, mechanical properties and tissue stresses. *Bone* **2010**, *47*, 248–255. [CrossRef] [PubMed]

45. Cao, J.J.; Wronski, T.J.; Iwaniec, U.; Phleger, L.; Kurimoto, P.; Boudignon, B.; Halloran, B.P. Aging increases stromal/osteoblastic cell-induced osteoclastogenesis and alters the osteoclast precursor pool in the mouse. *J. Bone Miner. Res.* **2005**, *20*, 1659–1668. [CrossRef] [PubMed]

46. Alwood, J.S.; Kumar, A.; Tran, L.H.; Wang, A.; Limoli, C.L.; Globus, R.K. Low-dose, ionizing radiation and age-related changes in skeletal microarchitecture. *J. Aging Res.* **2012**, *2012*, 481983. [CrossRef] [PubMed]

47. Oest, M.E.; Franken, V.; Kuchera, T.; Strauss, J.; Damron, T.A. Long-term loss of osteoclasts and unopposed cortical mineral apposition following limited field irradiation. *J. Orthop. Res.* **2015**, *33*, 334–342. [CrossRef] [PubMed]

48. Lloyd, S.A.; Bandstra, E.R.; Travis, N.D.; Nelson, G.A.; Bourland, J.D.; Pecaut, M.J.; Gridley, D.S.; Willey, J.S.; Bateman, T.A. Spaceflight-relevant types of ionizing radiation and cortical bone: Potential let effect? *Adv. Space Res.* **2008**, *42*, 1889–1897. [CrossRef] [PubMed]

49. Hada, M.; Zhang, Y.; Feiveson, A.; Cucinotta, F.A.; Wu, H. Association of inter- and intrachromosomal exchanges with the distribution of low- and high-let radiation-induced breaks in chromosomes. *Radiat. Res.* **2011**, *176*, 25–37. [CrossRef] [PubMed]

50. Mariotti, L.G.; Bertolotti, A.; Ranza, E.; Babini, G.; Ottolenghi, A. Investigation of the mechanisms underpinning il-6 cytokine release in bystander responses: The roles of radiation dose, radiation quality and specific ros/rns scavengers. *Int. J. Radiat. Biol.* **2012**, *88*, 751–762. [CrossRef] [PubMed]

51. Werner, E.; Wang, Y.; Doetsch, P.W. A single exposure to low- or high-let radiation induces persistent genomic damage in mouse epithelial cells in vitro and in lung tissue. *Radiat. Res.* **2017**, *188*, 373–380. [CrossRef] [PubMed]

52. Kim, K.A.; Kook, S.H.; Song, J.H.; Lee, J.C. A phenolic acid phenethyl urea derivative protects against irradiation-induced osteoblast damage by modulating intracellular redox state. *J. Cell. Biochem.* **2014**, *115*, 1877–1887. [PubMed]

53. Mi Choi, E.; Sik Suh, K.; Jung, W.W.; Young Park, S.; Ouk Chin, S.; Youl Rhee, S.; Kim Pak, Y.; Chon, S. Actein alleviates 2,3,7,8-tetrachlorodibenzo-p-dioxin-mediated cellular dysfunction in osteoblastic mc3t3-e1 cells. *Environ. Toxicol.* **2017**. [CrossRef] [PubMed]

54. Mao, X.W.; Nishiyama, N.C.; Campbell-Beachler, M.; Gifford, P.; Haynes, K.E.; Gridley, D.S.; Pecaut, M.J. Role of nadph oxidase as a mediator of oxidative damage in low-dose irradiated and hindlimb-unloaded mice. *Radiat. Res.* **2017**, *188*, 392–399. [CrossRef] [PubMed]

55. Krause, A.R.; Speacht, T.L.; Zhang, Y.; Lang, C.H.; Donahue, H.J. Simulated space radiation sensitizes bone but not muscle to the catabolic effects of mechanical unloading. *PLoS ONE* **2017**, *12*, e0182403. [CrossRef] [PubMed]

56. Shahnazari, M.; Dwyer, D.; Chu, V.; Asuncion, F.; Stolina, M.; Ominsky, M.; Kostenuik, P.; Halloran, B. Bone turnover markers in peripheral blood and marrow plasma reflect trabecular bone loss but not endocortical expansion in aging mice. *Bone* **2012**, *50*, 628–637. [CrossRef] [PubMed]

57. Bouxsein, M.L.; Boyd, S.K.; Christiansen, B.A.; Guldberg, R.E.; Jepsen, K.J.; Muller, R. Guidelines for assessment of bone microstructure in rodents using micro-computed tomography. *J. Bone Miner. Res.* **2010**, *25*, 1468–1486. [CrossRef] [PubMed]

International Journal of
Molecular Sciences

MDPI

Article

Combined Effects of Simulated Microgravity and Radiation Exposure on Osteoclast Cell Fusion

Srinivasan Shanmugarajan [1,2] **, Ye Zhang** [3]**, Maria Moreno-Villanueva** [1,4]**, Ryan Clanton** [5]**,
Larry H. Rohde** [2]**, Govindarajan T. Ramesh** [6]**, Jean D. Sibonga** [1] **and Honglu Wu** [1,*]

[1]   NASA Johnson Space Center, Houston, TX 77058, USA; srinimag@gmail.com (S.S.);
       maria.moreno-villanueva@uni-konstanz.de (M.M.-V.); jean.sibonga-1@nasa.gov (J.D.S.)
[2]   Department of Biological and Environmental Sciences, University of Houston Clear Lake,
       Houston, TX 77058, USA; Rohde@uhcl.edu
[3]   NASA Kennedy Space Center, Cape Canaveral, FL 32899, USA; ye.zhang-1@nasa.gov
[4]   Department of Biology, University of Konstanz, 78457 Konstanz, Germany
[5]   Department of Nuclear Engineering, Texas A & M University, College Station, TX 77843, USA;
       rc1025@tamu.edu
[6]   Department of Biology, Norfolk State University, Norfolk, VA 23504, USA; gtramesh@nsu.edu
[*]   Correspondence: honglu.wu-1@nasa.gov; Tel.: +281-483-6470; Fax: +281-483-3058

Received: 7 October 2017; Accepted: 15 November 2017; Published: 18 November 2017

Abstract: The loss of bone mass and alteration in bone physiology during space flight are one of the major health risks for astronauts. Although the lack of weight bearing in microgravity is considered a risk factor for bone loss and possible osteoporosis, organisms living in space are also exposed to cosmic radiation and other environmental stress factors. As such, it is still unclear as to whether and by how much radiation exposure contributes to bone loss during space travel, and whether the effects of microgravity and radiation exposure are additive or synergistic. Bone is continuously renewed through the resorption of old bone by osteoclast cells and the formation of new bone by osteoblast cells. In this study, we investigated the combined effects of microgravity and radiation by evaluating the maturation of a hematopoietic cell line to mature osteoclasts. RAW 264.7 monocyte/macrophage cells were cultured in rotating wall vessels that simulate microgravity on the ground. Cells under static 1g or simulated microgravity were exposed to γ rays of varying doses, and then cultured in receptor activator of nuclear factor-κB ligand (RANKL) for the formation of osteoclast giant multinucleated cells (GMCs) and for gene expression analysis. Results of the study showed that radiation alone at doses as low as 0.1 Gy may stimulate osteoclast cell fusion as assessed by GMCs and the expression of signature genes such as tartrate resistant acid phosphatase (*Trap*) and dendritic cell-specific transmembrane protein (*Dcstamp*). However, osteoclast cell fusion decreased for doses greater than 0.5 Gy. In comparison to radiation exposure, simulated microgravity induced higher levels of cell fusion, and the effects of these two environmental factors appeared additive. Interestingly, the microgravity effect on osteoclast stimulatory transmembrane protein (*Ocstamp*) and *Dcstamp* expressions was significantly higher than the radiation effect, suggesting that radiation may not increase the synthesis of adhesion molecules as much as microgravity.

Keywords: microgravity; radiation; osteoclast

1. Introduction

All living organisms on Earth undergo physiological changes in response to the space environment, microgravity in particular. In humans, spaceflight has resulted in complications such as cardiovascular deconditioning, reduced immune functions, and unbalanced bone and mineral turnover [1,2]. Of the health risks associated with space travel, alterations in the skeletal mass may be a risk factor for secondary osteoporosis [3,4]. Bone remodeling is a dynamic process with a balanced removal of old bone by osteoclasts followed with new bone formation by osteoblasts. In terms of osteoblasts, glucocorticoids are known inhibitors of osteoblast cell growth, and microgravity has been shown to systemically increase cortisol levels in osteoblast cultures in space [5]. In space, microgravity has been shown to promote osteoclast activities in vivo [6]. Enhanced differentiation of bone-resorbing osteoclasts has also been reported using in vitro cell models and simulated microgravity on the ground [7]. Such differentiations are associated with tumor necrosis factor-related apoptosis inducing ligand (TRAIL) expressions. Despite numerous in vivo and in vitro studies attempting to explain the bone loss phenomenon in space, a mechanism of these cellular changes has remained elusive [8]. Furthermore, the responses to microgravity in specific bone regions need to be investigated [9].

In addition to bone loss due to microgravity, cosmic radiation is another challenging factor in the spaceflight environment, and radiation-induced bone loss in astronauts might be an additional risk factor for osteoporosis. Space radiation consists of mostly high-energy protons and other heavier charged particles of high linear energy transfer (LET) [10]. Ionizing radiation has been shown to contribute significantly to bone homeostasis. For example, in women treated for a variety of pelvic tumors, ionizing radiation increased the 5-year incidence of hip fracture by 65% [11]. In mice, exposure to 2 Gy γ rays, protons, carbon, or iron radiation species caused a 30–40% loss of their trabecular bone volume fractions [12]. Furthermore, exposures of mouse bones to X-rays resulted in an increase in the osteoclast number and activity [11]. Irradiation of a single-limb in a murine model induced local and paradoxically systemic bone loss [13]. Even though these are clinically relevant doses, recent publications reported that spaceflight-relevant radiation doses also promote low bone turnover and osteoclast activity [14].

Whether exposures to microgravity and space radiation simultaneously produce additive or synergistic consequences has been investigated with a number of biological endpoints such as DNA damage response [15]. With regard to bone loss, low doses of high-LET radiation, in conjunction with partial-weight bearing, appeared to promote the induction of bone loss with an increase in sclerostin-positive osteocytes and wnt-signaling [16]. In a mouse model looking at the tibia bone surface, radiation caused a 46% increase in osteoclast number, hindlimb-unloading caused a 47% increase in osteoclast number, and the combination of radiation and hind-limb unloading caused a 64% increase in osteoclast number [17]. A possible mechanism of synergy between microgravity and radiation is the fact that hindlimb unloading and radiation both cause increases in oxidative stress [17]. Although animal studies in both microgravity environment and with radiation exposure have given selective evidence about changes in osteoclast functions, the fusion mechanism for the differentiation to osteoclasts, as influenced by microgravity and/or radiation exposure in the space environment is still poorly understood.

Osteoclasts are multinucleated giant cells formed by a group of mononuclear osteoclast precursor cells fusing in a differentiation process [18]. These monocyte-macrophage cells fuse to form multinucleated osteoclasts, which act upon bone surfaces for effective bone resorption. Cell fusion is a complex progress that involves several genes. The dendritic cell-specific transmembrane protein (DCSTAMP) and osteoclast stimulatory transmembrane protein (OCSTAMP) have been discovered as fusogens for osteoclast differentiation [19–21]. Along with fusion, OCSTAMP appears to be necessary for optimal bone resorption [20]. The connective tissue growth factor (CCN2/CTGF) protein is a matricellular protein involved in intercellular signaling and plays an important role in skeletal development [22]. CCN2/CTGF expression during osteoclastogenesis promotes osteoclast formation via induction of and interaction with DCSTAMP [23]. Osteoclasts precursors differentiate into mature multinucleated osteoclasts after stimulation with macrophage colony stimulating factor (M-CSF) and receptor activator of nuclear factor-κB ligand (RANKL). RANKL is produced by bone marrow stromal cells. Immune cells can bind to receptor activator of nuclear factor-κB (RANK) present on the surface of osteoclast precursors and trigger cascade of downstream signaling for osteoclast function [24]. RANKL is also responsible for the formation of heterogeneous population of DCSTAMP[lo] and DCSTAMP[hi], which allow for the fusion process to occur between multinucleated cells [19].

Our present study made use of the RAW 264.7 murine cell line, which is a pure macrophage/monocytic and has a pre-OC population. The cells can develop into highly bone-resorptive osteoclasts upon RANKL stimulation [25,26]. Microgravity on the ground was simulated by culturing the cells in rotating wall vessels (RWV) [27]. This microgravity-modeled system has been used for several cell types and the studies performed have revealed strikingly similar results to those obtained during spaceflight [28]. A Cs-137 γ source was used to deliver radiation to the cells.

This study was aimed specifically at investigating the combined effects of microgravity and radiation exposure on osteoclast fusion.

2. Results

2.1. Cell Growth

The number of cells/mL was determined after allowing 5-day cell culture with an initial concentration of 3×10^3 cells/mL. The cell concentration decreased significantly (one-way ANOVA: $p = 0.0091$ for radiation only and $p = 0.0027$ for microgravity and radiation conditions) as radiation dose increased. Furthermore, the assessed cell concentration was significantly (two way ANOVA: $p < 0.0001$) lower (0.7–1.0 $\times 10^4$ cells/mL) when cells were previously incubated in simulated microgravity, indicating that microgravity itself negatively affected cell growth (Figure 1). Although cells grew slower after 1 Gy, the cell concentration reached 2.1×10^4 cells/mL, which was 65.6% of the non-radiated cells.

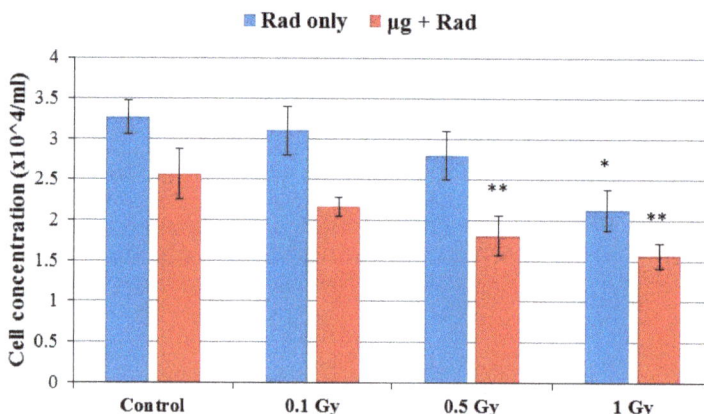

Figure 1. Concentration of RAW264.7 cells after exposure to different radiation doses. The number of cells decreased with increasing radiation doses. Cells previously cultured under simulated microgravity grew significantly slower (two-way ANOVA: $p < 0.0001$). Error bars mean standard deviation (SD) from three replicates. One-way ANOVA: $p = 0.0038$ for radiation only and $p = 0.0026$ for radiation + microgravity. Stars mean Dunnett's multiple comparison test: * $p < 0.05$ and ** $p < 0.01$, compared to the corresponding control.

2.2. Simulated Microgravity Increases Osteoclast Fusion

RAW 264.7 monocyte/macrophage cells were cultured in RWV or in static condition, and then treated with or without differentiation factor RANKL (25 or 50 ng/mL) for 5 days for the osteoclast fusion index. As shown in Figure 2, RANKL is needed for the multinucleated cell (MNC) formation. Although RANKL of 25 and 50 ng/mL induced similar numbers of MNCs containing 3–9 nuclei, the higher concentration apparently promoted more GMCs (Figure 2). For both concentrations, cells in the RWV culture condition demonstrated a significant three-fold increase (two-way ANOVA: $p < 0.0001$) in GMCs having 10 or more nuclei when compared to static cells. Furthermore, daily observation of the cells indicated that cell fusion in simulated microgravity started from 24 h of the culture, whereas in the static cells multinucleation began on Day 3. Our data support previous reports that microgravity stimulates increased osteoclast differentiation [26,29]. For studies of the combined radiation and microgravity effects, a RANKL concentration of 50 ng/mL was used.

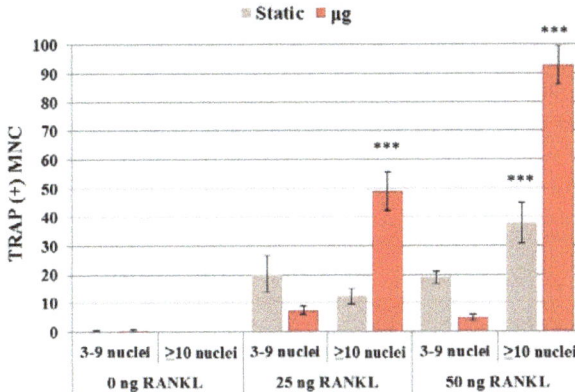

Figure 2. RANKL-dependent multinucleated cells formation under static and simulate microgravity. multinucleated cells (MNCs) with ≥ 10 nuclei significantly increases in microgravity compared to static condition (two-way ANOVA: $p < 0.0001$). The number of MNCs containing ≥ 10 nuclei increased significantly with increasing RANKL concentration in both static (one-way ANOVA: $p = 0.0002$) and microgravity (one-way ANOVA: $p = 0.0001$) conditions. Stars mean statistical significance compared to the corresponding control using Dunnett´s multiple comparison test (*** $p < 0.005$).

2.3. Radiation Exposure Increases Osteoclast Fusion

Without simulated microgravity culture, radiation alone increased osteoclast differentiation. As shown in Figure 3, γ rays of 0.1, 0.5, and 1.0 Gy had little impact on the number of MNCs containing 3–9 nuclei. However, the number of GMCs increased significantly even for doses as low as 0.1 Gy, The GMC number appeared to peak at doses of around 0.5 Gy and decreased at doses of 1 Gy.

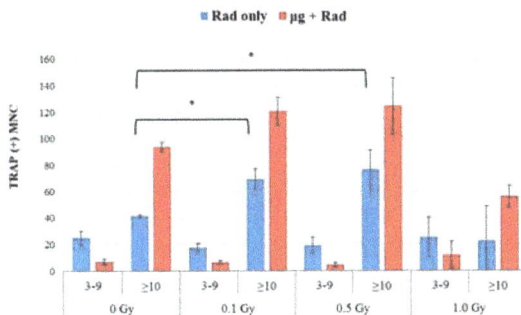

Figure 3. Induction of osteoclast fusion after radiation exposure static (blue bars) and under microgravity (red bars). 0.1 Gy and 0.5 Gy radiation significantly (* $p < 0.05$ *t*-test compared to 0 Gy) stimulated osteoclast fusion, but not 1 Gy. The number of multinucleated cells containing ≥ 10 nuclei increased significantly in radiation + simulated microgravity conditions compared with radiation alone (two-way ANOVA: $p = 0.0001$). Error bars mean SD from three independent experiments.

2.4. Effects of Combined Radiation Exposure and Simulated Microgravity

In RAW 264.7 cells cultured in simulated microgravity, radiation exposure had little impact on the number of MNCs containing 3–9 nuclei, similar to static culture controls (Figure 3). The number of GMCs, however, were higher in RWV cells in comparison to the static culture controls, for both the irradiated cells and non-irradiated controls. The GMC number after combined radiation and microgravity exposures also peaked at doses of around 0.1–0.5 Gy.

2.5. Osteoclast Fusion Genes Up-Regulated in Microgravity

RNAs collected 5 days after osteoclast culture were analyzed for *Trap*, *Dcstamp*, *Ocstamp*, and CGTF fusion gene expressions. Simulated microgravity alone increased the *Trap* gene expression in comparison to the static controls (Figure 4), in agreement with previous reports [26]. Radiation alone also activated *Trap* expressions, even at doses of 0.1 Gy, which is consistent with the GMC formation in the present study. The expression level of *Trap* was higher in cells after combined exposures to radiation and simulated microgravity in comparison to the cells exposed to radiation alone. *Trap* expressions peaked around 0.1–0.5 Gy for both gravity culture conditions.

Similarly, simulated microgravity alone demonstrated increased *Dcstamp* and *Ocstamp* fusion gene expressions (Figure 4). Radiation alone also upregulated the expression of both genes in a dose dependent manner similar to *Trap* expressions. However, expressions after combined exposure to microgravity and radiation showed that the contribution from microgravity was significantly greater than radiation, particularly for *Ocstamp*. Simulated microgravity and/or radiation exposure had little impact on CCN2/CTGF expression levels (Figure 4).

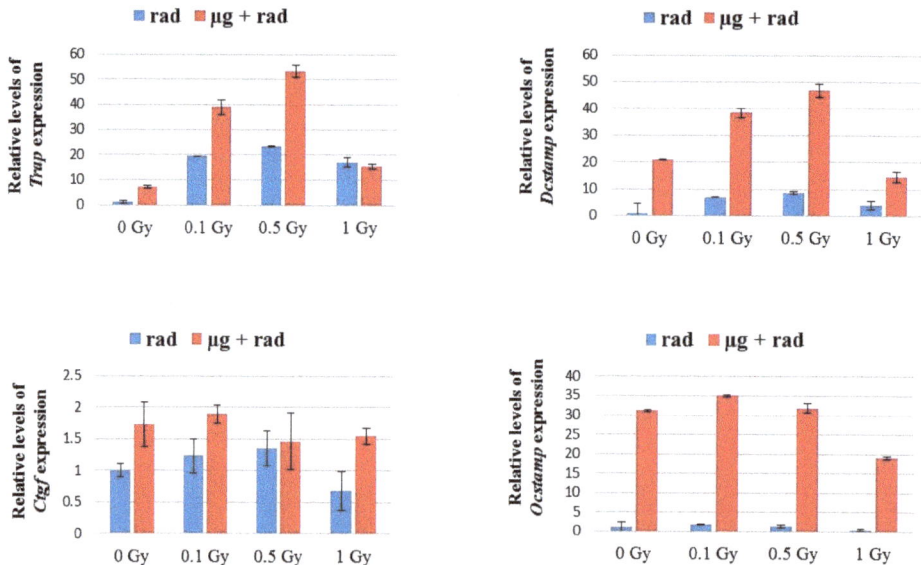

Figure 4. Expression of *Trap*, *Ocstamp*, *Dcstamp*, and *Ctgf* genes in response to different doses of gamma irradiation under the static (blue bars) and simulated microgravity (red bars) conditions. The increase in gene expression peaked at doses of 0.1–0.5 Gy. For all genes, except for *Ctgf*, expression was significantly higher in radiation + microgravity compared to radiation alone (two-way ANOVA of triplicates per each condition and dose; $p = 0.0001$).

3. Discussion

Microgravity and cosmic radiation are two of the most recognized environmental stress factors experienced during space travel. Microgravity is known to cause bone loss, but whether and how space radiation contributes to the potentially deleterious effects is still unclear. Loss of bone volume has been reported for low-LET radiation at doses of 1 Gy or above [30,31], and for high-LET radiation at doses below 1 Gy [16,32]. X-rays [33] and γ rays [34] have also been reported to increase osteoclast numbers in animals.

In this study, we investigated the formation of multinucleated osteoclast cells after γ irradiation at a range of doses between 0.1 and 1 Gy in RAW 264.7 cells. Analysis of survival of the cells indicated a typical dose response, consistent with the reported study showing that these cells are radiosensitive [35]. The cell concentration after 1 Gy γ irradiation was about 60% of the non-irradiated samples after a 5-day culture. However, correcting the number of quantified GMCs for radiation-induced decreased cell growth did not changed the output significantly, suggesting that the results presented were not primarily due to the cell killing effects (Figure 1). The dose response for the formation of MNCs was interesting. Radiation alone apparently stimulated MNC formation at doses as low as 0.1 Gy, particularly for GMCs that are relevant to bone resorption. GMC formation peaked at doses around 0.5 Gy and decreased for higher doses (Figure 2). The morphological data are consistent with the gene expressions for *Trap*, *Ocstamp* and *Dcstamp*, as shown in Figure 4. It should be noted that the window of doses for enhancement of MNCs between 0.1–0.5 Gy is relevant to space radiation exposures. In RAW 264.7 cells, γ rays have also been shown to promote osteoclast function, but at a dose of 2 Gy [36].

While the formation of MNCs under simulated microgravity for RAW 264.7 cells have been reported previously [7,26], our study was the first to investigate the combined effects of microgravity and radiation for this cell type. The present study confirmed that simulated microgravity alone stimulated the formation of MNCs, but the level of MNCs depended apparently on the concentration of RANKL (Figure 1) and the duration of osteoclast culture [26]. Osteoclasts in the present study were quantified by GMCs containing 10 or more nuclei. At 50 ng/mL concentration of RANKL, simulated microgravity alone would double the number of GMCs over the background (Figures 1 and 3). With combined exposure to radiation and microgravity, the number of GMCs increased as the dose of ionizing radiation increased, reaching a peak for doses around 0.1–0.5 Gy and decreased as the dose increased further beyond 0.5 Gy (Figure 3). Bone loss under partial weight bearing is another area of research interest [37,38] but is not addressed in the present study.

One of the fundamental questions in space biology research is whether the combined biological effects of microgravity and exposure to cosmic radiation are synergistic. While studies addressing this question have been carried out for half a century in space or using simulated microgravity on the ground, the reported results have been conflicting, at least for DNA damage response endpoints [15,39]. With regard to bone loss, the combined effects have been reported in studies using mostly rodents with hindlimbs elevated to simulate the effects of skeletal unloading (HU) while being exposed to radiation [1,16,30,32,40,41]. Some of these studies reported the radio-sensitizing effects of hindlimb unloading in some, but not all bones. To determine possible synergism for MNC formation, we present, in Figure 5, the number of GMCs by radiation alone, and combined radiation and microgravity, against the predicted number based on the additive effects. It is shown that microgravity alone induced about 52(=94 − 42) GMCs per dish in comparison to the background (Figure 5, 0 Gy). The predicted number based on additive effects (gray bar in Figure 5) was then the sum of this number and the number of GMCs for irradiated cells. For all three doses of 0.1, 0.5, and 1 Gy, the GMC number in cells after combined exposure to microgravity and radiation agreed well with the prediction, suggesting that the effects of these two factors were additive. It is interesting to note that microgravity enhanced GMCs (~52 MNCs/dish) at levels that are more than that of radiation-enhanced (~33 MNCs/dish), even at the peak radiation dose of 0.1 or 0.5 Gy, suggesting that microgravity would contribute more to osteoclast differentiation than radiation in the presence of these two factors in space.

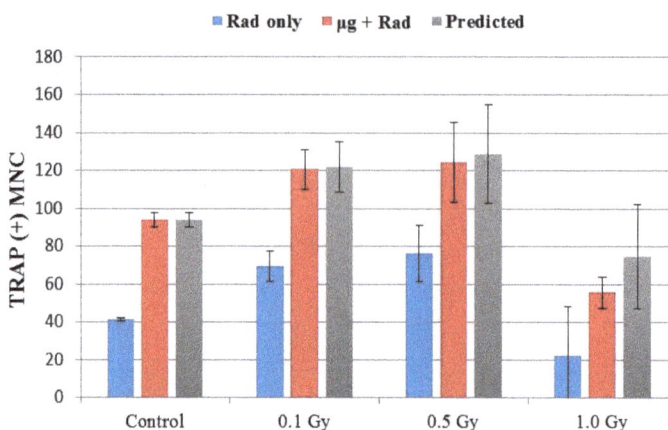

Figure 5. Osteoclast fusion prediction. Blue and red bars represent MNCs containing ≥ 10 nuclei analyzed in Figure 3 for radiation alone and combined radiation and microgravity exposures, respectively. Microgravity alone induced about 52 GMCs (=94 − 42) per dish in comparison to the background (0 Gy). The predicted number based on additive effects of radiation and microgravity (gray bars) is then the sum of this number and the number of GMCs after irradiation alone. For all three doses of 0.1, 0.5, and 1 Gy, the combined effects (red bars) agreed well with the prediction calculated based on the additive effects (gray bars).

In the present study, we assessed the expression of key genes involved in osteoclasts differentiation. No significant dysregulation of the *Ctfg* gene was observed, either for radiation alone or combined with simulated microgravity (Figure 4). Expressions of *Dcstamp* and *Ocstamp* peaked for doses of 0.1–0.5 Gy, but the fold changes for radiation alone were significantly lower than changes caused by microgravity (Figure 4). Such differences were particularly pronounced for *Ocstamp*. Expressions of *Ocstamp* and *Dcstamp* under simulated microgravity alone have been reported previously [7,26,42]. However, the observation that the microgravity effect on *Ocstamp* and *Dcstamp* expression is higher than the radiation effect suggests that radiation might not increase the synthesis of adhesion molecules as much as microgravity, and this could explain the lower number of GMCs formed after radiation when compared to microgravity. However, the protein levels and functionality of *Ocstamp* and *Dcstamp* need to be addressed in order to enforce these findings. Furthermore, we also analyzed expressions of *Trap*. TRAP is expressed in osteoclasts and is able to degrade skeletal phosphoproteins including osteopontin (OPN) [43]; therefore, TRAP has been associated with bone resorption [44]. In the present study, TRAP expression was induced by radiation, microgravity, and the combination of both (Figure 4). However, in contrast to *Ocstamp* and *Dcstamp*, upregulation of TRAP was higher in irradiated cells than in cells exposed to microgravity. Thus, it might be the case that radiation predominantly affects bone resorption while microgravity affects the formation of GMCs.

4. Materials and Methods

4.1. Cell Culture in Simulated Microgravity and γ Irradiation

RAW 264.7 murine macrophage cells were purchased from American Type Culture Collection (Manassas, VA, USA) and maintained in static condition with Delbecco's Modified Eagle's medium (DMEM) with 10% fetal bovine serum (FBS) and 1% penicillin/streptomycin (Invitrogen, Carlsbad, CA, USA) in a humidified incubator at 37 °C with 5% CO_2. NASA-developed ground-based rotating wall vessels (RWVs) were used to simulate the microgravity (μg) conditions. A Cs-137 γ source at NASA Johnson Space Center was used to deliver radiation of varying doses. Figure 6 shows the experimental timeline. RAW 264.7 cells suspended in complete media were cultured in rotation at 20 rpm or in a static condition for 48 h in a humidified incubator at 37 °C with 5% CO_2. The cells were then removed from the incubator, and exposed to γ rays at doses of 0.1, 0.5, or 1 Gy. After irradiation, the cells were cultured for the formation of multinucleated osteoclasts. This study focused on doses of 1 Gy or lower that are relevant to space radiation exposure.

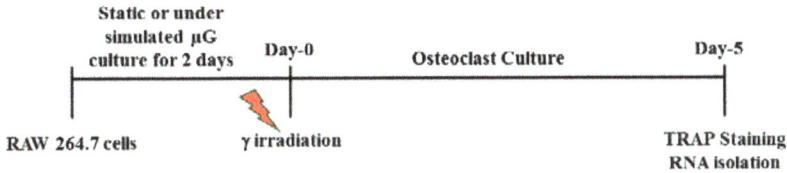

Figure 6. Timeline of the experiment. RAW 264.7 cells were cultured in static or under simulated microgravity for 48 h before exposed to varying doses of γ rays. Cells were then cultured in the presence of RANKL for 5 days for quantification of osteoclast differentiation and for gene expression analysis.

4.2. Cell Concentration and Viability

To determine viability of the cells after radiation exposure, irradiated and non-irradiated RAW 264.7 cells were seeded in the wells of 96-well plates at the density of 3×10^3 cells/mL. Cells were then incubated in a 37 °C for 5 days in DMEM supplemented with 10% FBS and 1% L-glutamine. At the end of the 5-day culture period, the number of cells in a well was measured with a Coulter counter. Viability of irradiated cells with prior culture in simulated microgravity was also analyzed.

4.3. Osteoclast Culture

RAW 264.7 cells were cultured under static or stimulated microgravity conditions, and then exposed to different doses of γ rays. After exposure, cells previously cultured under different gravity conditions were further incubated under a static condition in the presence of RANKL (50 ng/mL) and M-CSF (10 ng/mL) (R & D Systems, Minneapolis, MN, USA) in order to stimulate osteoclast differentiation. After 5 days, cells were fixed and stained for tartrate resistant acid phosphatase (TRAP) activities using an Acid Phosphatase, Leukocyte (TRAP) Kit (Sigma, St Luis, MO, USA). TRAP positive multinucleated osteoclasts were scored with a Zeiss microscope. Figure 7 shows the examples of TRAP positive multinucleated cells (MNCs). To quantify osteoclast differentiation, MNCs were grouped by the number of nuclei in a cell. Giant multinucleated cells (GMCs) were classified by those containing 10 or more nuclei [45].

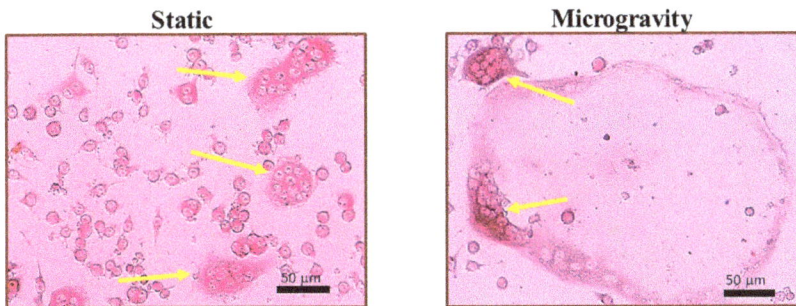

Figure 7. Multinucleated osteoclasts (pointed by arrows) formed in cells after a 5-day osteoclast culture in RANKL. Cells were maintained in static conditions or under simulated microgravity for 48 h immediately prior to osteoclast culture.

For the induction of MNCs, non-irradiated RAW 264.7 cells were incubated under static or rotating conditions with two different concentrations of RANKL: 25 ng/mL and 50 ng/mL.

4.4. Real-Time Polymerase Chain Reaction (RT-PCR) Analysis

At the end of 5 days of osteoclast culture, total RNA was isolated from cells previously cultured under the static or simulated microgravity conditions, with or without radiation exposure. Total RNA was reverse transcribed using random hexamers and Moloney murine leukemia virus reverse transcriptase (Invitrogen, Carlsbad, CA, USA). The resulting cDNAs were then subject to quantitative real-time reverse transcription polymerase chain reaction using specific primers for DCSTAMP, OCSTAMP, CCN2/CTGF, and TRAP as osteoclast markers. Relative levels of gene expressions were normalized in all the samples analyzed with respect to the levels of GAPDH amplification. Primers were designed and ordered from Invitrogen (Carlsbad, CA, USA), and the sequences were as follows: TRAP 5′-CGAC CATTGT TAGCCACATACG-3′ (sense) and 5′-TCGTCCTGAAGATACTGCAGGTT-3′ (anti-sense); CCN2/CTGF 5′-CCA CCCGAGTTACCAATGAC-3′ (sense) and 5′-GTGCAGCCAGAAAGCTCA-3′ (anti-sense); DCSTAMP 5′-CTAGCTGGCTGGACTTCATCC-3′ (sense) and 5′-TCATGCTGTCTAGG AGACCTC-3′ (anti-sense); GAPDH 5′-GCCAAA AGGGTCATCATCTC-3′ (sense) and 5′-GTCTTC TGGGTGGCAGTGAT-3′ (anti-sense); OCSTAMP 5′-GGCAGCCACGGAACAC-3′ (sense) and 5′-GCAGGGGGTCCCAAAG-3′ (anti-sense). RT-PCR analysis were performed at 94 °C for 4 min, followed by 35 cycles of amplification at 94 °C for 30 s, 58 °C for 1 min, 72 °C for 2 min, and 72 °C for 10 min as the final elongation step.

4.5. Statistical Analysis

The experiments were performed three times independently. In each of the experiments, cells were cultured in three RWV or dishes for each of the gravity and radiation dose conditions. Significant differences were analyzed by one-way ANOVA for comparing effects within one group and two-way ANOVA for comparing two groups. We used Prism version 6 as statistical software (GraphPad Software, La Jolla, CA, USA). For analyzing the effect of radiation on osteoclast fusion (Figure 3), a *t*-test was applied comparing each dose with 0 Gy.

Acknowledgments: This work was supported in part by the NASA Human Research Program and the University of Houston Institute for Space Systems Operations (ISSO) Program.

Author Contributions: Srinivasan Shanmugarajan and Honglu Wu conceived and designed the experiments; Srinivasan Shanmugarajan, Ye Zhang and Larry H. Rohde performed and supported the experiments; Maria Moreno-Villanueva and Ryan Clanton analyzed the data; Jean D. Sibong and Govindarajan T. Ramesh interpreted the data; Srinivasan Shanmugarajan, Maria Moreno-Villanueva, Ryan Clanton and Honglu Wu wrote and edited the paper.

Conflicts of Interest: The authors declare no conflict of interest.

References

1. Alwood, J.S.; Ronca, A.E.; Mains, R.C.; Shelhamer, M.J.; Smith, J.D.; Goodwin, T.J. From the bench to exploration medicine: NASA life sciences translational research for human exploration and habitation missions. *NPJ Microgravity* **2017**, *3*, 5. [CrossRef] [PubMed]
2. Zhang, Y.; Moreno-Villanueva, M.; Krieger, S.; Ramesh, G.T.; Neelam, S.; Wu, H. Transcriptomics, NF-κB Pathway, and Their Potential Spaceflight-Related Health Consequences. *Int. J. Mol. Sci.* **2017**, *18*, 1166. [CrossRef] [PubMed]
3. Lang, T.; LeBlanc, A.; Evans, H.; Lu, Y.; Genant, H.; Yu, A. Cortical and trabecular bone mineral loss from the spine and hip in long-duration spaceflight. *J. Bone Miner. Res.* **2004**, *19*, 1006–1012. [CrossRef] [PubMed]
4. Sibonga, J.D. Spaceflight-induce? Bone loss: Is there an osteoporosis risk? *Curr. Osteoporos. Rep.* **2013**, *11*, 92–98. [CrossRef] [PubMed]

5. Hughes-Fulford, M.; Tjandrawinata, R.; Fitzgerald, J.; Gasuad, K.; Gilbertson, V. Effects of microgravity on osteoblast growth. *Gravit. Space Biol. Bull.* **1998**, *11*, 51–60. [PubMed]

6. Chatani, M.; Mantoku, A.; Takeyama, K.; Abduweli, D.; Sugamori, Y.; Aoki, K.; Ohya, K.; Suzuki, H.; Uchida, S.; Sakimura, T.; et al. Microgravity promotes osteoclast activity in medaka fish reared at the international space station. *Sci. Rep.* **2015**, *5*, 14172. [CrossRef] [PubMed]

7. Sambandam, Y.; Baird, K.L.; Stroebel, M.; Kowal, E.; Balasubramanian, S.; Reddy, S.V. Microgravity Induction of TRAIL Expression in Preosteoclast Cells Enhances Osteoclast Differentiation. *Sci. Rep.* **2016**, *6*, 25143. [CrossRef] [PubMed]

8. Nabavi, N.; Khandani, A.; Camirand, A.; Harrison, R.E. Effects of microgravity on osteoclast bone resorption and osteoblast cytoskeletal organization and adhesion. *Bone* **2011**, *49*, 965–974. [CrossRef] [PubMed]

9. Vico, L.; Collet, P.; Guignandon, A.; Lafage-Proust, M.H.; Thomas, T.; Rehaillia, M.; Alexandre, C. Effects of long-term microgravity exposure on cancellous and cortical weight-bearing bones of cosmonauts. *Lancet* **2000**, *355*, 1607–1611. [CrossRef]

10. Durante, M.; Cucinotta, F.A. Heavy ion carcinogenesis and human space exploration. *Nat. Rev. Cancer* **2008**, *8*, 465–472. [CrossRef] [PubMed]

11. Willey, J.S.; Lloyd, S.A.J.; Nelson, G.A.; Bateman, T.A. Ionizing Radiation and Bone Loss: Space Exploration and Clinical Therapy Applications. *Clin. Rev. Bone Miner. Metab.* **2011**, *9*, 54–62. [CrossRef] [PubMed]

12. Hamilton, S.A.; Pecaut, M.J.; Gridley, D.S.; Travis, N.D.; Bandstra, E.R.; Willey, J.S.; Nelson, G.A.; Bateman, T.A. A murine model for bone loss from therapeutic and space-relevant sources of radiation. *J. Appl. Physiol.* **2006**, *101*, 789–793. [CrossRef] [PubMed]

13. Wright, L.E.; Buijs, J.T.; Kim, H.S.; Coats, L.E.; Scheidler, A.M.; John, S.K.; She, Y.; Murthy, S.; Ma, N.; Chin-Sinex, H.J.; et al. Single-Limb Irradiation Induces Local and Systemic Bone Loss in a Murine Model. *J. Bone Miner. Res.* **2015**, *30*, 1268–1279. [CrossRef] [PubMed]

14. Willey, J.S.; Lloyd, S.A.J.; Nelson, G.A.; Bateman, T.A. Space Radiation and Bone Loss. *Gravit. Space Biol. Bull.* **2011**, *25*, 14–21. [PubMed]

15. Moreno-Villanueva, M.; Wong, M.; Lu, T.; Zhang, Y.; Wu, H. Interplay of space radiation and microgravity in DNA damage and DNA damage response. *NPJ Microgravity* **2017**, *3*, 14. [CrossRef] [PubMed]

16. Macias, B.R.; Lima, F.; Swift, J.M.; Shirazi-Fard, Y.; Greene, E.S.; Allen, M.R.; Fluckey, J.; Hogan, H.A.; Braby, L.; Wang, S.; et al. Simulating the Lunar Environment: Partial Weightbearing and High-LET Radiation-Induce Bone Loss and Increase Sclerostin-Positive Osteocytes. *Radiat. Res.* **2016**, *186*, 254–263. [CrossRef] [PubMed]

17. Kondo, H.; Yumoto, K.; Alwood, J.S.; Mojarrab, R.; Wang, A.; Almeida, E.A.; Searby, N.D.; Limoli, C.L.; Globus, R.K. Oxidative stress and gamma radiation-induced cancellous bone loss with musculoskeletal disuse. *J. Appl. Physiol.* **2010**, *108*, 152–161. [CrossRef] [PubMed]

18. Takahashi, A.; Kukita, A.; Li, Y.J.; Zhang, J.Q.; Nomiyama, H.; Yamaza, T.; Ayukawa, Y.; Koyano, K.; Kukita, T. Tunneling nanotube formation is essential for the regulation of osteoclastogenesis. *J. Cell. Biochem.* **2013**, *114*, 1238–1247. [CrossRef] [PubMed]

19. Mensah, K.A.; Ritchlin, C.T.; Schwarz, E.M. RANKL induces heterogeneous DC-STAMP(lo) and DC-STAMP(hi) osteoclast precursors of which the DC-STAMP(lo) precursors are the master fusogens. *J. Cell. Physiol.* **2010**, *223*, 76–83. [CrossRef] [PubMed]

20. Wisitrasameewong, W.; Kajiya, M.; Movila, A.; Rittling, S.; Ishii, T.; Suzuki, M.; Matsuda, S.; Mazda, Y.; Torruella, M.R.; Azuma, M.M.; et al. DC-STAMP Is an Osteoclast Fusogen Engaged in Periodontal Bone Resorption. *J. Dent. Res.* **2017**, *96*, 685–693. [CrossRef] [PubMed]

21. Witwicka, H.; Hwang, S.Y.; Reyes-Gutierrez, P.; Jia, H.; Odgren, P.E.; Donahue, L.R.; Birnbaum, M.J.; Odgren, P.R. Studies of OC-STAMP in Osteoclast Fusion: A New Knockout Mouse Model, Rescue of Cell Fusion, and Transmembrane Topology. *PLoS ONE* **2015**, *10*, e0128275. [CrossRef] [PubMed]

22. Chen, C.C.; Lau, L.F. Functions and mechanisms of action of CCN matricellular proteins. *Int. J. Biochem. Cell Biol.* **2009**, *41*, 771–783. [CrossRef] [PubMed]

23. Nishida, T.; Emura, K.; Kubota, S.; Lyons, K.M.; Takigawa, M. CCN family 2/connective tissue growth factor (CCN2/CTGF) promotes osteoclastogenesis via induction of and interaction with dendritic cell-specific transmembrane protein (DC-STAMP). *J. Bone Miner. Res.* **2011**, *26*, 351–363. [CrossRef] [PubMed]

24. Takayanagi, H. New immune connections in osteoclast formation. *Ann. N. Y. Acad. Sci.* **2010**, *1192*, 117–123. [CrossRef] [PubMed]

25. Collin-Osdoby, P.; Osdoby, P. RANKL-mediated osteoclast formation from murine RAW 264.7 cells. *Methods Mol. Biol.* **2012**, *816*, 187–202. [CrossRef] [PubMed]
26. Sambandam, Y.; Blanchard, J.J.; Daughtridge, G.; Kolb, R.J.; Shanmugarajan, S.; Pandruvada, S.N.; Bateman, T.A.; Reddy, S.V. Microarray profile of gene expression during osteoclast differentiation in modelled microgravity. *J. Cell. Biochem.* **2010**, *111*, 1179–1187. [CrossRef] [PubMed]
27. Pellis, N.R.; Goodwin, T.J.; Risin, D.; McIntyre, B.W.; Pizzini, R.P.; Cooper, D.; Baker, T.L.; Spaulding, G.F. Changes in gravity inhibit lymphocyte locomotion through type I collagen. *In Vitro Cell. Dev. Biol. Anim.* **1997**, *33*, 398–405. [CrossRef] [PubMed]
28. Carmeliet, G. Bone Cell Biology in Microgravity. 1992. Avaliable online: https://lirias.kuleuven.be/handle/123456789/216494 (accessed on 7 October 2017).
29. Saxena, R.; Pan, G.; Dohm, E.D.; McDonald, J.M. Modeled microgravity and hindlimb unloading sensitize osteoclast precursors to RANKL-mediated osteoclastogenesis. *J. Bone Miner. Metab.* **2011**, *29*, 111–122. [CrossRef] [PubMed]
30. Lloyd, S.A.; Bandstra, E.R.; Willey, J.S.; Riffle, S.E.; Tirado-Lee, L.; Nelson, G.A.; Pecaut, M.J.; Bateman, T.A. Effect of proton irradiation followed by hindlimb unloading on bone in mature mice: A model of long-duration spaceflight. *Bone* **2012**, *51*, 756–764. [CrossRef] [PubMed]
31. Turner, R.T.; Iwaniec, U.T.; Wong, C.P.; Lindenmaier, L.B.; Wagner, L.A.; Branscum, A.J.; Menn, S.A.; Taylor, J.; Zhang, Y.; Wu, H.; et al. Acute exposure to high dose gamma-radiation results in transient activation of bone lining cells. *Bone* **2013**, *57*, 164–173. [CrossRef] [PubMed]
32. Yumoto, K.; Globus, R.K.; Mojarrab, R.; Arakaki, J.; Wang, A.; Searby, N.D.; Almeida, E.A.; Limoli, C.L. Short-term effects of whole-body exposure to (56)fe ions in combination with musculoskeletal disuse on bone cells. *Radiat. Res.* **2010**, *173*, 494–504. [CrossRef] [PubMed]
33. Willey, J.S.; Livingston, E.W.; Robbins, M.E.; Bourland, J.D.; Tirado-Lee, L.; Smith-Sielicki, H.; Bateman, T.A. Risedronate prevents early radiation-induced osteoporosis in mice at multiple skeletal locations. *Bone* **2010**, *46*, 101–111. [CrossRef] [PubMed]
34. Kondo, H.; Searby, N.D.; Mojarrab, R.; Phillips, J.; Alwood, J.; Yumoto, K.; Almeida, E.A.; Limoli, C.L.; Globus, R.K. Total-body irradiation of postpubertal mice with (137)Cs acutely compromises the microarchitecture of cancellous bone and increases osteoclasts. *Radiat. Res.* **2009**, *171*, 283–289. [CrossRef] [PubMed]
35. Zhang, J.; Wang, Z.; Wu, A.; Nie, J.; Pei, H.; Hu, W.; Wang, B.; Shang, P.; Li, B.; Zhou, G. Differences in responses to X-ray exposure between osteoclast and osteoblast cells. *J. Radiat. Res.* **2017**. [CrossRef] [PubMed]
36. Yang, B.; Zhou, H.; Zhang, X.D.; Liu, Z.; Fan, F.Y.; Sun, Y.M. Effect of radiation on the expression of osteoclast marker genes in RAW264.7 cells. *Mol. Med. Rep.* **2012**, *5*, 955–958. [CrossRef] [PubMed]
37. Ellman, R.; Spatz, J.; Cloutier, A.; Palme, R.; Christiansen, B.A.; Bouxsein, M.L. Partial reductions in mechanical loading yield proportional changes in bone density, bone architecture, and muscle mass. *J. Bone Miner. Res.* **2013**, *28*, 875–885. [CrossRef] [PubMed]
38. Swift, J.M.; Lima, F.; Macias, B.R.; Allen, M.R.; Greene, E.S.; Shirazi-Fard, Y.; Kupke, J.S.; Hogan, H.A.; Bloomfield, S.A. Partial weight bearing does not prevent musculoskeletal losses associated with disuse. *Med. Sci. Sports Exerc.* **2013**, *45*, 2052–2060. [CrossRef] [PubMed]
39. Lu, T.; Zhang, Y.; Kidane, Y.; Feiveson, A.; Stodieck, L.; Karouia, F.; Ramesh, G.; Rohde, L.; Wu, H. Cellular responses and gene expression profile changes due to bleomycin-induced DNA damage in human fibroblasts in space. *PLoS ONE* **2017**, *12*, e0170358. [CrossRef] [PubMed]
40. Ghosh, P.; Behnke, B.J.; Stabley, J.N.; Kilar, C.R.; Park, Y.; Narayanan, A.; Alwood, J.S.; Shirazi-Fard, Y.; Schreurs, A.S.; Globus, R.K.; et al. Effects of High-LET Radiation Exposure and Hindlimb Unloading on Skeletal Muscle Resistance Artery Vasomotor Properties and Cancellous Bone Microarchitecture in Mice. *Radiat. Res.* **2016**, *185*, 257–266. [CrossRef] [PubMed]
41. Krause, A.R.; Speacht, T.L.; Zhang, Y.; Lang, C.H.; Donahue, H.J. Simulated space radiation sensitizes bone but not muscle to the catabolic effects of mechanical unloading. *PLoS ONE* **2017**, *12*, e0182403. [CrossRef] [PubMed]
42. Makihira, S.; Kawahara, Y.; Yuge, L.; Mine, Y.; Nikawa, H. Impact of the microgravity environment in a 3-dimensional clinostat on osteoblast- and osteoclast-like cells. *Cell Biol. Int.* **2008**, *32*, 1176–1181. [CrossRef] [PubMed]

Int. J. Mol. Sci. **2017**, *18*, 2443

43. Hayman, A.R. Tartrate-resistant acid phosphatase (TRAP) and the osteoclast/immune cell dichotomy. *Autoimmunity* **2008**, *41*, 218–223. [CrossRef] [PubMed]

44. Kirstein, B.; Chambers, T.J.; Fuller, K. Secretion of tartrate-resistant acid phosphatase by osteoclasts correlates with resorptive behavior. *J. Cell. Biochem.* **2006**, *98*, 1085–1094. [CrossRef] [PubMed]

45. Kurihara, N.; Suda, T.; Miura, Y.; Nakauchi, H.; Kodama, H.; Hiura, K.; Hakeda, Y.; Kumegawa, M. Generation of osteoclasts from isolated hematopoietic progenitor cells. *Blood* **1989**, *74*, 1295–1302. [PubMed]

International Journal of
Molecular Sciences

|MDPI|

Review

The Impact of Oxidative Stress on the Bone System in Response to the Space Special Environment

Ye Tian, Xiaoli Ma, Chaofei Yang, Peihong Su, Chong Yin and Ai-Rong Qian *

Key Laboratory for Space Bioscience and Biotechnology, Bone Metabolism Lab, School of Life Sciences,
Northwestern Polytechnical University, Xi'an 710072, China; tianye@nwpu.edu.cn (Y.T.);
xiaoli225@mail.nwpu.edu.cn (X.M.); chaofei-yang2015@mail.nwpu.edu.cn (C.Y.); suph@mail.nwpu.edu.cn (P.S.);
yinchong42@mail.nwpu.edu.cn (C.Y.)
* Correspondence: qianair@nwpu.edu.cn

Received: 31 August 2017; Accepted: 9 October 2017; Published: 12 October 2017

Abstract: The space special environment mainly includes microgravity, radiation, vacuum and extreme temperature, which seriously threatens an astronaut's health. Bone loss is one of the most significant alterations in mammalians after long-duration habitation in space. In this review, we summarize the crucial roles of major factors—namely radiation and microgravity—in space in oxidative stress generation in living organisms, and the inhibitory effect of oxidative stress on bone formation. We discussed the possible mechanisms of oxidative stress-induced skeletal involution, and listed some countermeasures that have therapeutic potentials for bone loss via oxidative stress antagonism. Future research for better understanding the oxidative stress caused by space environment and the development of countermeasures against oxidative damage accordingly may facilitate human beings to live more safely in space and explore deeper into the universe.

Keywords: oxidative stress; bone loss; microgravity; radiation; countermeasure

---

1. Introduction

After the Moon landing in 1969, humankind never stop exploring the universe. For example, Shenzhou programs and the International Space Station (ISS) that are orbiting around the Earth recruit crew members continuously [1,2], and several research programs have been launched towards the Moon [3,4]. Even the Mars journey has been gradually industrialized [5,6]. We are standing in the space age now. Although space traveling sounds fascinating, it can cause dramatic changes of the human body, especially long-term spaceflights. More and more evidence proves that the space environment negatively affects human physiological functions with the extension of space stays. Gravitational unloading due to microgravity and cosmetic rays are conditions experienced by astronauts during space flight. The medical examinations conducted before, during and after spaceflight have revealed several health issues for space travelers, e.g., cardiovascular dysfunction, disruption in nervous system, and reduced immune function [7–11]. Bone loss induced by microgravity is also a well-documented alteration in astronauts [12–14]. It happens especially on weight-bearing bones and needs a very long duration to recover after returning to earth [15]. In the absence of countermeasures, this change can impact the performance and safety of crew members severely during extravehicular activities, and putting them at high risk of fracture [16]. Bone loss is one of the major obstacles to space exploration for human beings now.

Nowadays, researchers make great efforts to catch the mechanisms hidden behind the physiological alterations of bone during spaceflight and to develop countermeasures accordingly. Russian investigators found reductions in some blood antioxidants and increased lipid peroxidation in human after long-term space flight [17,18]. Urinary excretion of 8-iso-prostaglandin F2α and 8-oxo-7,8 dihydro-2 deoxyguanosine, which are markers of oxidative damage to lipids and DNA respectively, increased during and after long-duration space flight (90 to 180 days) [19]. It means that the balance between oxidant production and antioxidant defenses has been disturbed, and the excessive oxidants may attack DNA and membrane lipids resulting in oxidative damage. The pro-oxidative conditions caused by space environment may contribute to the bone alterations after long space habitation.

In this review, we summarized the oxidative effect to bone caused by microgravity and radiation, and expounded the relationship between oxidative stress and bone formation. The possible mechanisms will be discussed as well. Some prevention countermeasures of bone loss against oxidative injuries will be included too. This manuscript will help to capture the latest research progresses and inspire the possible direction of future studies.

2. Effects of Oxidative Stress on Bone Formation

The redox balance in the human body is maintained delicately, with the balance slightly inclined to oxidants [20]. Reactive oxygen species (ROS) are generated as normal by-products of aerobic metabolism, usually by leakage from the electron transport chain during oxidative phosphorylation in mitochondria [20,21]. The major forms of ROS include the superoxide anions (O_2^-, hydrogen peroxide (H_2O_2) and free radicals such as hydroxyl radicals (OH·). ROS at lower concentrations serve as signaling molecules to activate specific physiologic pathways that control several life processes [22,23]. Meanwhile, elevated levels of ROS can damage proteins, lipids, and DNA, eventually trigger oxidative stress and leading to cell death [24,25]. Oxidative damage to bio-macromolecule has been proved in the etiology of a wide variety of acute and chronic diseases, including osteoporosis [26].

It is reported that the increased level of ROS had opposite effects on osteoblast and osteoclast cells. ROS inhibits osteoblast function. It is believed that the increased level of ROS in osteoblast is one critical element of the pathophysiology of bone loss [27–30]. Almeida et al. reported that ROS inhibited osteoblast differentiation and promoted apoptosis [31–33]. ROS achieve this function by activating a small family of transcription factors known as Forkhead box O (FoxO), which contains four members: FoxO1, FoxO3a, FoxO4, and FoxO6 [34]. FoxOs defense ROS by up-regulating free radical scavenging enzymes such as Catalase, manganese superoxide dismutase (Mn-SOD), and glutathione peroxidase-1 (GPx-1) [35]. Importantly, FoxO-mediated transcription requires the binding of β-catenin that is also essential companion for T-cell factor (Tcf) family of transcription factors [36,37]. Without Tcf transcriptional activities, the downstream effects of the Wnt/β-catenin pathway cannot be conducted [37,38]. Thus, by competitive binding to β-catenin, FoxOs antagonizes Wnt/Tcf-mediated transcription after being activated by ROS (Scheme 1) [39]. Regarding the importance of Wnt/β-catenin/Tcf to bone formation, the attenuation of this pathway will inevitably lead to decreased osteogenesis [40]. Several researchers indicated that conditional deletion of FoxOs (FoxO1, FoxO3, FoxO4) in mice osteoblast resulted in a decrease in the number of osteoblasts, the rate of bone formation and bone mass but an increase of osteoblast apoptosis and oxidative stress in bone [41,42].

On the contrary, ROS play crucial roles in osteoclast differentiation and function. By increasing receptor activator of nuclear factor-kappa B ligand (RANKL) production and activating ERK/NF-κB/TNF/interleukin 6, ROS inhibit osteoclast apoptosis and promote osteoclastogenesis [29]. In addition, it is reported that RANKL could suppress the transcriptional activity of FoxOs, loss FoxOs' transcription factor function promoted osteoclast differentiation and survival, because intracellular H_2O_2 accumulation is pivotal for osteoclastogenesis and bone resorption [43]. Therefore, FoxOs are crucial regulators of both osteoblast and osteoclast physiology, and direct mechanistic links between oxidative stress and skeletal involution.

Another oxidative stress-related pathway includes Nrf2/HO-1, which also can adjust cellular ROS via a switch on gene transcription of several antioxidative enzymes such as SOD, Catalase, GPx, etc. [44]. Mitochondrial dynamics [45], endoplasmic reticulum stress pathway [46], and autophagy [47] are also participated in the bone loss induced by oxidative stress.

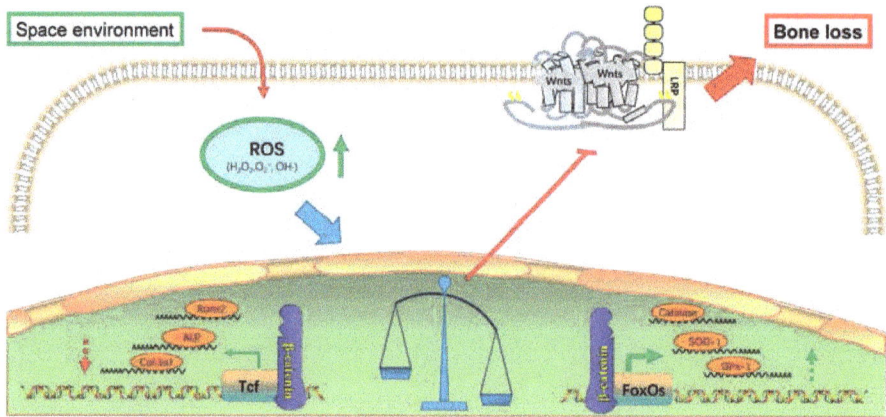

Scheme 1. The scheme of possible mechanism of space environment-induced bone loss: the increment of ROS caused by Space environment antagonizes the skeletal effects of Wnt/β-catenin/Tcf by diverting β-catenin from Tcf—to FoxOs-mediated transcription. LRP: LDL receptor-related proteins.

3. Microgravity Increases Oxidative Stress in Bone System

Microgravity conditions in space cause an imbalance between bone formation and resorption [48–51]. The average rate of aBMD loss is 1–1.5% per month evaluated by dual-energy X-ray absorptionmetry (DXA) scans from preflight and postflight [52]. Bone loss under microgravity conditions is relevant to oxidative injury. Microgravity is considered to increase free radical formation and causes oxidative stress [53–56]. Findings either from the real spaceflight missions or ground-based models (head-down bed rest model and hind-limb unloading rodents) all demonstrated elevated oxidative damage markers and attenuated total antioxidant capacity [53,57,58]. Hind-limb unloading (HLU) rodents, rotary wall vessel bioreactor (RWVB) and Random Positioning Machines (RPMs) are commonly used microgravity models in vivo and in vitro. Xin et al. and Sun et al. both observed that malondialdehyde levels (oxidant marker) were raised but total sulfhydryl content (anti-oxidant marker) descended in femurs of HLU Sprague-Dawley (SD) rats [59,60]. MC3T3-E1 cells that were exposed to RWVB had higher cellular ROS levels but lower differential abilities [59,60]. On the contrary, RWVB treatment-induced ROS generation facilitated osteoclastogenesis of RAW264.7 cells [59,60]. Their findings illustrated that the generation of ROS increased in response to microgravity. The excessive ROS destroyed normal function of osteoblasts but enhanced the osteoclasts' capabilities, which lead to insufficient bone formation and massive bone absorption.

It is believed that the oxidative damage caused by the space environment is related to insufficient nutrition intake and disturbed iron metabolism as well [54,55]. By analyzing blood and urine samples from 23 crew members who participated in missions lasting 50 to 247 days on the ISS, Zwart et al. found serum ferritin was positively correlated with 8-hydroxy-2′-deoxyguanosine ($r = 0.53$, $p < 0.001$) and prostaglandin F2α ($r = 0.26$, $p < 0.001$), which are oxidative damage makers [55]. In addition, they revealed that greater amount of ferritin during flight is accompanied by greater loss in bone mineral density in the total hip ($p = 0.031$), trochanter ($p = 0.006$), hip neck ($p = 0.044$), and ($p = 0.049$) after flight [55]. Their research inspired us that microgravity-induced bone loss may be associated with oxidative stress caused by increased iron store.

Besides iron metabolism, the downregulation of anti-oxidative enzymes like Mn-SOD are also key reasons for oxidative stress-induced bone loss in response to microgravity [61–63]. The deficiency of anti-oxidative enzymes can cause distinct weakness in bone and bone fragility [64], and dysfunctional oxidative defense system will exacerbate bone loss via suppressed osteoblastic abilities during mechanical unloading [65].

In brief, microgravity affects oxidative status of bone in many aspects. Mechanical unloading-induced bone loss is closely associated with increased ROS level in different types of bone cell in response to microgravity. Through disturbing oxidative-antioxidative defense systems, microgravity breaks the equilibrium between bone formation and bone absorption leading to skeletal fragility.

4. Radiation Induces Oxidative Stress in Bone System

In addition to microgravity, cosmic radiation is another predominant feature of the space environment, and it is a strong incentive to oxidative stress [66–69]. To date, direct research in spaceflight about the connection between oxidative stress caused by radiation and bone involution is rare. However, some ground-based study suggested the inhibitory effect of radiation to bone formation. Irradiation suppressed bone-like nodule formation, alkaline phosphatase (ALP) activity and expression of osteoblast markers in MC3T3-E1 cells [70]. Meanwhile, the depletion of antioxidant defense enzymes and accumulation of cellular ROS were observed [70]. A similar phenomenon was also exhibited in bone marrow-derived skeletal cell progenitors after a single dose (1–5 Gy) irradiation (137Cs Gy/min) exposure [71]. In an HLU mouse model, total body gamma irradiation (1 or 2 Gy of 137Cs) to C57BL/6 mice decreased cancellous bone volume fractions in the proximal tibiae and lumbar vertebrae significantly, but increased osteoclast surface 47% in the tibiae [72]. Irradiation to total body also stimulated generation of ROS in marrow cells and promoted cell apoptosis [72]. These results inferred that irradiation may cause oxidative stress and inhibit the osteoblasts' growth and differentiation, but encourage bone absorption. Thus, cosmic radiation may affect critical bone cell functions by stimulating production of ROS, and its suppressive effect to osteoblast involved oxidative stress-mediated activation of Nrf2/HO-1 pathway [70].

5. Countermeasures against Bone Loss Caused by Oxidative Stress in Spaceflight

The development of effective countermeasures against oxidative damage in bone during long-term spaceflight is essential. Expanded investments in ground-based or in-flight studies revealed some approaches for antagonism of oxidative stress triggered by microgravity and cosmic radiation.

Adequate intake of antioxidant vitamins (e.g., vitamins C and E and carotenoids) can reduce oxidative damages in bones [73,74]. Some research indicated drinking of hydrogen water could relieve microgravity-induced reduction of bone mineral density and augmentation of malondialdehyde in bone tissue [60]. In addition, consuming a diet that provides other naturally occurring antioxidants, such as carotenoids and flavonoids is also effective to reverse microgravity-induced skeletal involution. For example, curcumin, a phenolic natural product isolated from the rhizome of Curcuma Longa (turmeric), could attenuate HLU-induced bone loss by suppressing oxidative stress [59].

Some natural products have exhibited skeletal benefits against oxidative stress. Tanshinol, extracted from Salvia miltiorrhiza Bunge, rescued the decrease of osteoblastic differentiation via down-regulation of FoxO3a signaling and upregulation of Wnt signal under oxidative stress [75]. Some extracts from teas also have osteogenic benefits against oxidative stress [76,77]. Other antioxidants, e.g., α-lipoic acid and N-acetyl cysteine, could restore the changes induced by oxidative stress in bone as well [70,72]. Although these data are not from ground-based models or in-flight studies, they can still enlighten the development of countermeasures against bone loss induced by oxidative stress during space flight.

Humans are embarking on the adjustment of diet in long-duration space flight. It seems that intake of rich antioxidants will prevent oxidative damage caused by space environment. With the progress of research, certain medical approaches will be created and we will conquer oxidative stress-induced bone loss eventually.

6. Conclusions and Perspectives

Space is a stressful environment. Microgravity and cosmic rays are main adverse factors challenging the survival of organisms. The oxidative stress triggered by spaceflight causes a variety of damages to the human body including skeletal involution. In this review, we summarized the stimulation of oxidative stress by radiation and microgravity in space, and its inhibitory effect on bone formation. We discussed the possible mechanisms of oxidative injury induced by space environment to the bone system. Presently, it is mainly believed that the attributed factors include inadequate nutrition intake, increased iron store, and an impaired oxidative defense system. Some countermeasures that have therapeutic potentials for bone loss via oxidative stress antagonism are also mentioned in this manuscript.

Although some progress has been made, the mechanisms of oxidative injuries induced by space habitation are not fully understood. For example, how is gravity sensed and transduced in the bone system and how does it cause the elevation of ROS correspondingly? How do ROS cause bone loss under the special space environment? Further steps need to be taken to thoroughly clarify the whole predisposing process of oxidative stress in space flight and mechanotransduction of gravity, and to develop countermeasures accordingly. It is understandable that resources are limited for spaceflight itself, because the crew time and sample return are restricted and the subject pools are small. Therefore, some ground-based models can be used as vital experimental platforms that allow researchers to examine the effects of the special space environment on bone system. To date, researchers suggest the potential application of antioxidants as a useful dietary source in astronauts' lifestyles. Perhaps it is one solution for oxidative injury during long-term space habitation. The development of countermeasures against oxidative damage will facilitate human beings residing longer in space and truly entering the space era.

Acknowledgments: This study was supported by China Postdoctoral Science Foundation, No. 2017M613210.

Author Contributions: Ye Tian, Xiaoli Ma, Chaofei Yang, Peihong Su, and Chong Yin drafted the manuscript. Ai-Rong Qian designed the project.

Conflicts of Interest: The authors declare no conflict of interest.

Abbreviations

ROS	Reactive Oxygen Species
HLU	Hind-limb Unloadings
ISS	International Space Station
FoxO	Forkhead box O
Mn-SOD	Manganese superoxide dismutase
GPx-1	Glutathione peroxidase-1
Tcf	T-cell factor
LRP	LDL receptor-related proteins
aBMD	Areal bone mineral density
DXA	Dual-energy X-ray absorptionmetry
RWVB	Rotary wall vessel bioreactor
SD	Sprague-Dawley
RANKL	Receptor activator of nuclear factor-kappa B ligand

References

1. Williams, D.R.; Turnock, M. Human space exploration the next fifty years. *Mcgill. J. Med.* **2011**, *13*, 76. [PubMed]
2. Zhao, L.; Gao, Y.; Mi, D.; Sun, Y. Mining potential biomarkers associated with space flight in Caenorhabditis elegans experienced Shenzhou-8 mission with multiple feature selection techniques. *Mutat. Res.* **2016**, *791–792*, 27–34. [CrossRef] [PubMed]

3. Fu, Y.; Li, L.; Xie, B.; Dong, C.; Wang, M.; Jia, B.; Shao, L.; Dong, Y.; Deng, S.; Liu, H. How to Establish a Bioregenerative Life Support System for Long-Term Crewed Missions to the Moon or Mars. *Astrobiology* **2016**, *16*, 925–936. [CrossRef] [PubMed]

4. Byloos, B.; Coninx, I.; Van Hoey, O.; Cockell, C.; Nicholson, N.; Ilyin, V.; Van Houdt, R.; Boon, N.; Leys, N. The Impact of Space Flight on Survival and Interaction of Cupriavidus metallidurans CH34 with Basalt, a Volcanic Moon Analog Rock. *Front. Microbiol.* **2017**, *8*, 671. [CrossRef] [PubMed]

5. Witze, A. NASA rethinks approach to Mars exploration. *Nature* **2016**, *538*, 149–150. [CrossRef] [PubMed]

6. Caceres, M. Creating a space exploration industry. *Aerosp. Am.* **2005**, *43*, 10–12. [PubMed]

7. Demontis, G.C.; Germani, M.M.; Caiani, E.G.; Barravecchia, I.; Passino, C.; Angeloni, D. Human Pathophysiological Adaptations to the Space Environment. *Front. Physiol.* **2017**, *8*, 547. [CrossRef] [PubMed]

8. Otsuka, K.; Cornelissen, G.; Furukawa, S.; Kubo, Y.; Hayashi, M.; Shibata, K.; Mizuno, K.; Aiba, T.; Ohshima, H.; Mukai, C. Long-term exposure to space's microgravity alters the time structure of heart rate variability of astronauts. *Heliyon* **2016**, *2*, e00211. [CrossRef] [PubMed]

9. Mao, X.W.; Nishiyama, N.C.; Pecaut, M.J.; Campbell-Beachler, M.; Gifford, P.; Haynes, K.E.; Becronis, C.; Gridley, D.S. Simulated Microgravity and Low-Dose/Low-Dose-Rate Radiation Induces Oxidative Damage in the Mouse Brain. *Radiat. Res.* **2016**, *185*, 647–657. [CrossRef] [PubMed]

10. Luo, H.; Wang, C.; Feng, M.; Zhao, Y. Microgravity inhibits resting T cell immunity in an exposure time-dependent manner. *Int. J. Med. Sci.* **2014**, *11*, 87–96. [CrossRef] [PubMed]

11. Sanzari, J.K.; Romeroweaver, A.L.; James, G.; Krigsfeld, G.; Lin, L.; Diffenderfer, E.S.; Kennedy, A.R. Leukocyte Activity Is Altered in a Ground Based Murine Model of Microgravity and Proton Radiation Exposure. *PLoS ONE* **2013**, *8*, e71757. [CrossRef] [PubMed]

12. Cazzaniga, A.; Maier, J.A.M.; Castiglioni, S. Impact of simulated microgravity on human bone stem cells: New hints for space medicine. *Biochem. Biophys. Res. Commun.* **2016**, *473*, 181–186. [CrossRef] [PubMed]

13. Grimm, D.; Grosse, J.; Wehland, M.; Mann, V.; Reseland, J.E.; Sundaresan, A.; Corydon, T.J. The impact of microgravity on bone in humans. *Bone* **2016**, *87*, 44–56. [CrossRef] [PubMed]

14. Cappellesso, R.; Nicole, L.; Guido, A.; Pizzol, D. Spaceflight osteoporosis: Current state and future perspective. *Endocr. Regul.* **2015**, *49*, 231–239. [CrossRef] [PubMed]

15. Sibonga, J.D. Spaceflight-induced bone loss: Is there an osteoporosis risk? *Curr. Osteoporos. Rep.* **2013**, *11*, 92–98. [CrossRef] [PubMed]

16. Lang, T.; Van Loon, J.; Bloomfield, S.; Vico, L.; Chopard, A.; Rittweger, J.; Kyparos, A.; Blottner, D.; Vuori, I.; Gerzer, R.; et al. Towards human exploration of space: The THESEUS review series on muscle and bone research priorities. *NPJ Microgravity* **2017**, *3*, 8. [CrossRef] [PubMed]

17. Markin, A.A.; Zhuravlëva, O.A. Lipid peroxidation and antioxidant defense system in rats after a 14-day space flight in the "Space-2044" spacecraft. *Aviakosm. Ekolog. Med.* **1993**, *27*, 47–50. [PubMed]

18. Markin, A.A.; Popova, I.A.; Vetrova, E.G.; Zhuravleva, O.A.; Balashov, O.I. Lipid peroxidation and activity of diagnostically significant enzymes in cosmonauts after flights of various durations. *Aviakosm. Ekolog. Med.* **1997**, *31*, 14–18. [PubMed]

19. Stein, T.P.; Leskiw, M.J. Oxidant damage during and after spaceflight. *Am. J. Physiol. Endocrinol. Metab.* **2000**, *278*, E375–E382. [PubMed]

20. Balaban, R.S.; Nemoto, S.; Finkel, T. Mitochondria, oxidants, and aging. *Cell* **2005**, *120*, 483–495. [CrossRef] [PubMed]

21. Giorgio, M.; Migliaccio, E.; Orsini, F.; Paolucci, D.; Moroni, M.; Contursi, C.; Pelliccia, G.; Luzi, L.; Minucci, S.; Marcaccio, M.; et al. Electron transfer between cytochrome c and p66Shc generates reactive oxygen species that trigger mitochondrial apoptosis. *Cell* **2005**, *122*, 221–233. [CrossRef] [PubMed]

22. Quarrie, J.K.; Riabowol, K.T. Murine models of life span extension. *Sci. Aging Knowl. Environ.* **2004**, *2004*, re5. [CrossRef] [PubMed]

23. Finkel, T.; Holbrook, N.J. Oxidants, oxidative stress and the biology of ageing. *Nature* **2000**, *408*, 239–247. [CrossRef] [PubMed]

24. Glasauer, A.; Chandel, N.S. Ros. *Curr. Biol.* **2013**, *23*, R100–R102. [CrossRef] [PubMed]

25. Valko, M.; Leibfritz, D.; Moncol, J.; Cronin, M.T.; Mazur, M.; Telser, J. Free radicals and antioxidants in normal physiological functions and human disease. *Int. J. Biochem. Cell Biol.* **2007**, *39*, 44–84. [CrossRef] [PubMed]

26. De Boer, J.; Andressoo, J.O.; de Wit, J.; Huijmans, J.; Beems, R.B.; van Steeg, H.; Weeda, G.; van der Horst, G.T.; van Leeuwen, W.; Themmen, A.P.; et al. Premature aging in mice deficient in DNA repair and transcription. *Science* **2002**, *296*, 1276–1279. [CrossRef] [PubMed]

27. Bai, X.C.; Lu, D.; Bai, J.; Zheng, H.; Ke, Z.Y.; Li, X.M.; Luo, S.Q. Oxidative stress inhibits osteoblastic differentiation of bone cells by ERK and NF-kappaB. *Biochem. Biophys. Res. Commun.* **2004**, *314*, 197–207. [CrossRef] [PubMed]

28. Lean, J.M.; Davies, J.T.; Fuller, K.; Jagger, C.J.; Kirstein, B.; Partington, G.A.; Urry, Z.L.; Chambers, T.J. A crucial role for thiol antioxidants in estrogen-deficiency bone loss. *J. Clin. Investig.* **2003**, *112*, 915–923. [CrossRef] [PubMed]

29. Almeida, M.; Han, L.; Martin-Millan, M.; Plotkin, L.I.; Stewart, S.A.; Roberson, P.K.; Kousteni, S.; O'Brien, C.A.; Bellido, T.; Parfitt, A.M.; et al. Skeletal involution by age-associated oxidative stress and its acceleration by loss of sex steroids. *J. Biol. Chem.* **2007**, *282*, 27285–27297. [CrossRef] [PubMed]

30. Manolagas, S.C. From estrogen-centric to aging and oxidative stress: A revised perspective of the pathogenesis of osteoporosis. *Endocr. Rev.* **2010**, *31*, 266–300. [CrossRef] [PubMed]

31. Almeida, M.; Han, L.; Martin-Millan, M.; O'Brien, C.A.; Manolagas, S.C. Oxidative stress antagonizes Wnt signaling in osteoblast precursors by diverting beta-catenin from T cell factor- to forkhead box O-mediated transcription. *J. Biol. Chem.* **2007**, *282*, 27298–27305. [CrossRef] [PubMed]

32. Almeida, M.; Ambrogini, E.; Han, L.; Manolagas, S.C.; Jilka, R.L. Increased lipid oxidation causes oxidative stress, increased peroxisome proliferator-activated receptor-gamma expression, and diminished pro-osteogenic Wnt signaling in the skeleton. *J. Biol. Chem.* **2009**, *284*, 27438–27448. [CrossRef] [PubMed]

33. Almeida, M.; Martin-Millan, M.; Ambrogini, E.; Bradsher, R., 3rd; Han, L.; Chen, X.D.; Roberson, P.K.; Weinstein, R.S.; O'Brien, C.A.; Jilka, R.L.; et al. Estrogens attenuate oxidative stress and the differentiation and apoptosis of osteoblasts by DNA-binding-independent actions of the ERalpha. *J. Bone. Miner. Res.* **2010**, *25*, 769–781. [PubMed]

34. Katoh, M.; Katoh, M. Human FOX gene family (Review). *Int. J. Oncol.* **2004**, *25*, 1495–1500. [CrossRef] [PubMed]

35. Klotz, L.O.; Sanchez-Ramos, C.; Prieto-Arroyo, I.; Urbanek, P.; Steinbrenner, H.; Monsalve, M. Redox regulation of FoxO transcription factors. *Redox. Biol.* **2015**, *6*, 51–72. [CrossRef] [PubMed]

36. Essers, M.A.; de Vries-Smits, L.M.; Barker, N.; Polderman, P.E.; Burgering, B.M.; Korswagen, H.C. Functional interaction between beta-catenin and FOXO in oxidative stress signaling. *Science* **2005**, *308*, 1181–1184. [CrossRef] [PubMed]

37. Staal, F.J.; Clevers, H. Tcf/Lef transcription factors during T-cell development: Unique and overlapping functions. *Hematol. J.* **2000**, *1*, 3–6. [CrossRef] [PubMed]

38. Moon, R.T.; Bowerman, B.; Boutros, M.; Perrimon, N. The promise and perils of Wnt signaling through beta-catenin. *Science* **2002**, *296*, 1644–1646. [CrossRef] [PubMed]

39. Iyer, S.; Ambrogini, E.; Bartell, S.M.; Han, L.; Roberson, P.K.; de Cabo, R.; Jilka, R.L.; Weinstein, R.S.; O'Brien, C.A.; Manolagas, S.C.; et al. FOXOs attenuate bone formation by suppressing Wnt signaling. *J. Clin. Investig.* **2013**, *123*, 3409–3419. [CrossRef] [PubMed]

40. Manolagas, S.C.; Almeida, M. Gone with the Wnts: Beta-catenin, T-cell factor, forkhead box O, and oxidative stress in age-dependent diseases of bone, lipid, and glucose metabolism. *Mol. Endocrinol.* **2007**, *21*, 2605–2614. [CrossRef] [PubMed]

41. Rached, M.T.; Kode, A.; Xu, L.; Yoshikawa, Y.; Paik, J.H.; Depinho, R.A.; Kousteni, S. FoxO1 is a positive regulator of bone formation by favoring protein synthesis and resistance to oxidative stress in osteoblasts. *Cell Metab.* **2010**, *11*, 147–160. [CrossRef] [PubMed]

42. Ambrogini, E.; Almeida, M.; Martin, M. FoxO-Mediated Defense against Oxidative Stress in Osteoblasts Is Indispensable for Skeletal Homeostasis in Mice. *Cell Metab.* **2010**, *11*, 136. [CrossRef] [PubMed]

43. Bartell, S.M.; Kim, H.N.; Ambrogini, E.; Han, L.; Iyer, S.; Serra Ucer, S.; Rabinovitch, P.; Jilka, R.L.; Weinstein, R.S.; Zhao, H.; et al. FoxO proteins restrain osteoclastogenesis and bone resorption by attenuating H_2O_2 accumulation. *Nat. Commun.* **2014**, *5*, 3773. [CrossRef] [PubMed]

44. Zhu, H.; Zhang, L.; Itoh, K.; Yamamoto, M.; Ross, D.; Trush, M.A.; Zweier, J.L.; Li, Y. Nrf2 controls bone marrow stromal cell susceptibility to oxidative and electrophilic stress. *Free. Radic. Biol. Med.* **2006**, *41*, 132–143. [CrossRef] [PubMed]

45. Gan, X.; Huang, S.; Yu, Q.; Yu, H.; Yan, S.S. Blockade of Drp1 rescues oxidative stress-induced osteoblast dysfunction. *Biochem. Biophys. Res. Commun.* **2015**, *468*, 719–725. [CrossRef] [PubMed]

46. Yang, Y.H.; Li, B.; Zheng, X.F.; Chen, J.W.; Chen, K.; Jiang, S.D.; Jiang, L.S. Oxidative damage to osteoblasts can be alleviated by early autophagy through the endoplasmic reticulum stress pathway–implications for the treatment of osteoporosis. *Free Radic. Biol. Med.* **2014**, *77*, 10–20. [CrossRef] [PubMed]

47. Almeida, M.; O'Brien, C.A. Basic biology of skeletal aging: Role of stress response pathways. *J. Gerontol. A Biol. Sci. Med. Sci.* **2013**, *68*, 1197–1208. [CrossRef] [PubMed]

48. Smith, S.M.; Heer, M. Calcium and bone metabolism during space flight. *Nutrition* **2002**, *18*, 849–852. [CrossRef]

49. Smith, S.M.; Heer, M.A.; Shackelford, L.C.; Sibonga, J.D.; Ploutz-Snyder, L.; Zwart, S.R. Benefits for bone from resistance exercise and nutrition in long-duration spaceflight: Evidence from biochemistry and densitometry. *J. Bone Miner. Res.* **2012**, *27*, 1896–1906. [CrossRef] [PubMed]

50. Smith, S.M.; Wastney, M.E.; O'Brien, K.O.; Morukov, B.V.; Larina, I.M.; Abrams, S.A.; Davis-Street, J.E.; Oganov, V.; Shackelford, L.C. Bone Markers, Calcium Metabolism, and Calcium Kinetics During Extended-Duration Space Flight on the Mir Space Station. *J. Bone Miner. Res. Off. J. Am. Soc. Bone Miner. Res.* **2005**, *20*, 208–218. [CrossRef] [PubMed]

51. Morgan, J.L.; Zwart, S.R.; Heer, M.; Ploutz-Snyder, R.; Ericson, K.; Smith, S.M. Bone metabolism and nutritional status during 30-day head-down-tilt bed rest. *J. Appl. Physiol. (1985)* **2012**, *113*, 1519–1529. [CrossRef] [PubMed]

52. LeBlanc, A.; Schneider, V.; Shackelford, L.; West, S.; Oganov, V.; Bakulin, A.; Voronin, L. Bone mineral and lean tissue loss after long duration space flight. *J. Musculoskelet. Neuronal Interact.* **2000**, *1*, 157–160. [PubMed]

53. Zwart, S.R.; Oliver, S.A.; Fesperman, J.V.; Kala, G.; Krauhs, J.; Ericson, K.; Smith, S.M. Nutritional status assessment before, during, and after long-duration head-down bed rest. *Aviat. Space Environ. Med.* **2009**, *80*, A15–A22. [CrossRef] [PubMed]

54. Smith, S.M.; Zwart, S.R.; Block, G.; Rice, B.L.; Davis-Street, J.E. The nutritional status of astronauts is altered after long-term space flight aboard the International Space Station. *J. Nutr.* **2005**, *135*, 437–443. [PubMed]

55. Zwart, S.R.; Morgan, J.L.; Smith, S.M. Iron status and its relations with oxidative damage and bone loss during long-duration space flight on the International Space Station. *Am. J. Clin. Nutr.* **2013**, *98*, 217–223. [CrossRef] [PubMed]

56. Rizzo, A.M.; Corsetto, P.A.; Montorfano, G.; Milani, S.; Zava, S.; Tavella, S.; Cancedda, R.; Berra, B. Effects of long-term space flight on erythrocytes and oxidative stress of rodents. *PLoS ONE* **2012**, *7*, e32361. [CrossRef] [PubMed]

57. Lawler, J.M.; Song, W.; Demaree, S.R. Hindlimb unloading increases oxidative stress and disrupts antioxidant capacity in skeletal muscle. *Free Radic. Biol. Med.* **2003**, *35*, 9–16. [CrossRef]

58. Chowdhury, P.; Soulsby, M.; Kim, K. L-carnitine influence on oxidative stress induced by hind limb unloading in adult rats. *Aviat. Space Environ. Med.* **2007**, *78*, 554–556. [PubMed]

59. Xin, M.; Yang, Y.; Zhang, D.; Wang, J.; Chen, S.; Zhou, D. Attenuation of hind-limb suspension-induced bone loss by curcumin is associated with reduced oxidative stress and increased vitamin D receptor expression. *Osteoporos. Int.* **2015**, *26*, 1–12. [CrossRef] [PubMed]

60. Sun, Y.; Shuang, F.; Chen, D.M.; Zhou, R.B. Treatment of hydrogen molecule abates oxidative stress and alleviates bone loss induced by modeled microgravity in rats. *Osteoporos. Int.* **2013**, *24*, 969–978. [CrossRef] [PubMed]

61. Takahashi, K.; Okumura, H.; Guo, R.; Naruse, K. Effect of Oxidative Stress on Cardiovascular System in Response to Gravity. *Int. J. Mol. Sci.* **2017**, *18*, 1426. [CrossRef] [PubMed]

62. Smith, S.M.; Davis-Street, J.E.; Fesperman, J.V.; Smith, M.D.; Rice, B.L.; Zwart, S.R. Nutritional status changes in humans during a 14-day saturation dive: The NASA Extreme Environment Mission Operations V project. *J. Nutr.* **2004**, *134*, 1765–1771. [PubMed]

63. Hollander, J.; Gore, M.; Fiebig, R.; Mazzeo, R.; Ohishi, S.; Ohno, H.; Ji, L. Spaceflight downregulates antioxidant defense systems in rat liver. *Free Radic. Biol. Med.* **1998**, *24*, 385–390. [CrossRef]

64. Nojiri, H.; Saita, Y.; Morikawa, D.; Kobayashi, K.; Tsuda, C.; Miyazaki, T.; Saito, M.; Marumo, K.; Yonezawa, I.; Kaneko, K.; Shirasawa, T.; Shimizu, T. Cytoplasmic superoxide causes bone fragility owing to low-turnover osteoporosis and impaired collagen cross-linking. *J. Bone Miner. Res.* **2011**, *26*, 2682–2694. [CrossRef] [PubMed]

65. Morikawa, D.; Nojiri, H.; Saita, Y.; Kobayashi, K.; Watanabe, K.; Ozawa, Y.; Koike, M.; Asou, Y.; Takaku, T.; Kaneko, K.; Shimizu, T. Cytoplasmic reactive oxygen species and SOD1 regulate bone mass during mechanical unloading. *J. Bone Miner. Res.* **2013**, *28*, 2368–2380. [CrossRef] [PubMed]

66. Beck, M.; Moreels, M.; Quintens, R.; Abou-El-Ardat, K.; El-Saghire, H.; Tabury, K.; Michaux, A.; Janssen, A.; Neefs, M.; Van Oostveldt, P.; et al. Chronic exposure to simulated space conditions predominantly affects cytoskeleton remodeling and oxidative stress response in mouse fetal fibroblasts. *Int. J. Mol. Med.* **2014**, *34*, 606–615. [CrossRef] [PubMed]

67. Suman, S.; Rodriguez, O.C.; Winters, T.A.; Fornace, A.J., Jr.; Albanese, C.; Datta, K. Therapeutic and space radiation exposure of mouse brain causes impaired DNA repair response and premature senescence by chronic oxidant production. *Aging* **2013**, *5*, 607–622. [CrossRef] [PubMed]

68. Christofidou-Solomidou, M.; Pietrofesa, R.A.; Arguiri, E.; Schweitzer, K.S.; Berdyshev, E.V.; McCarthy, M.; Corbitt, A.; Alwood, J.S.; Yu, Y.; Globus, R.K.; et al. Space radiation-associated lung injury in a murine model. *Am. J. Physiol. Lung Cell. Mol. Physiol.* **2015**, *308*, L416–L428. [CrossRef] [PubMed]

69. Guan, J.; Stewart, J.; Ware, J.H.; Zhou, Z.; Donahue, J.J.; Kennedy, A.R. Effects of dietary supplements on the space radiation-induced reduction in total antioxidant status in CBA mice. *Radiat. Res.* **2006**, *165*, 373–378. [CrossRef] [PubMed]

70. Kook, S.H.; Kim, K.A.; Ji, H.; Lee, D.; Lee, J.C. Irradiation inhibits the maturation and mineralization of osteoblasts via the activation of Nrf2/HO-1 pathway. *Mol. Cell Biochem.* **2015**, *410*, 255–266. [CrossRef] [PubMed]

71. Kondo, H.; Limoli, C.; Searby, N.D.; Almeida, E.A.; Loftus, D.J.; Vercoutere, W.; Morey-Holton, E.; Giedzinski, E.; Mojarrab, R.; Hilton, D.; et al. Shared oxidative pathways in response to gravity-dependent loading and gamma-irradiation of bone marrow-derived skeletal cell progenitors. *Radiatsionnaia Biol. Radioecol.* **2007**, *47*, 281–285. [PubMed]

72. Kondo, H.; Yumoto, K.; Alwood, J.S.; Mojarrab, R.; Wang, A.; Almeida, E.A.; Searby, N.D.; Limoli, C.L.; Globus, R.K. Oxidative stress and gamma radiation-induced cancellous bone loss with musculoskeletal disuse. *J. Appl. Physiol. (1985)* **2010**, *108*, 152–161. [CrossRef] [PubMed]

73. Maillet, A.; Beaufrere, B.; Di, N.P.; Elia, M.; Pichard, C. Weightlessness as an accelerated model of nutritional disturbances. *Curr. Opin. Clin. Nutr. Metab. Care* **2001**, *4*, 301–306. [CrossRef] [PubMed]

74. Bergouignan, A.; Stein, T.P.; Habold, C.; Coxam, V.; O'Gorman, D.; Blanc, S. Towards human exploration of space: The THESEUS review series on nutrition and metabolism research priorities. *NPJ Microgravity* **2016**, *2*, 16029. [CrossRef] [PubMed]

75. Yang, Y.; Su, Y.; Wang, D.; Chen, Y.; Wu, T.; Li, G.; Sun, X.; Cui, L. Tanshinol attenuates the deleterious effects of oxidative stress on osteoblastic differentiation via Wnt/FoxO3a signaling. *Oxid. Med. Cell. Longev.* **2013**, *2013*, 351895. [CrossRef] [PubMed]

76. Zeng, X.; Tian, J.; Cai, K.; Wu, X.; Wang, Y.; Zheng, Y.; Su, Y.; Cui, L. Promoting osteoblast differentiation by the flavanes from Huangshan Maofeng tea is linked to a reduction of oxidative stress. *Phytomed. Int. J. Phytother. Phytopharmacol.* **2014**, *21*, 217–224. [CrossRef] [PubMed]

77. Vester, H.; Holzer, N.; Neumaier, M.; Lilianna, S.; Nussler, A.K.; Seeliger, C. Green Tea Extract (GTE) improves differentiation in human osteoblasts during oxidative stress. *J. Inflamm.* **2014**, *11*, 15. [CrossRef] [PubMed]

International Journal of
Molecular Sciences

MDPI

Review

Redox Signaling and Its Impact on Skeletal and Vascular Responses to Spaceflight

Candice G. T. Tahimic [1,2] and Ruth K. Globus [1,*]

[1]   Space Biosciences Division, NASA Ames Research Center, Moffett Field, CA 94035, USA;
      candiceginn.t.tahimic@nasa.gov
[2]   KBRWyle, Moffett Field, CA 94035, USA
*   Correspondence: Ruth.K.Globus@nasa.gov; Tel.: +1-650-604-1743

Received: 2 September 2017; Accepted: 10 October 2017; Published: 16 October 2017

Abstract: Spaceflight entails exposure to numerous environmental challenges with the potential to contribute to both musculoskeletal and vascular dysfunction. The purpose of this review is to describe current understanding of microgravity and radiation impacts on the mammalian skeleton and associated vasculature at the level of the whole organism. Recent experiments from spaceflight and ground-based models have provided fresh insights into how these environmental stresses influence mechanisms that are related to redox signaling, oxidative stress, and tissue dysfunction. Emerging mechanistic knowledge on cellular defenses to radiation and other environmental stressors, including microgravity, are useful for both screening and developing interventions against spaceflight-induced deficits in bone and vascular function.

Keywords: spaceflight; bone; vasculature; oxidative stress; microgravity; hindlimb unloading; radiation; reactive oxygen species; antioxidant

1. The Spaceflight Environment and Its Impact on Skeletal and Vascular Health

Microgravity and radiation are two unique elements of the spaceflight environment that pose challenges to the health of an organism. Microgravity leads to a cephalad fluid shift and profound reductions in mechanical loading of bone and muscle. Spaceflight causes perturbations in calcium homeostasis and site-specific reductions in bone mass (osteopenia), and thus may pose long-term risks for skeletal health and tissue repair [1–4]. In rodent models of weightlessness such as hindlimb unloading (HU), the onset of osteopenia correlates with reductions in skeletal perfusion, vascular density (rarefication), and vasodilation responses similar to that observed in aging [5–7]. These are serious risks for long-duration, exploration-class missions when astronauts will face the challenges of increased exposure to space radiation and abrupt transitions between different gravitational states upon return to Earth.

Beyond the Earth's protective magnetosphere, astronauts will be exposed to a complex combination of ionizing radiation from galactic cosmic radiation (GCR) and intermittent solar particle events (SPEs). GCR is comprised of α particles, protons and a small percentage of high-charge and high-energy (HZE) nuclei while SPEs generate highly energetic protons and heavy ions. HZE particles are of particular concern during exploration-class missions [8] (reviewed in [9]). HZE easily penetrate spacecraft shielding and biological tissue, deposit very large amounts of energy along linear tracks (high linear energy transfer, high LET) that cause DNA strand breakage, and generate secondary radiations that may have additional detrimental biological effects [10–14]. Other responses to ionizing radiation include oxidative stress, damage to proteins, lipids, and DNA, adversely affecting functions of membranes, the extracellular matrix, cell cycle and survival [15–17].

2. Oxidative Stress and Its Link to Spaceflight-Induced Tissue Dysfunction

2.1. Oxidative Damage Associated with Spaceflight and Its Analogs

2.1.1. Evidence from Spaceflight

Spaceflight and the return to Earth may lead to oxidative damage in blood and various tissues as a result of excessive reactive oxygen species/reactive nitrogen species (ROS/RNS) in both humans and animals [18–23]. Collectively, a number of studies indicate that various tissues undergo altered redox status during and/or after spaceflight. Urinary excretion of the oxidative damage markers, 8-iso-prostaglandin F2α (8-iso-PGF2α) and 8-hydroxydeoxyguanosine (8-OHdG) were measured in-flight (88 to 186 days in orbit) and post-flight (up to 14 days) in Mir mission crew [19]. The isoprostane, 8-iso-PGF2α, is a marker for oxidative damage to membrane lipids and is produced by the peroxidation of arachidonic acid in membrane phospholipids while 8-OHdG is an oxidized derivative of the nucleoside, deoxyguanosine, and is therefore used to assess oxidative damage to DNA. 8-OHdG excretion was unchanged during spaceflight and increased postflight. No changes in 8-OHdG levels were observed in Earth-based individuals that underwent bed rest although isoprostane was increased in the ensuing recovery period. Changes in isoprostane production were attributed to decreased generation of oxygen radicals from the electron transport chain due to reduced caloric intake in-flight, whereas the post-flight increases in the excretion of oxidative damage markers may be partly caused by a combination of increased metabolic activity following flight and the loss of some antioxidant defenses during flight. The downregulation of antioxidant defenses as a potential mechanism for the increased levels of oxidative damage post-flight is supported by the observation that hair follicle samples from International Space Station (ISS) crew members at post-flight display decreased expression of endogenous antioxidant genes, including Mn superoxide dismutase (MnSOD), CuZnSOD, glutathione peroxidase 4 (*GPX4*) and kelch-like ECH-associated protein 1 (*KEAP1*), the regulator of nuclear factor erythroid 2-related factor 2 (*NFE2L2* a.k.a. *NRF2*), a master transcription factor that regulates hundreds of oxidative defense-related genes [24].

Consistent with the findings in spaceflight crew members, rodents subjected to spaceflight also exhibit alterations in redox signaling relative to ground controls. In rats, short-duration spaceflight (6 days) increases cardiac gene expression of mitochondrial redox-related enzymes [25], suggesting a possible stress response and/or change in energy metabolism. Although short duration spaceflight (7 days) does not alter heart mass in rats [26], differences between flight and ground controls were observed in the vasculature of mice [27,28]. Spaceflight induces similar effects in the liver of both rats and mice, which includes elevated expression levels of antioxidant genes and markers of oxidative damage [18,29]. Male rats flown in STS-63 for eight days showed reduced total glutathione (GSH) content and activities of CuZnSOD (*SOD1*), catalase, GSH reductase, and GSH sulfur-transferase in liver [18]. In ocular tissue, astronauts are susceptible to optic disc edema, globe flattening, and choroidal fold formation [30]. The mechanisms are unknown, although mice flown in STS-135 also showed ocular tissue damage as well as alterations in several critical genes involved in regulating vascular endothelial cell response to oxidative stress and apoptosis [31].

2.1.2. Evidence from Ground-Based Models for Spaceflight

Ground-based simulations of microgravity and space radiation also demonstrate a link between oxidative stress and tissue impairments. Rodents exposed to proton (50 cGy) or ^{56}Fe (15 cGy) display increased oxidative damage in the heart and decrements in clinical measures of cardiac function [32] while γ-radiation exposure (^{137}Cs, 1–2 Gy) increases ROS generation and lipid peroxidation in bone marrow, and decreases cancellous bone volume fraction [33]. Furthermore, bone marrow from rodents exposed to radiation display increased expression of inflammatory cytokines interleukin-6 (*IL6*), tumor necrosis factor alpha (*TNFα*), monocyte chemoattractant protein-1 (*MCP1*), and the pro-osteoclastogenic signal, receptor activator of nuclear factor kappa-B ligand (*RANKL*), as well as upregulation of *NRF2* [34,35].

Hindlimb unloading (HU) is a widely accepted rodent model to simulate weightlessness for many different tissues (reviewed in [36–38]). Marrow cells of the osteoblast lineage from HU mice show elevated ROS within a week of HU [39]. Marrow and bone show parallel increases in gene expression of the cytosolic fraction of the free radical scavenger, superoxide dismutase (*SOD*), changes that coincide with HU-induced reductions in osteoblast activity and bone loss [39]. HU also causes adaptation-related changes in cardiovascular function [40–44], and several studies have linked HU-induced vascular responses and oxidative stress [45–47]. Taken together, observations from both spaceflight and ground-based rodent models that simulate weightlessness show that oxidative stress and damage may occur in response to microgravity and radiation across multiple, if not all, tissues.

3. The Role of Nitric Oxide (NO) and Reactive Oxygen Species (ROS) Signaling in Skeletal and Vascular Disease

3.1. NO and ROS Signaling: Mechanisms and Impact on Tissue Function

Excess reactive oxygen species/reactive nitrogen species (ROS/RNS) and inflammation are implicated both in age-related diseases, such as osteoporosis and atherosclerosis, and following insults such as radiation exposure [15]. ROS can directly stimulate bone resorption by osteoclasts [48,49], and the bone loss due to aging and estrogen deficiency may be partly attributed to oxidative stress [50,51]. Furthermore, treatment with the potent antioxidant, α-lipoic acid, can prevent inflammation- and acute radiation-induced bone loss [33,52]. Oxidative damage from the production of excess ROS/RNS in skeletal tissues during spaceflight may lead to delayed deficits in bone structural integrity and reduced mechanical strength during recovery and the aging process.

Under physiological conditions, ROS and RNS such as nitric oxide (NO) can function as signaling molecules to regulate important processes such as mechanotransduction [53–59] and vascular function [60]. However, at high levels, ROS are potent inducers of apoptosis in response to a vast array of cellular insults [61,62]. Superoxide (O_2^-) in particular is cytotoxic and is produced when molecular oxygen reacts with an aqueous electron. O_2^- can be generated by exposure of cells to ionizing irradiation and as a by-product of metabolism, primarily from mitochondria. O_2^- is an upstream component of many oxidative pathways and can be rapidly converted into hydrogen peroxide (H_2O_2) by members of the superoxide dismutase (*SOD*) family of antioxidants [63,64]. Catalase, glutathione peroxidase (*GPX*), and other peroxidases such as peroxiredoxin then convert H_2O_2 into H_2O [65]. H_2O_2 can react with endogenous Fe^{2+} or other transition metals via a fenton mechanism to generate the highly reactive hydroxyl radical (OH•) [66] which is one of the most damaging free radical species produced by exposure to ionizing radiation [67].

Normal vascular function requires both basal and stimulated production of NO in the endothelium by endothelial nitric oxide synthase (*eNOS*). The *eNOS* enzyme utilizes L-arginine to produce NO and L-citrulline. NO then activates soluble guanylate cyclase (*sGC*) by binding to it. This leads to increased production of cyclic guanosine monophosphate (cGMP) which in turn mediates vascular smooth muscle relaxation [68].

Genetic ablation studies of *eNOS* in rodents demonstrate the critical role of endothelial NO signaling in vascular health [69]. *eNOS* knockout (KO) mice lack endothelium-derived relaxing factor (*EDRF*) activity in response to endothelium-dependent vasodilators (e.g., acetylcholine) [70,71]. Furthermore, they are hypertensive [71], display increased vascular smooth muscle cell proliferation [72], and platelet aggregation [73], as well as a higher predisposition to atherosclerosis [74], thrombosis [75], and stroke [69,76,77].

The activity of *eNOS* is regulated via complex and concerted processes, including transcriptional control, substrate availability, interactions with other proteins and co-factors and post-translational modification [69]. Under conditions of oxidative stress, the biological activity of NO may decline in resistance arteries. For example, in the presence of O_2^-, NO rapidly combines to form peroxynitrite.

SOD neutralizes O_2^- by converting it to H_2O_2, thus preventing its reaction with NO and increasing the half-life and bioavailability of NO [64,78].

The generation of excess ROS is thought to be of one of the primary mechanisms by which ionizing radiation causes tissue damage. Dr. M. Delp [79] and others propose a model wherein microgravity and radiation promote the generation of ROS in the resistance vasculature of bone that creates an imbalance favoring peroxynitrite over NO production. This lowers the bioavailability of NO, diminishes endothelium-dependent vasodilation, and disrupts the coupling of bone circulation with bone remodeling. While more studies are required to test the validity of this model, results from our group and others, as described in the succeeding sections, appear to be consistent with such a model.

3.2. Bone and Vascular Function during Development Are Intimately Associated

Vasculature is critical for normal skeletal development, postnatal growth, remodeling, and fracture repair. The embryonic skeleton undergoes endochondral ossification (reviewed in [80,81]) where cartilage, an avascular tissue, is gradually converted into bone, one of the most vascularized tissues in vertebrates [80]. This process occurs through a series of complex and coordinated signals that induce terminal differentiation of chondrocytes and their subsequent death. The extracellular matrix produced by the chondrocytes is then mineralized and partly degraded by chondroclasts and preosteoclasts, thus promoting the invasion of blood vessels. Following vascular invasion, osteogenic progenitors are recruited to the site, where they form mineralized bone [81]. A similar mechanism occurs during postnatal growth of the long bones, fracture repair and bone remodeling, although smaller in scale than what occurs during development. There is evidence that re-vascularization in response to bone injury is impaired in spaceflight. Rats that underwent osteotomy and then exposed to microgravity display reduced angiogenesis at the site of injury [82]. Since proper repair of skeletal tissue requires competent vascular invasion, this raises concerns of a higher risk for impaired fracture healing in individuals as a consequence of spaceflight.

The coupling of vascular perfusion and skeletal function has been demonstrated in a number of studies [83–85]. In addition, administration of anti-osteoporotic bisphosphonates improved blood flow and angiogenesis in aged rodents [86] while endochondral bone formation and angiogenesis are impaired in mice lacking certain isoforms of the pro-angiogenic signal, vascular endothelial growth factor (*VEGF*) [87]. Increased *VEGF* levels from over-expression of hypoxia-inducible factor alpha (*HIFα*) promotes both angiogenesis and osteogenesis, while loss of *HIF1a* in osteoblasts resulted in thinner and less vascularized bones [88].

3.3. Vascular-Bone Coupling Occurs via Redox-Dependent Mechanisms: Implications for Tissue Responses to Spaceflight

Cardiovascular deconditioning is one of the potential health risks associated with spaceflight [89,90]. A majority of astronauts experience orthostatic intolerance upon return from long-duration flight (129–190 days) [89,91–93] which is attributed to impairments in raising peripheral resistance [79]. Muscle vasculature plays an integral role in elevating peripheral resistance and, therefore, has been the subject of a number of studies using spaceflight models. In rodents, both spaceflight and hindlimb unloading impair vasoconstriction of gastrocnemius resistance arteries [94,95]. In addition, resistance arteries of mouse gastrocnemius [96] and rat soleus [97] both display defects in endothelium-dependent vasodilation. In rats, the soleus resistance arteries exhibit diminished expression levels of *eNOS* and *SOD1*, although the gastrocnemius resistance arteries do not show such alterations [97,98]. The basis for the site-specific differences in redox status of vascular resistance arteries is not fully understood. However, there is speculation that this is partly due to region-specific differences in perfusion and/or differences in metabolic rates of slow versus fast twitch muscle fibers and the relative abundance of these fiber types in the two muscle sites.

Similar to the observations in muscle resistance arteries, the vasculature of bone also exhibits functional impairments under spaceflight conditions. Two weeks of HU in rats results in deficits in vasoconstrictor and vasodilator properties of the femoral principal nutrient artery (PNA) [99], the primary route for blood circulation to the femur. HU animals display decrements in bone and marrow perfusion and increases in vascular resistance. These changes are not attributed to enhanced vasoconstrictor responsiveness of the bone resistance arteries, but are associated with decreased endothelium-dependent vasodilation and vascular remodeling that reduces PNA maximal diameter.

The skeleton, like muscle, undergoes deconditioning during spaceflight (and its analogs) which manifests as a reduction in bone mass due to a transient net increase in resorption [100–102] as well as diminished mechanical strength [4]. These coincide with functional impairments and changes in redox status in associated vasculature [93]. Greater decrements in percent cancellous bone volume were observed after 13 to 16 days of HU and exposure to simulated space radiation when treatments were combined, compared to untreated controls. HU reduced trabecular thickness, whereas irradiation reduced trabecular number but not thickness, accounting for the sometimes greater deficit in percent cancellous bone volume when HU and radiation are combined.

In the gastrocnemius feed artery, which served as a surrogate artery for the PNA, the early effects of HU and IR (^{56}Fe) are to each impair peak endothelium-dependent vasodilation, with the combination of HU and IR exacerbating this deficit. These group differences are abolished in the presence of nitric oxide synthase (*NOS*) inhibitors, indicating that the impairment induced by HU and IR was mediated through the *NOS* signaling pathway. Vasodilation response to the NO donor DEA-NONOate is also impaired by HU and IR, indicating that the depressed endothelium-dependent vasodilation could be mediated in part through a reduced smooth muscle responsiveness to NO. Further, peak vasodilation correlates positively with percent cancellous bone volume which suggests a coupling of bone and vasculature responses to stressors associated with spaceflight. In contrast to vasodilator responses, HU and IR as single treatments or combined have little effect on vasoconstrictor properties or pressure-diameter responses. HU and HU with IR results in decreased levels of *eNOS* protein, while IR and HU with IR leads to diminished superoxide dismutase-1 (*SOD1*) and higher xanthine oxidase (*XO*) protein content. Decrements in the bioavailability of NO via reduction in *eNOS* protein levels, lower anti-oxidant capacity (*SOD1*) and higher pro-oxidant capacity (*XO*) may contribute to the deficits in *NOS* signaling in resistance arteries.

When the vascular effects of long-term (6–7 months) recovery from HU, IR (^{56}Fe), and the combination of HU and IR are examined, only IR sustains an impairment in peak endothelium-dependent vasodilation [103]. The IR-induced deficit in endothelial vasodilator function is abolished by chemical inhibition of *NOS*, indicating that the sustained dysfunction is mediated through the *NOS* signaling pathway. Similar to the findings from an early timepoint (two weeks of HU with or without IR) [93], HU and IR have little effect on vasoconstrictor properties. These findings indicate that although both simulated weightlessness and irradiation produce early effects of impaired vascular endothelial function, only those produced by IR are sustained.

IR and HU causes differential effects on loss and thinning of trabeculae, contributing to lower percent cancellous bone volume when the treatments were combined. Both treatments impair endothelium-dependent vasodilation of skeletal muscle resistance arteries via a NO-dependent signaling pathway. The impairment of *NOS* signaling, however, seems to be differentially affected by unloading and irradiation; HU diminishes vascular *eNOS* protein expression whereas IR reduces *SOD* expression and increases pro-oxidant *XO* protein levels. If these findings in skeletal muscle resistance arteries hold for bone vasculature, impairments in endothelium-dependent vasodilation may lead to reductions in skeletal perfusion and perturbations in the coupling of bone cell and vascular endothelium activity.

3.4. Cellular Defenses to Oxidative Damage Are Important for Preserving Skeletal and Vascular Health

Most aerobic organisms have multiple defenses against the damaging effects of ROS, which include enzymatic and non-enzymatic antioxidants. Studies on perturbation of key antioxidant molecules and signaling pathways highlight the importance of antioxidant defenses in preserving the functional integrity of tissues and cells. In the following section, we summarize the state of knowledge on the role of endogenous antioxidant proteins in maintaining skeletal and vascular health with an emphasis on studies using genetic models of gain or loss of protein function. The cytoprotective effect of these proteins on Earth has, therefore, formed the rationale for investigating their importance in tissue and cellular defenses against the stressors associated with spaceflight [18,24] and its Earth-based analogs [34,35]. In this section, we also cite studies that illustrate the emerging role of these antioxidant proteins in modulating skeletal and vascular responses to spaceflight.

3.4.1. Nuclear Factor Erythroid 2-Related Factor 2 (*NRF2*)

NRF2 is a master transcription factor that regulates cellular redox balance and protective antioxidant and detoxification responses across multiple species, from *Drosophila* to humans [104–106]. It is a member of the Cap-N-Collar family of regulatory proteins composed of *NRF1*, *NRF3* and *BACH1* and *BACH2* [106] and is ubiquitously expressed with the highest concentrations occuring in heart, muscle and brain [107]. Under normal conditions, *NRF2* is sequestered to the cytosol via binding to its inhibitor, *KEAP1*, and is therefore subject to proteosomal degradation. During conditions of excess ROS, *NRF2* dissociates from *KEAP1* and undergoes translocation into the nucleus. *NRF2* then binds to antioxidant response-element sequences in the genome to promote transcription of hundreds of antioxidant genes [108] including *SOD1* as well as phase II detoxification enzymes [106].

In *Drosophila*, loss-of-function mutations in *dKEAP1*, an endogenous inhibitor of the *NRF2* fly homolog *CncC*, extends lifespan and increases resistance to the oxidizer paraquat [104]. Mice in which *NRF2* is globally deleted are viable, fertile and exhibit no apparent phenotypic defects [109,110]. However, consistent with its cytoprotective function, *NRF2* deletion in mice leads to increased sensitivity to radiation, oxidative damage and age-related disease pathologies [105,111]. Specifically, *NRF2* knockouts display accelerated heart failure in a myocardial infarct model [112], enhanced retinal degeneration [113,114], bone loss [110,115,116], and deficits in bone and endothelial stem and progenitor cell populations [115,117].

A number of studies have characterized the skeletal phenotype of *NRF2* knockouts and show that the effects of *NRF2* appear to be sex- and age-dependent. Cultured osteoclasts from *NRF2* knockout mice display elevated ROS levels, increased maturation and defective production of antioxidant enzymes. On the other hand, pre-treatment with *NRF2* agonists sulforaphane and curcumin, inhibit osteoclast maturation via activation of *NRF2* [118]. In female rodents, *NRF2* deletion results in decrements in percent cancellous bone volume and trabecular number as well as increased trabecular separation, which persists up to eight months of age but are no longer apparent thereafter. In addition, primary cultures of bone marrow-derived stromal cells from these KOs show diminished colony formation with no differences in osteoblast proliferation in vivo. Similarly, ex vivo osteoclast cultures from *NRF2* KOs mature faster than those from wild-type animals, although in vivo osteoclast counts are unchanged [115]. In contrast, *NRF2* deletion in male rodents leads to higher bone mass, mineral apposition rate and osteoblast number compared to wild-type controls [119]. These sex-dependent differences were corroborated in a more rigorous study comparing male and female *NRF2* KOs [110]. *NRF2* deficiency in females results in reduced femoral and spinal bone mineral density, while loss of *NRF2* in males leads to an improvement in the said bone structural parameters compared to sex-matched controls. Furthermore, both young (3-month old) and old (15-month old) KO females display decreased expression of *NRF2* target antioxidant enzymes; in contrast, these deficits are only observed in aging males [110].

Collectively, the abovementioned studies demonstrate that *NRF2* plays a critical role in maintaining the functional and structural integrity of bone and vasculature here on Earth, and that loss of the *NRF2* gene mimics some of the features of spaceflight-induced bone loss. However, how *NRF2* impacts tissue responses to elements of the space environment (radiation or microgravity) or its analogs is less clear. One study determined the consequence of *NRF2* loss in a rodent radiotherapy model at high doses (20 Gy) of ionizing radiation [116] where it was found that *NRF2* ablation exacerbates bone loss and impairs colony-forming capacity of bone marrow-derived progenitors [116]. Likewise, the role of *NRF2* in tissue responses to microgravity has not been thoroughly examined, although there is some evidence of its function in mechanical load-driven bone formation. *NRF2* KOs are less responsive to the anabolic effects of mechanical loading of the ulna compared to wild-type controls, as indicated by blunted bone formation rate and decreased relative mineralizing surface. In addition, cultured primary osteoblasts from mechanically loaded ulna of *NRF2* KOs display reductions in the expression levels of antioxidant enzymes [120].

Microgravity causes a cephalad fluid shift [121] as well as changes in ion homeostasis [122–124] and there is indirect evidence that *NRF2* may play a role in the endothelial response to perturbations in ionic balance. *NRF2* KO and wild-type controls were fed either a low or high salt diet and vascular function was analyzed. Endothelium-dependent dilation to acetylcholine is unchanged in the middle cerebral arteries (MCA) in both groups fed a low-salt diet. High-salt diet eliminates endothelium-dependent dilation to acetylcholine in both genotypes. However, unlike wild-type controls, *NRF2* KO rats fail to respond to the rescuing effects of angiotensin II infusion on high-salt diet-induced endothelial dysfunction and microvessel rarefication [125].

3.4.2. CuZn Superoxide Dismutase (*SOD1*)

SOD1 gene expression is downregulated in hair follicles of spacefight crew members [24] while its protein levels are decreased in livers of spaceflown rats [18] although its role in protecting vasculature and bone from spaceflight stressors needs to be further elucidated. *SOD*s are the major antioxidant defense systems against O_2^-, catalyzing the conversion of O_2^- to H_2O_2, which is then further reduced to water by catalase, peroxiredoxins (PRx), or glutathione peroxidases (*GPX*) [64]. In mammals, the *SOD* family is comprised of three isoforms: cytoplasmic CuZnSOD (*SOD1*), mitochondrial MnSOD (*SOD2*), and the extracellular CuZnSOD (*SOD3*), and as their names indicate, require a metal co-factor (Cu, Mn, or Zn) for their activation. Both male and female *SOD1* knockout mice exhibit increased oxidative stress, and decreased bone mineral density and mechanical strength (bending stiffness) as assessed by three-point bending compared to sex-matched wild type controls [126]. Consistent with the observations in bone, *SOD1* deletion also leads to increased oxidative damage in skeletal muscle, accelerates age-dependent skeletal muscle atrophy and motor function deficits as assessed by voluntary wheel running and rotating rod test [127]. Similar to results from deletion of its transcriptional regulator *NRF2*, loss of the *SOD1* gene leads to musculoskeletal deficits, also resembling some of tissue deficits observed in spaceflight and its Earth-based analogs.

3.4.3. Catalase

Catalase is an anti-oxidant that converts the reactive species, H_2O_2, into water and molecular oxygen. Transgenic mice over-expressing the human catalase (*hCAT*) gene targeted to the mitochondria (*mCAT* mice) display extended lifespan [128] and reductions in numerous pathologies [129–131] including cardiovascular disease [131–133] and Alzheimer-related amyloid deposition [130]. Further, the transgene effectively quenches mitochondrial oxidative stress in macrophages which are precursors to bone-resorbing osteoclasts [134]. Consistent with the latter finding, over-expression of catalase targeted to mitochondria of bone-resorbing osteoclasts protects from bone loss caused by the loss of estrogens [135]. In addition, the transgene rescues cardiomyopathy including impaired systolic and diastolic function in mice that carry homozygous mutations in the exonuclease encoding domain of mitochondrial DNA polymerase γ (Polg(m/m)) [131].

Catalase protein levels and activity are decreased in livers of spaceflown rats [18]. Furthermore, a number of studies involving the mitochondrial *hCAT* transgene demonstrate the importance of mitochondrial ROS quenching in the cellular defense against HZE particles. *mCAT* animals display protection from proton-induced deficits in synaptic signaling, dendritic complexity, memory and cognition [136–138]. The impact of mitochondrial ROS quenching on skeletal and cardiovascular responses to microgravity and radiation is less understood and is currently under investigation by our team and others.

4. Implications for the Development of Spaceflight Countermeasures

Whole body irradiation increases ROS generation by bone marrow cells [33,139], damages skeletal lipids [33], and upregulates the activity of the antioxidant enzyme *NRF2* in both bone marrow and mineralized tissue [35,140]. On the other hand, *NRF2* deficiency can exacerbate radiation-induced bone loss [116]. Collectively, these data suggest that pharmacologic interventions directed toward limiting excess ROS may moderate the radiation stress response of bone cells.

Within a period of three days to a month, relatively low doses of radiation (≤ 2 Gy) lead to progressive loss of cancellous bone, as shown by our group and others [34,35,141–144], which is preceded by a rapid increase in expression levels of inflammatory cytokines such as *IL1* and *TNFα* and pro-osteoclastogenic signals *RANKL* and *MCP1* as well as the oxidative defense gene, *NRF2* [34,35]. On the basis of these findings, we hypothesized that diets or drugs capable of preventing early increases in these molecular signals can mitigate cancellous bone loss caused by both low LET and high LET radiation. The hypothesis was tested by evaluating a number of candidate interventions including: (1) an antioxidant diet cocktail (AOX) composed of *N*-acetyl cysteine, ascorbic acid, L-selenomethionine, dihydrolipoic acid and vitamin E which in combination has been shown to protect a variety of tissues from ionizing radiation [145–147]; (2) dihydrolipoic acid (DHLA), which possesses antioxidant properties [148,149]; (3) the non-steroidal anti-inflammatory drug, Ibuprofen [150,151]; and (4) Dried Plum (DP, 25% by weight), previously reported to inhibit osteoclast activity and protect from age-related bone loss [152–155]. Findings from this study indicate that high levels of pro-resorptive, pro-inflammatory, and oxidative stress-related genes in bone marrow strongly correlate with cancellous bone loss. DP diet completely prevents radiation-induced increases in these molecular signals and the ensuing cancellous bone loss [35] which appears to strengthen the rationale for using antioxidants to mitigate radiation-induced bone loss. However, seemingly paradoxical is that the other two antioxidant-based interventions, DHLA and AOX, failed to protect from bone loss induced by radiation. These findings do not nullify the feasibility of using antioxidants as a countermeasure, yet highlight very important considerations in countermeasure design for radiation-induced bone loss. Firstly, an effective strategy to mitigate radiation-induced bone loss must at least include components that prevent the early rise in expression of pro-resorptive, pro-inflammatory, and oxidative stress-related genes. This knowledge is gained from the observation that treatments which fail to mitigate the changes in all these three molecular signals (such as AOX and Ibuprofen) ultimately were unsuccessful in preventing radiation-induced bone loss. Secondly, there are additional, equally important signaling molecules likely to mediate radiation-induced bone loss. This is a lesson learned from the finding that DHLA, which appeared nearly as effective as DP in preventing radiation-induced increases in expression of these markers, did not protect skeletal microarchitecture. Lastly, an assessment of the efficacy of an intervention on multiple organ systems must be incorporated in countermeasure development efforts as antioxidants appear to confer various levels of protection in different tissues.

5. Concluding Statements

Studies overwhelmingly demonstrate a link between oxidative damage and tissue dysfunction that ensues from exposure to spaceflight factors (Figure 1). One of the outstanding questions that needs to be addressed is whether oxidative damage mediates or results from progressive tissue degeneration in the course of spaceflight [156]. More studies using genetic models for altered redox status will contribute to resolving this question. Future interplanetary travel will increase the duration of exposure of humans to spaceflight. Therefore, additional comprehensive investigations involving extended timepoints that simulate the duration of long-term missions and the ensuing recovery must be conducted to better estimate health risks to crew. Furthermore, the latent effects of spaceflight on the cardiovascular system and on stem and progenitor cells of the bone marrow must be examined for their potential to exacerbate the effects of aging. The use of antioxidant-based countermeasures must be guided by the understanding that low levels of ROS signaling are important for normal cellular function. Therefore, future studies must also assess the potential of any antioxidant-based countermeasure to perturb these essential signaling events and their consequences on the functional integrity of tissues. Research progress on the aformentioned areas will help in the development of effective approaches for maintaining crew health during and after interplanetary missions.

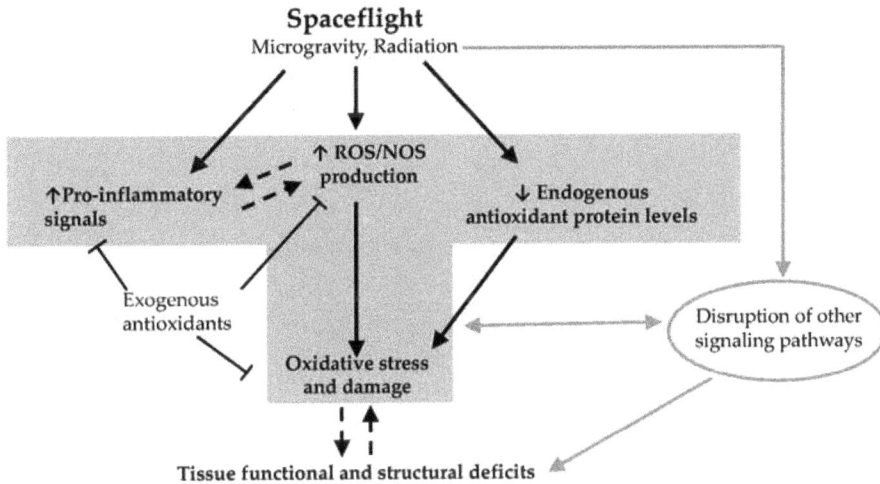

Figure 1. Hypothetical model on how spaceflight leads to deficits in tissue function and structural integrity. Exposure of tissues to elements of the spaceflight environment such as microgravity and radiation (and potentially other unknown factors) leads to enhanced production of reactive oxygen species (ROS) and reactive nitrogen species (NOS), increased levels of pro-inflammatory signals, and downregulation of endogenous antioxidant defenses. This leads to excess ROS/NOS due to an imbalance between endogenous antioxidant protein levels and ROS/NOS production. Excess ROS/NOS leads to oxidative damage of proteins, lipids and DNA which in turn result in deficits in tissue function and structural integrity. Other non-redox signaling processes may also contribute to these deficits. Some exogenous antioxidants found in the diet may block the increases in ROS/NOS levels and inflammatory signals, thereby preventing oxidative damage. It remains to be elucidated whether inflammation causes ROS production and/or vice versa in the context of spaceflight, and whether oxidative damage mediates or results from progressive tissue degeneration as a consequence of spaceflight. T-bar arrow: inhibitory effect; dotted line arrow: cause and effect needs further elucidation. Gray arrows depict the contribution of non-redox related processes in spaceflight-induced deficits in tissue structure and function.

Acknowledgments: We express gratitude to our valued collaborators, Michael Delp (Florida State University, Tallahassee, FL, USA) and Charles Limoli (University of California, Irvine, CA, USA), as well as Joshua Alwood, Ann-Sofie Schreurs, Yasaman Shirazi-Fard, Mohammed Shahnazari and the many other motivated postdoctoral scholars and students in the Bone and Signaling Lab who performed some of the research described in this review. We thank Marianne Sowa for helpful input during the preparation of this article. RKG is supported by grants from the NASA Space Biology Program and the NASA Human Health and Countermeasures Program; in particular we are grateful for research support from the National Space Biomedical Research Institute grant #MA02501 under NASA cooperative agreement NCC 9-58.

Conflicts of Interest: The authors declare no conflict of interest.

References

1. Lang, T.; LeBlanc, A.; Evans, H.; Lu, Y.; Genant, H.; Yu, A. Cortical and trabecular bone mineral loss from the spine and hip in long-duration spaceflight. *J. Bone Miner. Res.* **2004**, *19*, 1006–1012. [CrossRef] [PubMed]
2. LeBlanc, A.; Schneider, V.; Shackelford, L.; West, S.; Oganov, V.; Bakulin, A.; Voronin, L. Bone mineral and lean tissue loss after long duration space flight. *J. Musculoskelet. Neuronal Interact.* **2000**, *1*, 157–176. [PubMed]
3. Sibonga, J.D.; Evans, H.J.; Sung, H.G.; Spector, E.R.; Lang, T.F.; Oganov, V.S.; Bakulin, A.V.; Shackelford, L.C.; LeBlanc, A.D. Recovery of spaceflight-induced bone loss: Bone mineral density after long-duration missions as fitted with an exponential function. *Bone* **2007**, *41*, 973–978. [CrossRef] [PubMed]
4. Keyak, J.H.; Koyama, A.K.; LeBlanc, A.; Lu, Y.; Lang, T.F. Reduction in proximal femoral strength due to long-duration spaceflight. *Bone* **2009**, *44*, 449–495. [CrossRef] [PubMed]
5. Bloomfield, S.A.; Hogan, H.A.; Delp, M.D. Decreases in bone blood flow and bone material properties in aging Fischer-344 rats. *Clin. Orthop. Relat. Res.* **2002**, 248–285. [CrossRef]
6. Prisby, R.D.; Ramsey, M.W.; Behnke, B.J.; Dominguez, J.M., 2nd; Donato, A.J.; Allen, M.R.; Delp, M.D. Aging reduces skeletal blood flow, endothelium-dependent vasodilation, and NO bioavailability in rats. *J. Bone Miner. Res.* **2007**, *22*, 1280–1288. [CrossRef] [PubMed]
7. Burkhardt, R.; Kettner, G.; Bohm, W.; Schmidmeier, M.; Schlag, R.; Frisch, B.; Mallmann, B.; Eisenmenger, W.; Gilg, T. Changes in trabecular bone, hematopoiesis and bone marrow vessels in aplastic anemia, primary osteoporosis, and old age: A comparative histomorphometric study. *Bone* **1987**, *8*, 157–176. [CrossRef]
8. National Council on Radiation Protection and Measurements (NCRP). *Information Needed to Make Recommendations for Space Missions Beyond Low-Earth Orbit*; US National Council for Radiation Protection and Measurements: Bethesda, MD, USA, 2006.
9. Chancellor, J.C.; Scott, G.B.; Sutton, J.P. Space radiation: The number one risk to astronaut health beyond low earth orbit. *Life* **2014**, *4*, 491–510. [CrossRef] [PubMed]
10. Silberberg, R.; Tsao, C.H.; Adams, J.H., Jr.; Letaw, J.R. Radiation doses and LET distributions of cosmic rays. *Radiat. Res.* **1984**, *98*, 209–292. [CrossRef] [PubMed]
11. Cucinotta, F.A.; Katz, R.; Wilson, J.W.; Townsend, L.W.; Shinn, J.; Hajnal, F. Biological effectiveness of high-energy protons: Target fragmentation. *Radiat. Res.* **1991**, *127*, 130–137. [CrossRef] [PubMed]
12. Wilson, J.W.; Townsend, L.W.; Badavi, F.F. Galactic HZE propagation through the Earth's atmosphere. *Radiat. Res.* **1987**, *109*, 173–183. [CrossRef] [PubMed]
13. Sridharan, D.M.; Asaithamby, A.; Bailey, S.M.; Costes, S.V.; Doetsch, P.W.; Dynan, W.S.; Kronenberg, A.; Rithidech, K.N.; Saha, J.; Snijders, A.M.; et al. Understanding cancer development processes after HZE-particle exposure: Roles of ROS, DNA damage repair and inflammation. *Radiat. Res.* **2015**, *183*, 1–26. [CrossRef] [PubMed]
14. Saha, J.; Wilson, P.; Thieberger, P.; Lowenstein, D.; Wang, M.; Cucinotta, F.A. Biological characterization of low-energy ions with high-energy deposition on human cells. *Radiat. Res.* **2014**, *182*, 282–291. [CrossRef] [PubMed]
15. Limoli, C.L.; Giedzinski, E.; Rola, R.; Otsuka, S.; Palmer, T.D.; Fike, J.R. Radiation response of neural precursor cells: Linking cellular sensitivity to cell cycle checkpoints, apoptosis and oxidative stress. *Radiat. Res.* **2004**, *161*, 17–27. [CrossRef] [PubMed]
16. Sancar, A.; Lindsey-Boltz, L.A.; Unsal-Kacmaz, K.; Linn, S. Molecular mechanisms of mammalian DNA repair and the DNA damage checkpoints. *Annu. Rev. Biochem* **2004**, *73*, 39–85. [CrossRef] [PubMed]
17. Mavragani, I.V.; Nikitaki, Z.; Souli, M.P.; Aziz, A.; Nowsheen, S.; Aziz, K.; Rogakou, E.; Georgakilas, A.G. Complex DNA Damage: A Route to Radiation-Induced Genomic Instability and Carcinogenesis. *Cancers* **2017**, *9*. [CrossRef] [PubMed]
18. Hollander, J.; Gore, M.; Fiebig, R.; Mazzeo, R.; Ohishi, S.; Ohno, H.; Ji, L.L. Spaceflight downregulates antioxidant defense systems in rat liver. *Free Radic. Biol. Med.* **1998**, *24*, 385–390. [CrossRef]
19. Stein, T.P.; Leskiw, M.J. Oxidant damage during and after spaceflight. *Am. J. Physiol. Endocrinol. Metab.* **2000**, *278*, E375–E382. [PubMed]
20. Dinkova-Kostova, A.T.; Holtzclaw, W.D.; Cole, R.N.; Itoh, K.; Wakabayashi, N.; Katoh, Y.; Yamamoto, M.; Talalay, P. Direct evidence that sulfhydryl groups of *Keap1* are the sensors regulating induction of phase 2 enzymes that protect against carcinogens and oxidants. *Proc. Natl. Acad. Sci. USA* **2002**, *99*, 11908–11913. [CrossRef] [PubMed]

21. De Luca, C.; Deeva, I.; Mariani, S.; Maiani, G.; Stancato, A.; Korkina, L. Monitoring antioxidant defenses and free radical production in space-flight, aviation and railway engine operators, for the prevention and treatment of oxidative stress, immunological impairment, and pre-mature cell aging. *Toxicol. Ind. Health* **2009**, *25*, 259–267. [CrossRef] [PubMed]

22. Smith, S.M.; Davis-Street, J.E.; Rice, B.L.; Nillen, J.L.; Gillman, P.L.; Block, G. Nutritional status assessment in semiclosed environments: Ground-based and space flight studies in humans. *J. Nutr.* **2001**, *131*, 2053–2061. [PubMed]

23. Smith, S.M.; Zwart, S.R.; Block, G.; Rice, B.L.; Davis-Street, J.E. The nutritional status of astronauts is altered after long-term space flight aboard the International Space Station. *J. Nutr.* **2005**, *135*, 437–474. [PubMed]

24. Indo, H.P.; Majima, H.J.; Terada, M.; Suenaga, S.; Tomita, K.; Yamada, S.; Higashibata, A.; Ishioka, N.; Kanekura, T.; Nonaka, I.; et al. Changes in mitochondrial homeostasis and redox status in astronauts following long stays in space. *Sci. Rep.* **2016**, *6*. [CrossRef] [PubMed]

25. Connor, M.K.; Hood, D.A. Effect of microgravity on the expression of mitochondrial enzymes in rat cardiac and skeletal muscles. *J. Appl. Physiol.* **1998**, *84*, 593–598. [PubMed]

26. Ray, C.A.; Vasques, M.; Miller, T.A.; Wilkerson, M.K.; Delp, M.D. Effect of short-term microgravity and long-term hindlimb unloading on rat cardiac mass and function. *J. Appl. Physiol.* **2001**, *91*, 1207–1213. [PubMed]

27. Taylor, C.R.; Hanna, M.; Behnke, B.J.; Stabley, J.N.; McCullough, D.J.; Davis, R.T., 3rd; Ghosh, P.; Papadopoulos, A.; Muller-Delp, J.M.; Delp, M.D. Spaceflight-induced alterations in cerebral artery vasoconstrictor, mechanical, and structural properties: Implications for elevated cerebral perfusion and intracranial pressure. *FASEB J.* **2013**, *27*, 2282–2292. [CrossRef] [PubMed]

28. Dabertrand, F.; Porte, Y.; Macrez, N.; Morel, J.L. Spaceflight regulates ryanodine receptor subtype 1 in portal vein myocytes in the opposite way of hypertension. *J. Appl. Physiol.* **2012**, *112*, 471–480. [CrossRef] [PubMed]

29. Baqai, F.P.; Gridley, D.S.; Slater, J.M.; Luo-Owen, X.; Stodieck, L.S.; Ferguson, V.; Chapes, S.K.; Pecaut, M.J. Effects of spaceflight on innate immune function and antioxidant gene expression. *J. Appl. Physiol.* **2009**, *106*, 1935–1942. [CrossRef] [PubMed]

30. Mader, T.H.; Gibson, C.R.; Pass, A.F.; Kramer, L.A.; Lee, A.G.; Fogarty, J.; Tarver, W.J.; Dervay, J.P.; Hamilton, D.R.; Sargsyan, A.; et al. Optic disc edema, globe flattening, choroidal folds, and hyperopic shifts observed in astronauts after long-duration space flight. *Ophthalmology* **2011**, *118*, 2058–2069. [CrossRef] [PubMed]

31. Mao, X.W.; Pecaut, M.J.; Stodieck, L.S.; Ferguson, V.L.; Bateman, T.A.; Bouxsein, M.; Jones, T.A.; Moldovan, M.; Cunningham, C.E.; Chieu, J.; et al. Spaceflight environment induces mitochondrial oxidative damage in ocular tissue. *Radiat. Res.* **2013**, *180*, 340–350. [CrossRef] [PubMed]

32. Yan, X.; Sasi, S.P.; Gee, H.; Lee, J.; Yang, Y.; Mehrzad, R.; Onufrak, J.; Song, J.; Enderling, H.; Agarwal, A.; et al. Cardiovascular risks associated with low dose ionizing particle radiation. *PLoS ONE* **2014**, *9*. [CrossRef] [PubMed]

33. Kondo, H.; Yumoto, K.; Alwood, J.S.; Mojarrab, R.; Wang, A.; Almeida, E.A.; Searby, N.D.; Limoli, C.L.; Globus, R.K. Oxidative stress and γ radiation-induced cancellous bone loss with musculoskeletal disuse. *J. Appl. Physiol.* **2010**, *108*, 126–152. [CrossRef] [PubMed]

34. Alwood, J.S.; Shahnazari, M.; Chicana, B.; Schreurs, A.S.; Kumar, A.; Bartolini, A.; Shirazi-Fard, Y.; Globus, R.K. Ionizing radiation stimulates expression of pro-osteoclastogenic genes in marrow and skeletal tissue. *J. Interferon Cytokine Res.* **2015**, *35*, 480–487. [CrossRef] [PubMed]

35. Schreurs, A.S.; Shirazi-Fard, Y.; Shahnazari, M.; Alwood, J.S.; Truong, T.A.; Tahimic, C.G.; Limoli, C.L.; Turner, N.D.; Halloran, B.; Globus, R.K. Dried plum diet protects from bone loss caused by ionizing radiation. *Sci. Rep.* **2016**, *6*. [CrossRef] [PubMed]

36. Morey-Holton, E.R.; Globus, R.K. Hindlimb unloading rodent model: Technical aspects. *J. Appl. Physiol.* **2002**, *92*, 1367–1377. [CrossRef] [PubMed]

37. Globus, R.K.; Morey-Holton, E. Hindlimb unloading: Rodent analog for microgravity. *J. Appl. Physiol.* **2016**, *120*, 1196–1206. [CrossRef] [PubMed]

38. Bloomfield, S.A.; Martinez, D.A.; Boudreaux, R.D.; Mantri, A.V. Microgravity stress: Bone and connective tissue. *Compr. Physiol.* **2016**, *6*, 645–658. [CrossRef] [PubMed]

39. Morikawa, D.; Nojiri, H.; Saita, Y.; Kobayashi, K.; Watanabe, K.; Ozawa, Y.; Koike, M.; Asou, Y.; Takaku, T.; Kaneko, K.; et al. Cytoplasmic reactive oxygen species and *SOD1* regulate bone mass during mechanical unloading. *J. Bone Miner. Res.* **2013**, *28*, 2368–2380. [CrossRef] [PubMed]
40. Moffitt, J.A.; Henry, M.K.; Welliver, K.C.; Jepson, A.J.; Garnett, E.R. Hindlimb unloading results in increased predisposition to cardiac arrhythmias and alters left ventricular connexin 43 expression. *Am. J. Physiol. Regul. Integr. Comp. Physiol.* **2013**, *304*, R362–R373. [CrossRef] [PubMed]
41. Kang, H.; Fan, Y.; Sun, A.; Jia, X.; Deng, X. Simulated microgravity exposure modulates the phenotype of cultured vascular smooth muscle cells. *Cell Biochem. Biophys.* **2013**, *66*, 121–130. [CrossRef] [PubMed]
42. Tsvirkun, D.; Bourreau, J.; Mieuset, A.; Garo, F.; Vinogradova, O.; Larina, I.; Navasiolava, N.; Gauquelin-Koch, G.; Gharib, C.; Custaud, M.A. Contribution of social isolation, restraint, and hindlimb unloading to changes in hemodynamic parameters and motion activity in rats. *PLoS ONE* **2012**, *7*. [CrossRef] [PubMed]
43. Jung, A.S.; Harrison, R.; Lee, K.H.; Genut, J.; Nyhan, D.; Brooks-Asplund, E.M.; Shoukas, A.A.; Hare, J.M.; Berkowitz, D.E. Simulated microgravity produces attenuated baroreflex-mediated pressor, chronotropic, and inotropic responses in mice. *Am. J. Physiol Heart Circ. Physiol.* **2005**, *289*, H600–H607. [CrossRef] [PubMed]
44. Powers, J.; Bernstein, D. The mouse as a model of cardiovascular adaptations to microgravity. *J. Appl. Physiol.* **2004**, *97*, 1686–1692. [CrossRef] [PubMed]
45. Zhang, R.; Bai, Y.G.; Lin, L.J.; Bao, J.X.; Zhang, Y.Y.; Tang, H.; Cheng, J.H.; Jia, G.L.; Ren, X.L.; Ma, J. Blockade of AT1 receptor partially restores vasoreactivity, *NOS* expression, and superoxide levels in cerebral and carotid arteries of hindlimb unweighting rats. *J. Appl. Physiol.* **2009**, *106*, 251–258. [CrossRef] [PubMed]
46. Zhang, R.; Ran, H.H.; Ma, J.; Bai, Y.G.; Lin, L.J. NAD(P)H oxidase inhibiting with apocynin improved vascular reactivity in tail-suspended hindlimb unweighting rat. *J. Physiol. Biochem.* **2012**, *68*, 99–105. [CrossRef] [PubMed]
47. Kanazashi, M.; Okumura, Y.; Al-Nassan, S.; Murakami, S.; Kondo, H.; Nagatomo, F.; Fujita, N.; Ishihara, A.; Roy, R.R.; Fujino, H. Protective effects of astaxanthin on capillary regression in atrophied soleus muscle of rats. *Acta Physiol.* **2013**, *207*, 405–415. [CrossRef] [PubMed]
48. Lean, J.M.; Davies, J.T.; Fuller, K.; Jagger, C.J.; Kirstein, B.; Partington, G.A.; Urry, Z.L.; Chambers, T.J. A crucial role for thiol antioxidants in estrogen-deficiency bone loss. *J. Clin. Investig.* **2003**, *112*, 915–923. [CrossRef] [PubMed]
49. Garrett, I.R.; Boyce, B.F.; Oreffo, R.O.; Bonewald, L.; Poser, J.; Mundy, G.R. Oxygen-derived free radicals stimulate osteoclastic bone resorption in rodent bone in vitro and in vivo. *J. Clin. Investig.* **1990**, *85*, 632–639. [CrossRef] [PubMed]
50. Almeida, M.; Han, L.; Martin-Millan, M.; Plotkin, L.I.; Stewart, S.A.; Roberson, P.K.; Kousteni, S.; O'Brien, C.A.; Bellido, T.; Parfitt, A.M.; et al. Skeletal involution by age-associated oxidative stress and its acceleration by loss of sex steroids. *J. Biol. Chem.* **2007**, *282*, 27285–27297. [CrossRef] [PubMed]
51. Manolagas, S.C.; Almeida, M. Gone with the Wnts: β-Catenin, T-cell factor, forkhead box O, and oxidative stress in age-dependent diseases of bone, lipid, and glucose metabolism. *Mol. Endocrinol.* **2007**, *21*, 2605–2614. [CrossRef] [PubMed]
52. Ha, H.; Lee, J.H.; Kim, H.N.; Kim, H.M.; Kwak, H.B.; Lee, S.; Kim, H.H.; Lee, Z.H. α-Lipoic acid inhibits inflammatory bone resorption by suppressing prostaglandin E2 synthesis. *J. Immunol.* **2006**, *176*, 111–117. [CrossRef] [PubMed]
53. Chatterjee, S.; Fisher, A.B. Mechanotransduction: Forces, sensors, and redox signaling. *Antioxid. Redox Signal.* **2014**, *20*, 868–887. [CrossRef] [PubMed]
54. Hsieh, H.J.; Liu, C.A.; Huang, B.; Tseng, A.H.; Wang, D.L. Shear-induced endothelial mechanotransduction: The interplay between reactive oxygen species (ROS) and nitric oxide (NO) and the pathophysiological implications. *J. Biomed. Sci.* **2014**, *21*. [CrossRef] [PubMed]
55. Noris, M.; Morigi, M.; Donadelli, R.; Aiello, S.; Foppolo, M.; Todeschini, M.; Orisio, S.; Remuzzi, G.; Remuzzi, A. Nitric oxide synthesis by cultured endothelial cells is modulated by flow conditions. *Circ. Res.* **1995**, *76*, 536–564. [CrossRef] [PubMed]
56. Fukaya, Y.; Ohhashi, T. Acetylcholine- and flow-induced production and release of nitric oxide in arterial and venous endothelial cells. *Am. J. Physiol.* **1996**, *270*, H99–H106. [PubMed]

57. Korenaga, R.; Ando, J.; Tsuboi, H.; Yang, W.; Sakuma, I.; Toyo-oka, T.; Kamiya, A. Laminar flow stimulates ATP- and shear stress-dependent nitric oxide production in cultured bovine endothelial cells. *Biochem. Biophys. Res. Commun.* **1994**, *198*, 213–219. [CrossRef] [PubMed]

58. Klein-Nulend, J.; Helfrich, M.H.; Sterck, J.G.; MacPherson, H.; Joldersma, M.; Ralston, S.H.; Semeins, C.M.; Burger, E.H. Nitric oxide response to shear stress by human bone cell cultures is endothelial nitric oxide synthase dependent. *Biochem. Biophys. Res. Commun.* **1998**, *250*, 108–181. [CrossRef] [PubMed]

59. Bakker, A.D.; Soejima, K.; Klein-Nulend, J.; Burger, E.H. The production of nitric oxide and prostaglandin E(2) by primary bone cells is shear stress dependent. *J. Biomech.* **2001**, *34*, 671–677. [CrossRef]

60. Godo, S.; Shimokawa, H. Divergent roles of endothelial nitric oxide synthases system in maintaining cardiovascular homeostasis. *Free Radic. Biol. Med.* **2017**, *109*, 4–10. [CrossRef] [PubMed]

61. Polyak, K.; Xia, Y.; Zweier, J.L.; Kinzler, K.W.; Vogelstein, B. A model for *p53*-induced apoptosis. *Nature* **1997**, *389*, 300–305. [CrossRef] [PubMed]

62. Lotem, J.; Peled-Kamar, M.; Groner, Y.; Sachs, L. Cellular oxidative stress and the control of apoptosis by wild-type *p53*, cytotoxic compounds, and cytokines. *Proc. Natl. Acad. Sci. USA* **1996**, *93*, 9166–9171. [CrossRef] [PubMed]

63. McCord, J.M.; Fridovich, I. Superoxide dismutase. An enzymic function for erythrocuprein (hemocuprein). *J. Biol. Chem.* **1969**, *244*, 6049–6055. [PubMed]

64. Fukai, T.; Ushio-Fukai, M. Superoxide dismutases: Role in redox signaling, vascular function, and diseases. *Antioxid. Redox Signal.* **2011**, *15*, 1583–1606. [CrossRef] [PubMed]

65. Rhee, S.G.; Yang, K.S.; Kang, S.W.; Woo, H.A.; Chang, T.S. Controlled elimination of intracellular H_2O_2: Regulation of peroxiredoxin, catalase, and glutathione peroxidase via post-translational modification. *Antioxid. Redox Signal.* **2005**, *7*, 619–626. [CrossRef] [PubMed]

66. Sharpe, M.A.; Robb, S.J.; Clark, J.B. Nitric oxide and Fenton/Haber-Weiss chemistry: Nitric oxide is a potent antioxidant at physiological concentrations. *J. Neurochem.* **2003**, *87*, 386–394. [CrossRef] [PubMed]

67. Balasubramanian, B.; Pogozelski, W.K.; Tullius, T.D. DNA strand breaking by the hydroxyl radical is governed by the accessible surface areas of the hydrogen atoms of the DNA backbone. *Proc. Natl. Acad. Sci. USA* **1998**, *95*, 9738–9743. [CrossRef] [PubMed]

68. Mergia, E.; Friebe, A.; Dangel, O.; Russwurm, M.; Koesling, D. Spare guanylyl cyclase NO receptors ensure high NO sensitivity in the vascular system. *J. Clin. Investig.* **2006**, *116*, 1731–1737. [CrossRef] [PubMed]

69. Atochin, D.N.; Huang, P.L. Endothelial nitric oxide synthase transgenic models of endothelial dysfunction. *Pflugers Arch.* **2010**, *460*, 965–974. [CrossRef] [PubMed]

70. Atochin, D.N.; Demchenko, I.T.; Astern, J.; Boso, A.E.; Piantadosi, C.A.; Huang, P.L. Contributions of endothelial and neuronal nitric oxide synthases to cerebrovascular responses to hyperoxia. *J. Cereb. Blood Flow Metab.* **2003**, *23*, 1219–1226. [CrossRef] [PubMed]

71. Huang, P.L.; Huang, Z.; Mashimo, H.; Bloch, K.D.; Moskowitz, M.A.; Bevan, J.A.; Fishman, M.C. Hypertension in mice lacking the gene for endothelial nitric oxide synthase. *Nature* **1995**, *377*, 239–242. [CrossRef] [PubMed]

72. Huang, P.L. Lessons learned from nitric oxide synthase knockout animals. *Semin. Perinatol.* **2000**, *24*, 87–90. [CrossRef]

73. Freedman, J.E.; Sauter, R.; Battinelli, E.M.; Ault, K.; Knowles, C.; Huang, P.L.; Loscalzo, J. Deficient platelet-derived nitric oxide and enhanced hemostasis in mice lacking the NOSIII gene. *Circ. Res.* **1999**, *84*, 1416–1421. [CrossRef] [PubMed]

74. Kuhlencordt, P.J.; Rosel, E.; Gerszten, R.E.; Morales-Ruiz, M.; Dombkowski, D.; Atkinson, W.J.; Han, F.; Preffer, F.; Rosenzweig, A.; Sessa, W.C.; et al. Role of endothelial nitric oxide synthase in endothelial activation: Insights from eNOS knockout endothelial cells. *Am. J. Physiol. Cell Physiol.* **2004**, *286*, C1195–C1202. [CrossRef] [PubMed]

75. Lefer, D.J.; Jones, S.P.; Girod, W.G.; Baines, A.; Grisham, M.B.; Cockrell, A.S.; Huang, P.L.; Scalia, R. Leukocyte-endothelial cell interactions in nitric oxide synthase-deficient mice. *Am. J. Physiol.* **1999**, *276*, H1943–H1950. [PubMed]

76. Huang, Z.; Huang, P.L.; Ma, J.; Meng, W.; Ayata, C.; Fishman, M.C.; Moskowitz, M.A. Enlarged infarcts in endothelial nitric oxide synthase knockout mice are attenuated by nitro-L-arginine. *J. Cereb. Blood Flow Metab.* **1996**, *16*, 981–987. [CrossRef] [PubMed]

77. Atochin, D.N.; Wang, A.; Liu, V.W.; Critchlow, J.D.; Dantas, A.P.; Looft-Wilson, R.; Murata, T.; Salomone, S.; Shin, H.K.; Ayata, C.; et al. The phosphorylation state of *eNOS* modulates vascular reactivity and outcome of cerebral ischemia in vivo. *J. Clin. Investig.* **2007**, *117*, 1961–1967. [CrossRef] [PubMed]

78. Pacher, P.; Beckman, J.S.; Liaudet, L. Nitric oxide and peroxynitrite in health and disease. *Physiol. Rev.* **2007**, *87*, 315–424. [CrossRef] [PubMed]

79. Ghosh, P.; Behnke, B.J.; Stabley, J.N.; Kilar, C.R.; Park, Y.; Narayanan, A.; Alwood, J.S.; Shirazi-Fard, Y.; Schreurs, A.S.; Globus, R.K.; et al. Effects of high-let radiation exposure and hindlimb unloading on skeletal muscle resistance artery vasomotor properties and cancellous bone microarchitecture in mice. *Radiat. Res.* **2016**, *185*, 257–276. [CrossRef] [PubMed]

80. Ortega, N.; Behonick, D.J.; Werb, Z. Matrix remodeling during endochondral ossification. *Trends Cell Biol.* **2004**, *14*, 86–93. [CrossRef] [PubMed]

81. Mackie, E.J.; Ahmed, Y.A.; Tatarczuch, L.; Chen, K.S.; Mirams, M. Endochondral ossification: How cartilage is converted into bone in the developing skeleton. *Int. J. Biochem. Cell Biol.* **2008**, *40*, 46–62. [CrossRef] [PubMed]

82. Kirchen, M.E.; O'Connor, K.M.; Gruber, H.E.; Sweeney, J.R.; Fras, I.A.; Stover, S.J.; Sarmiento, A.; Marshall, G.J. Effects of microgravity on bone healing in a rat fibular osteotomy model. *Clin. Orthop. Relat. Res.* **1995**, 231–242.

83. Vandamme, K.; Holy, X.; Bensidhoum, M.; Deschepper, M.; Logeart-Avramoglou, D.; Naert, I.; Duyck, J.; Petite, H. Impaired osteoblastogenesis potential of progenitor cells in skeletal unloading is associated with alterations in angiogenic and energy metabolism profile. *Biomed. Mater. Eng.* **2012**, *22*, 219–226. [CrossRef] [PubMed]

84. Fei, J.; Peyrin, F.; Malaval, L.; Vico, L.; Lafage-Proust, M.H. Imaging and quantitative assessment of long bone vascularization in the adult rat using microcomputed tomography. *Anat. Rec.* **2010**, *293*, 215–252. [CrossRef]

85. Colleran, P.N.; Wilkerson, M.K.; Bloomfield, S.A.; Suva, L.J.; Turner, R.T.; Delp, M.D. Alterations in skeletal perfusion with simulated microgravity: A possible mechanism for bone remodeling. *J. Appl. Physiol.* **2000**, *89*, 1046–1054. [PubMed]

86. Ramasamy, S.K.; Kusumbe, A.P.; Schiller, M.; Zeuschner, D.; Bixel, M.G.; Milia, C.; Gamrekelashvili, J.; Limbourg, A.; Medvinsky, A.; Santoro, M.M.; et al. Blood flow controls bone vascular function and osteogenesis. *Nat. Commun.* **2016**, *7*. [CrossRef] [PubMed]

87. Maes, C.; Carmeliet, P.; Moermans, K.; Stockmans, I.; Smets, N.; Collen, D.; Bouillon, R.; Carmeliet, G. Impaired angiogenesis and endochondral bone formation in mice lacking the vascular endothelial growth factor isoforms *VEGF164* and *VEGF188*. *Mech. Dev.* **2002**, *111*, 61–73. [CrossRef]

88. Wang, Y.; Wan, C.; Deng, L.; Liu, X.; Cao, X.; Gilbert, S.R.; Bouxsein, M.L.; Faugere, M.C.; Guldberg, R.E.; Gerstenfeld, L.C.; et al. The hypoxia-inducible factor α pathway couples angiogenesis to osteogenesis during skeletal development. *J. Clin. Investig.* **2007**, *117*, 1616–1626. [CrossRef] [PubMed]

89. Bungo, M.W.; Johnson, P.C., Jr. Cardiovascular examinations and observations of deconditioning during the space shuttle orbital flight test program. *Aviat. Space Environ. Med.* **1983**, *54*, 1001–1004. [PubMed]

90. Mulvagh, S.L.; Charles, J.B.; Riddle, J.M.; Rehbein, T.L.; Bungo, M.W. Echocardiographic evaluation of the cardiovascular effects of short-duration spaceflight. *J. Clin. Pharmacol.* **1991**, *31*, 1024–1026. [CrossRef] [PubMed]

91. Meck, J.V.; Reyes, C.J.; Perez, S.A.; Goldberger, A.L.; Ziegler, M.G. Marked exacerbation of orthostatic intolerance after long- vs. short-duration spaceflight in veteran astronauts. *Psychosom. Med.* **2001**, *63*, 865–873. [PubMed]

92. Buckey, J.C., Jr.; Lane, L.D.; Levine, B.D.; Watenpaugh, D.E.; Wright, S.J.; Moore, W.E.; Gaffney, F.A.; Blomqvist, C.G. Orthostatic intolerance after spaceflight. *J. Appl. Physiol.* **1996**, *81*, 7–18. [PubMed]

93. Lee, S.M.; Feiveson, A.H.; Stein, S.; Stenger, M.B.; Platts, S.H. Orthostatic intolerance after ISS and space shuttle missions. *Aerosp. Med. Hum. Perform.* **2015**, *86* (Suppl. S12), 54–67. [CrossRef] [PubMed]

94. Stabley, J.N.; Dominguez, J.M., 2nd; Dominguez, C.E.; Mora Solis, F.R.; Ahlgren, J.; Behnke, B.J.; Muller-Delp, J.M.; Delp, M.D. Spaceflight reduces vasoconstrictor responsiveness of skeletal muscle resistance arteries in mice. *J. Appl. Physiol.* **2012**, *113*, 1439–1445. [CrossRef] [PubMed]

95. Delp, M.D. Myogenic and vasoconstrictor responsiveness of skeletal muscle arterioles is diminished by hindlimb unloading. *J. Appl. Physiol.* **1999**, *86*, 1178–1184. [PubMed]

96. Prisby, R.D.; Alwood, J.S.; Behnke, B.J.; Stabley, J.N.; McCullough, D.J.; Ghosh, P.; Globus, R.K.; Delp, M.D. Effects of hindlimb unloading and ionizing radiation on skeletal muscle resistance artery vasodilation and its relation to cancellous bone in mice. *J. Appl. Physiol.* **2016**, *120*, 97–106. [CrossRef] [PubMed]

97. Jasperse, J.L.; Woodman, C.R.; Price, E.M.; Hasser, E.M.; Laughlin, M.H. Hindlimb unweighting decreases ecNOS gene expression and endothelium-dependent dilation in rat soleus feed arteries. *J. Appl. Physiol.* **1999**, *87*, 1476–1482. [PubMed]

98. Woodman, C.R.; Schrage, W.G.; Rush, J.W.; Ray, C.A.; Price, E.M.; Hasser, E.M.; Laughlin, M.H. Hindlimb unweighting decreases endothelium-dependent dilation and *eNOS* expression in soleus not gastrocnemius. *J. Appl. Physiol.* **2001**, *91*, 1091–1098. [PubMed]

99. Prisby, R.D.; Behnke, B.J.; Allen, M.R.; Delp, M.D. Effects of skeletal unloading on the vasomotor properties of the rat femur principal nutrient artery. *J. Appl. Physiol.* **2015**, *118*, 980–988. [CrossRef] [PubMed]

100. Kaplanskii, A.S.; Durnova, G.N.; Sakharova, Z.F.; Il'ina-Kakueva, E.I. [Histomorphometric analysis of the bones of rats on board the Kosmos 1667 biosatellite]. *Kosm. Biol. Aviakosm. Med.* **1987**, *21*, 25–31. [PubMed]

101. Lueken, S.A.; Arnaud, S.B.; Taylor, A.K.; Baylink, D.J. Changes in markers of bone formation and resorption in a bed rest model of weightlessness. *J. Bone Miner. Res.* **1993**, *8*, 1433–1438. [CrossRef] [PubMed]

102. Shahnazari, M.; Kurimoto, P.; Boudignon, B.M.; Orwoll, B.E.; Bikle, D.D.; Halloran, B.P. Simulated spaceflight produces a rapid and sustained loss of osteoprogenitors and an acute but transitory rise of osteoclast precursors in two genetic strains of mice. *Am. J. Physiol. Endocrinol. Metab.* **2012**, *303*, E1354–E1362. [CrossRef] [PubMed]

103. Delp, M.D.; Charvat, J.M.; Limoli, C.L.; Globus, R.K.; Ghosh, P. apollo lunar astronauts show higher cardiovascular disease mortality: Possible deep space radiation effects on the vascular endothelium. *Sci Rep.* **2016**, *6*. [CrossRef] [PubMed]

104. Sykiotis, G.P.; Bohmann, D. *Keap1/Nrf2* signaling regulates oxidative stress tolerance and lifespan in Drosophila. *Dev. Cell* **2008**, *14*, 76–85. [CrossRef] [PubMed]

105. Ma, Q. Role of *nrf2* in oxidative stress and toxicity. *Annu. Rev. Pharmacol. Toxicol.* **2013**, *53*, 401–412. [CrossRef] [PubMed]

106. Loboda, A.; Damulewicz, M.; Pyza, E.; Jozkowicz, A.; Dulak, J. Role of Nrf2/HO-1 system in development, oxidative stress response and diseases: An evolutionarily conserved mechanism. *Cell. Mol. Life Sci.* **2016**, *73*, 3221–3247. [CrossRef] [PubMed]

107. Moi, P.; Chan, K.; Asunis, I.; Cao, A.; Kan, Y.W. Isolation of NF-E2-related factor 2 (*Nrf2*), a NF-E2-like basic leucine zipper transcriptional activator that binds to the tandem NF-E2/AP1 repeat of the β-globin locus control region. *Proc. Natl. Acad. Sci. USA* **1994**, *91*, 9926–9930. [CrossRef] [PubMed]

108. Zhang, H.; Davies, K.J.A.; Forman, H.J. Oxidative stress response and *Nrf2* signaling in aging. *Free Radic. Biol. Med.* **2015**, *88*, 314–336. [CrossRef] [PubMed]

109. Chan, K.; Lu, R.; Chang, J.C.; Kan, Y.W. NRF2, a member of the NFE2 family of transcription factors, is not essential for murine erythropoiesis, growth, and development. *Proc. Natl. Acad. Sci. USA* **1996**, *93*, 13943–13948. [CrossRef] [PubMed]

110. Pellegrini, G.G.; Cregor, M.; McAndrews, K.; Morales, C.C.; McCabe, L.D.; McCabe, G.P.; Peacock, M.; Burr, D.; Weaver, C.; Bellido, T. Nrf2 regulates mass accrual and the antioxidant endogenous response in bone differently depending on the sex and age. *PLoS ONE* **2017**, *12*. [CrossRef] [PubMed]

111. McDonald, J.T.; Kim, K.; Norris, A.J.; Vlashi, E.; Phillips, T.M.; Lagadec, C.; della Donna, L.; Ratikan, J.; Szelag, H.; Hlatky, L.; et al. Ionizing radiation activates the *Nrf2* antioxidant response. *Cancer Res.* **2010**, *70*, 8886–8895. [CrossRef] [PubMed]

112. Strom, J.; Chen, Q.M. Loss of Nrf2 promotes rapid progression to heart failure following myocardial infarction. *Toxicol. Appl. Pharmacol.* **2017**, *327*, 52–58. [CrossRef] [PubMed]

113. Larabee, C.M.; Desai, S.; Agasing, A.; Georgescu, C.; Wren, J.D.; Axtell, R.C.; Plafker, S.M. Loss of *Nrf2* exacerbates the visual deficits and optic neuritis elicited by experimental autoimmune encephalomyelitis. *Mol. Vis.* **2016**, *22*, 1503–1513. [PubMed]

114. Zhao, Z.; Chen, Y.; Wang, J.; Sternberg, P.; Freeman, M.L.; Grossniklaus, H.E.; Cai, J. Age-related retinopathy in *NRF2*-deficient mice. *PLoS ONE* **2011**, *6*. [CrossRef] [PubMed]

115. Kim, J.H.; Singhal, V.; Biswal, S.; Thimmulappa, R.K.; DiGirolamo, D.J. Nrf2 is required for normal postnatal bone acquisition in mice. *Bone Res.* **2014**, *2*. [CrossRef] [PubMed]

116. Rana, T.; Schultz, M.A.; Freeman, M.L.; Biswas, S. Loss of Nrf2 accelerates ionizing radiation-induced bone loss by upregulating *RANKL*. *Free Radic. Biol Med.* **2012**, *53*, 2298–2307. [CrossRef] [PubMed]

117. Gremmels, H.; de Jong, O.G.; Hazenbrink, D.H.; Fledderus, J.O.; Verhaar, M.C. The transcription factor *Nrf2* protects angiogenic capacity of endothelial colony-forming cells in high-oxygen radical stress conditions. *Stem Cells Int.* **2017**, *2017*. [CrossRef] [PubMed]

118. Hyeon, S.; Lee, H.; Yang, Y.; Jeong, W. Nrf2 deficiency induces oxidative stress and promotes *RANKL*-induced osteoclast differentiation. *Free Radic. Biol. Med.* **2013**, *65*, 789–799. [CrossRef] [PubMed]

119. Park, C.K.; Lee, Y.; Kim, K.H.; Lee, Z.H.; Joo, M.; Kim, H.H. Nrf2 is a novel regulator of bone acquisition. *Bone* **2014**, *63*, 36–46. [CrossRef] [PubMed]

120. Sun, Y.X.; Li, L.; Corry, K.A.; Zhang, P.; Yang, Y.; Himes, E.; Mihuti, C.L.; Nelson, C.; Dai, G.; Li, J. Deletion of *Nrf2* reduces skeletal mechanical properties and decreases load-driven bone formation. *Bone* **2015**, *74*, 1–9. [CrossRef] [PubMed]

121. Leach, C.S. A review of the consequences of fluid and electrolyte shifts in weightlessness. *Acta Astronaut.* **1979**, *6*, 1123–1135. [CrossRef]

122. Natochin, Y.V.; Kozyrevskaya, G.I.; Grigor'yev, A.I. Study of water-salt metabolism and renal function in cosmonauts. *Acta Astronaut.* **1975**, *2*, 175–188. [CrossRef]

123. Grigor'ev, A.I.; Dorokhova, B.P.; Semenov, V.; Morukov, B.V.; Baichorov, E.O. [Water-electrolyte balance and functional state of the kidneys in cosmonauts after a 185-day space flight]. *Kosm. Biol. Aviakosm. Med.* **1985**, *19*, 17–21.

124. Grigoriev, A.I.; Morukov, B.V.; Vorobiev, D.V. Water and electrolyte studies during long-term missions onboard the space stations SALYUT and MIR. *Clin. Investig.* **1994**, *72*, 169–198. [CrossRef] [PubMed]

125. Priestley, J.R.; Kautenburg, K.E.; Casati, M.C.; Endres, B.T.; Geurts, A.M.; Lombard, J.H. The *NRF2* knockout rat: A new animal model to study endothelial dysfunction, oxidant stress, and microvascular rarefaction. *Am. J. Physiol. Heart Circ. Physiol.* **2016**, *310*, H478–H488. [CrossRef] [PubMed]

126. Smietana, M.J.; Arruda, E.M.; Faulkner, J.A.; Brooks, S.V.; Larkin, L.M. Reactive oxygen species on bone mineral density and mechanics in Cu,Zn superoxide dismutase (*Sod1*) knockout mice. *Biochem. Biophys. Res. Commun.* **2010**, *403*, 149–153. [CrossRef] [PubMed]

127. Muller, F.L.; Song, W.; Liu, Y.; Chaudhuri, A.; Pieke-Dahl, S.; Strong, R.; Huang, T.T.; Epstein, C.J.; Roberts, L.J., 2nd; Csete, M.; et al. Absence of CuZn superoxide dismutase leads to elevated oxidative stress and acceleration of age-dependent skeletal muscle atrophy. *Free Radic. Biol. Med.* **2006**, *40*, 1993–2004. [CrossRef] [PubMed]

128. Schriner, S.E.; Linford, N.J.; Martin, G.M.; Treuting, P.; Ogburn, C.E.; Emond, M.; Coskun, P.E.; Ladiges, W.; Wolf, N.; van Remmen, H.; et al. Extension of murine life span by overexpression of catalase targeted to mitochondria. *Science* **2005**, *308*, 1909–1911. [CrossRef] [PubMed]

129. Treuting, P.M.; Linford, N.J.; Knoblaugh, S.E.; Emond, M.J.; Morton, J.F.; Martin, G.M.; Rabinovitch, P.S.; Ladiges, W.C. Reduction of age-associated pathology in old mice by overexpression of catalase in mitochondria. *J. Gerontol. A Biol. Sci. Med. Sci.* **2008**, *63*, 813–832. [CrossRef] [PubMed]

130. Mao, P.; Manczak, M.; Calkins, M.J.; Truong, Q.; Reddy, T.P.; Reddy, A.P.; Shirendeb, U.; Lo, H.H.; Rabinovitch, P.S.; Reddy, P.H. Mitochondria-targeted catalase reduces abnormal APP processing, amyloid β production and *BACE1* in a mouse model of Alzheimer's disease: Implications for neuroprotection and lifespan extension. *Hum. Mol. Genet.* **2012**, *21*, 2973–2990. [CrossRef] [PubMed]

131. Dai, D.F.; Chen, T.; Wanagat, J.; Laflamme, M.; Marcinek, D.J.; Emond, M.J.; Ngo, C.P.; Prolla, T.A.; Rabinovitch, P.S. Age-dependent cardiomyopathy in mitochondrial mutator mice is attenuated by overexpression of catalase targeted to mitochondria. *Aging Cell* **2010**, *9*, 536–564. [CrossRef] [PubMed]

132. Dai, D.F.; Santana, L.F.; Vermulst, M.; Tomazela, D.M.; Emond, M.J.; MacCoss, M.J.; Gollahon, K.; Martin, G.M.; Loeb, L.A.; Ladiges, W.C.; et al. Overexpression of catalase targeted to mitochondria attenuates murine cardiac aging. *Circulation* **2009**, *119*, 2789–2797. [CrossRef] [PubMed]

133. Dai, D.F.; Rabinovitch, P. Mitochondrial oxidative stress mediates induction of autophagy and hypertrophy in angiotensin-II treated mouse hearts. *Autophagy* **2011**, *7*, 917–918. [CrossRef] [PubMed]

134. Wang, Y.; Wang, G.Z.; Rabinovitch, P.S.; Tabas, I. Macrophage mitochondrial oxidative stress promotes atherosclerosis and nuclear factor-κB-mediated inflammation in macrophages. *Circ. Res.* **2014**, *114*, 421–433. [CrossRef] [PubMed]

135. Ucer, S.; Iyer, S.; Kim, H.N.; Han, L.; Rutlen, C.; Allison, K.; Thostenson, J.D.; de Cabo, R.; Jilka, R.L.; O'Brien, C.; et al. The effects of aging and sex steroid deficiency on the murine skeleton are independent and mechanistically distinct. *J. Bone Miner. Res.* **2017**, *32*, 560–574. [CrossRef] [PubMed]

136. Liao, A.C.; Craver, B.M.; Tseng, B.P.; Tran, K.K.; Parihar, V.K.; Acharya, M.M.; Limoli, C.L. Mitochondrial-targeted human catalase affords neuroprotection from proton irradiation. *Radiat. Res.* **2013**, *180*, 1–6. [CrossRef] [PubMed]

137. Parihar, V.K.; Allen, B.D.; Tran, K.K.; Chmielewski, N.N.; Craver, B.M.; Martirosian, V.; Morganti, J.M.; Rosi, S.; Vlkolinsky, R.; Acharya, M.M.; et al. Targeted overexpression of mitochondrial catalase prevents radiation-induced cognitive dysfunction. *Antioxid. Redox Signal.* **2015**, *22*, 78–91. [CrossRef] [PubMed]

138. Chmielewski, N.N.; Caressi, C.; Giedzinski, E.; Parihar, V.K.; Limoli, C.L. Contrasting the effects of proton irradiation on dendritic complexity of subiculum neurons in wild type and MCAT mice. *Environ. Mol. Mutagen.* **2016**, *57*, 364–371. [CrossRef] [PubMed]

139. Chang, J.; Feng, W.; Wang, Y.; Luo, Y.; Allen, A.R.; Koturbash, I.; Turner, J.; Stewart, B.; Raber, J.; Hauer-Jensen, M.; et al. Whole-body proton irradiation causes long-term damage to hematopoietic stem cells in mice. *Radiat. Res.* **2015**, *183*, 240–248. [CrossRef] [PubMed]

140. Chen, L.; Ran, Q.; Xiang, Y.; Xiang, L.; Chen, L.; Li, F.; Wu, J.; Wu, C.; Li, Z. Co-activation of *PKC-Δ* by *CRIF1* modulates oxidative stress in bone marrow multipotent mesenchymal stromal cells after Irradiation by phosphorylating NRF2 ser40. *Theranostics* **2017**, *7*, 2634–2648. [CrossRef] [PubMed]

141. Alwood, J.S.; Yumoto, K.; Mojarrab, R.; Limoli, C.L.; Almeida, E.A.; Searby, N.D.; Globus, R.K. Heavy ion irradiation and unloading effects on mouse lumbar vertebral microarchitecture, mechanical properties and tissue stresses. *Bone* **2010**, *47*, 248–285. [CrossRef] [PubMed]

142. Yumoto, K.; Globus, R.K.; Mojarrab, R.; Arakaki, J.; Wang, A.; Searby, N.D.; Almeida, E.A.; Limoli, C.L. Short-term effects of whole-body exposure to (56)fe ions in combination with musculoskeletal disuse on bone cells. *Radiat. Res.* **2010**, *173*, 494–504. [CrossRef] [PubMed]

143. Lloyd, S.A.; Bandstra, E.R.; Willey, J.S.; Riffle, S.E.; Tirado-Lee, L.; Nelson, G.A.; Pecaut, M.J.; Bateman, T.A. Effect of proton irradiation followed by hindlimb unloading on bone in mature mice: A model of long-duration spaceflight. *Bone* **2012**, *51*, 756–766. [CrossRef] [PubMed]

144. Alwood, J.S.; Kumar, A.; Tran, L.H.; Wang, A.; Limoli, C.L.; Globus, R.K. Low-dose, ionizing radiation and age-related changes in skeletal microarchitecture. *J. Aging Res.* **2012**, *2012*. [CrossRef] [PubMed]

145. Guan, J.; Stewart, J.; Ware, J.H.; Zhou, Z.; Donahue, J.J.; Kennedy, A.R. Effects of dietary supplements on the space radiation-induced reduction in total antioxidant status in CBA mice. *Radiat. Res.* **2006**, *165*, 373–378. [CrossRef] [PubMed]

146. Kennedy, A.R.; Davis, J.G.; Carlton, W.; Ware, J.H. Effects of dietary antioxidant supplementation on the development of malignant lymphoma and other neoplastic lesions in mice exposed to proton or iron-ion radiation. *Radiat. Res.* **2008**, *169*, 615–625. [CrossRef] [PubMed]

147. Wambi, C.; Sanzari, J.; Wan, X.S.; Nuth, M.; Davis, J.; Ko, Y.H.; Sayers, C.M.; Baran, M.; Ware, J.H.; Kennedy, A.R. Dietary antioxidants protect hematopoietic cells and improve animal survival after total-body irradiation. *Radiat. Res.* **2008**, *169*, 384–396. [CrossRef] [PubMed]

148. Matsugo, S.; Yan, L.J.; Han, D.; Trischler, H.J.; Packer, L. Elucidation of antioxidant activity of α lipoic acid toward hydroxyl radical. *Biochem. Biophys. Res. Commun.* **1995**, *208*, 161–167. [CrossRef] [PubMed]

149. Brown, S.L.; Kolozsvary, A.; Liu, J.; Jenrow, K.A.; Ryu, S.; Kim, J.H. Antioxidant diet supplementation starting 24 hours after exposure reduces radiation lethality. *Radiat. Res.* **2010**, *173*, 462–468. [CrossRef] [PubMed]

150. Espirito Santo, A.I.; Ersek, A.; Freidin, A.; Feldmann, M.; Stoop, A.A.; Horwood, N.J. Selective inhibition of *TNFR1* reduces osteoclast numbers and is differentiated from anti-TNF in a LPS-driven model of inflammatory bone loss. *Biochem. Biophys. Res. Commun.* **2015**, *464*, 1145–1150. [CrossRef] [PubMed]

151. Goldring, S.R.; Purdue, P.E.; Crotti, T.N.; Shen, Z.; Flannery, M.R.; Binder, N.B.; Ross, F.P.; McHugh, K.P. Bone remodelling in inflammatory arthritis. *Ann. Rheum. Dis.* **2013**, *72* (Suppl. S2). [CrossRef] [PubMed]

152. Vinson, J.A.; Zubik, L.; Bose, P.; Samman, N.; Proch, J. Dried fruits: Excellent in vitro and in vivo antioxidants. *J. Am. Coll. Nutr.* **2005**, *24*, 44–50. [CrossRef] [PubMed]

153. Halloran, B.P.; Wronski, T.J.; VonHerzen, D.C.; Chu, V.; Xia, X.; Pingel, J.E.; Williams, A.A.; Smith, B.J. Dietary dried plum increases bone mass in adult and aged male mice. *J. Nutr.* **2010**, *140*, 1781–1787. [CrossRef] [PubMed]

154. Johnson, C.D.; Lucas, E.A.; Hooshmand, S.; Campbell, S.; Akhter, M.P.; Arjmandi, B.H. Addition of fructooligosaccharides and dried plum to soy-based diets reverses bone loss in the ovariectomized rat. *Evid. Based Complement. Alternat. Med.* **2011**, *2011*. [CrossRef] [PubMed]

155. Rendina, E.; Lim, Y.F.; Marlow, D.; Wang, Y.; Clarke, S.L.; Kuvibidila, S.; Lucas, E.A.; Smith, B.J. Dietary supplementation with dried plum prevents ovariectomy-induced bone loss while modulating the immune response in C57BL/6J mice. *J. Nutr. Biochem.* **2012**, *23*, 60–68. [CrossRef] [PubMed]

156. Takahashi, K.; Okumura, H.; Guo, R.; Naruse, K. Effect of oxidative stress on cardiovascular system in response to gravity. *Int. J. Mol. Sci.* **2017**, *18*. [CrossRef] [PubMed]

International Journal of
Molecular Sciences

MDPI

Review

Effect of Oxidative Stress on Cardiovascular System in Response to Gravity

Ken Takahashi [1,*], Hiroki Okumura [2], Rui Guo [1,3] and Keiji Naruse [1]

[1] Department of Cardiovascular Physiology, Graduate School of Medicine, Dentistry and Pharmaceutical
    Sciences, Okayama University, Okayama 700-8558, Japan; kakuei0830@gmail.com (R.G.);
    knaruse@md.okayama-u.ac.jp (K.N.)
[2] Department of Medicine, Okayama University, Okayama 700-8558, Japan; pbuu59vj@s.okayama-u.ac.jp
[3] Department of Cardiovascular Surgery, Harbin Medical University, Harbin 150001, China
*   Correspondence: takah-k2@okayama-u.ac.jp; Tel.: +81-86-235-7119

Received: 1 June 2017; Accepted: 27 June 2017; Published: 4 July 2017

Abstract: Long-term habitation in space leads to physiological alterations such as bone loss, muscle atrophy, and cardiovascular deconditioning. Two predominant factors—namely space radiation and microgravity—have a crucial impact on oxidative stress in living organisms. Oxidative stress is also involved in the aging process, and plays important roles in the development of cardiovascular diseases including hypertension, left ventricular hypertrophy, and myocardial infarction. Here, we discuss the effects of space radiation, microgravity, and a combination of these two factors on oxidative stress. Future research may facilitate safer living in space by reducing the adverse effects of oxidative stress.

Keywords: oxidative stress; reactive oxygen species; radiation; microgravity

1. Introduction

Five million years after the birth of humankind, we are living in the space age, with the International Space Station continuously accommodating crew members orbiting around the Earth, planning commercial flights to the Moon, and even discussing Mars exploration realistically [1–3]. These frontiers excite humanity. We can pursue it and are destined to do so. However, as the duration of stay in space extends to months and years, it has gradually become evident that the space environment affects our physiological functions. A few obvious alterations were identified in the earlier days of space exploration: bone loss [4–6], muscle atrophy [5–7], and cardiovascular deconditioning, of which orthostatic intolerance is one of the symptoms [6,8–10]. Analysis of the long-term effects of spaceflight on human health requires several decades.

The Apollo program was a magnificent project that embodied science technology and exploration, sending 24 astronauts from the Earth to the lunar orbit. While the program represented a significant and unshakable milestone in human history, an alarming fact regarding health risks was reported 40 years later; this report indicated that the Apollo lunar astronauts show higher cardiovascular disease mortality rate [11], caused by heart failure, myocardial infarction, stroke, brain aneurysm, or blood clots than their counterparts who experienced the space environment only at low Earth orbit (LEO) and who did not experience space travel [11]. The authors of this report, which include a researcher from the National Aeronautics and Space Administration (NASA), assumed that the reason for the higher mortality is space radiation, based on an experiment in mice. Meanwhile, a serious opposing opinion on this report was expressed in terms of the method of data collection and analysis [12]. Therefore, adequate care should be taken to consider the cardiovascular disease mortality rate in response to space radiation. However, it is of great importance to scrutinize the possible effect of the space environment on human health.

One of the alterations caused by long-term space stay is the development of pro-oxidative conditions, including elevated expression of oxidative enzymes (e.g., nicotinamide adenine dinucleotide phosphate (NADP$^+$) oxidase (NOX)) and decreased expression of anti-oxidative enzymes (e.g., superoxide dismutase, SOD, and glutathione peroxidase, GPx). Pro-oxidative conditions are observed in spaceflight and simulated space environments (radiation and microgravity) in various types of organs and cells, including erythrocytes [13,14], endothelial cells [15], retina [16], skin [17], brain [18,19], neuronal cells [20], liver [21,22], and skeletal muscles [23,24]. In several studies, the direct detection of increased reactive oxygen species (ROS) or the detection of substances produced by oxidative reactions were reported [13,16,18,20,21,23], implying that the production of oxidative substances increases in the space environment. In this review, we first provide an overview of oxidative stress; then, we discuss the generation of oxidative stress in response to radiation, microgravity, and a combination of these two factors.

2. What Are Reactive Oxygen Species (ROS)?

ROS are oxidizing agents produced by both endogenous (mitochondria, peroxisomes, lipoxygenases, NOX, and cytochrome P450) and exogenous (ultraviolet light, ionizing radiation, chemotherapeutics, inflammatory cytokines, and environmental toxins) factors [25]. Superoxide anion ($O_2{}^-$) [26], hydrogen peroxide (H_2O_2), and hydroxyl radicals (OH·) are the major types of ROS—each of which has preferential biological targets based on its chemical properties [27]. ROS have two different actions. First, with their unstable and highly reactive chemical properties, ROS react with lipids, proteins, and DNA [27], resulting in aging, disease, and cell death [25,28]. Second, in contrast to the first destructive action, ROS are involved in cellular homeostatic functions such as proliferation [29,30] via heat-shock transcription factor 1, nuclear factor-κB, p53, phosphoinositide 3-kinase, and mitogen-activated protein kinase pathways [25]. In contrast to the pro-oxidative process and enzymes described above, SOD, peroxiredoxin, glutathione reductase, GPx, and catalase (CAT) are anti-oxidative enzymes that reduce the levels of ROS.

ROS in the Cardiovascular System

As described by Sugamura and Keaney [31], biological processes in the mitochondrial respiratory chain and subsequent enzymatic processes cause ROS generation in the cardiovascular system. The enzymes involved in these processes are NOX, xanthine oxidase (XO), lipoxygenase, nitric oxide synthase (NOS), and myeloperoxidase (Figure 1). Subtypes of NOX proteins are widely expressed in the cardiovascular system—specifically NOX1 (vascular smooth muscle cells), NOX2 (endothelium, vascular smooth muscle cells, adventitia, and cardiomyocytes), NOX4 (endothelium, vascular smooth muscle cells, cardiomyocytes, and cardiac stem cells), and NOX5 (vascular smooth muscle cells) [32]. NOX is involved in the development of cardiovascular diseases such as hypertension, left ventricular hypertrophy, and myocardial infarction [32]. Besides, NOX is also involved in cardiovascular physiology including angiogenesis [33] and blood pressure regulation [34].

Endothelial dysfunction is a hallmark of cardiovascular diseases [35]. An increase in oxidative stress leads to monomerization of the endothelial isoform of NOS (eNOS), which in turn causes further production of superoxide anion rather than nitric oxide [36]. Insufficiency of the nitric oxide production contributes to endothelial dysfunction and the resultant cardiovascular disorders, including hypertension [37].

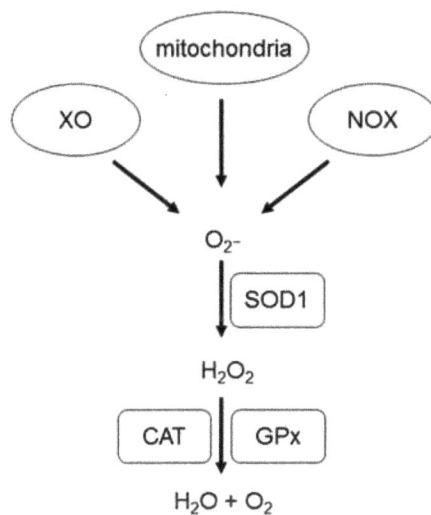

Figure 1. Generation of reactive oxygen species. Superoxide anion is produced by xanthine oxidase (XO), mitochondria, and NADP$^+$ oxidase (NOX). Superoxide anion is converted to hydrogen peroxide (H_2O_2) by superoxide dismutase 1 (SOD1), and then to H_2O and O_2 by catalase (CAT) and glutathione peroxidase (GPx).

3. ROS Generation in Response to Radiation

Exposure to hazardous radiation from galactic cosmic rays and solar particle events such as solar flares significantly decreases at the altitude of LEO and below due to shielding by the Earth's atmosphere and magnetosphere [38]. Therefore, the cause of higher cardiovascular risk in the Apollo lunar astronauts is inferred to be severe deep space radiation [11]. Indeed, space radiation causes adverse effects such as DNA damage [39,40] and cell senescence [41]. The adverse effects of radiation are firstly due to direct damage to cellular structures such as DNA, which are exposed to radiation. Secondly, radiation decomposes water molecules to ROS such as O_2^-, $OH\cdot$, and H_2O_2 [42]. Thirdly, high-charge and high-energy (HZE) ion particle radiation—which is a component of galactic cosmic rays—generates secondary radiation around its initial location [43]. Lastly, ROS may spread to nearby cells and cause long-term damage [44]. It is reported that long-term space stay affects neuronal function [45–47], possibly due to the effects of space radiation on the nervous system [48,49].

Cardiovascular ROS Generation in Response to Radiation

In the cardiovascular system, radiation causes ischemic heart disease [50], cardiomyopathy, and stroke [51,52]. Supporting this fact is the observation that exposure to HZE radiation facilitates the activation of XO (Figure 1) and a resultant increase of ROS in vascular endothelial cells in rats [53]. Furthermore, XO activity was elevated, and aortic stiffness was higher, even 4 and 6 months after a single radiation exposure, respectively. Increased XO expression in response to HZE radiation was confirmed in mouse endothelial cells as well [11]. In terms of the long-term effects of HZE radiation, elevated ROS and mitochondrial superoxide were observed in intestinal epithelial cells, along with increased NOX expression and decreased SOD and CAT expressions, 1 year after exposure [54].

4. ROS Generation in Response to Microgravity

The predominant mechanism of ROS generation in microgravity conditions seems to be the upregulation of oxidative enzymes and downregulation of anti-oxidative enzymes. For example, simulated microgravity has been reported to induce a decrease in the antioxidant enzymes SOD, GPx, and CAT and an increase in the amount of ROS in rat neuronal PC12 cells 96 h after the onset [20]. A similar increase in ROS production was observed in another neuronal cell line, SH-SY5Y [55]. Decreased expression of the anti-oxidative enzyme CAT in the soleus muscle was reported in mice habituated in space for 30 days [56]. Wise et al. reported increased ROS and decreased glutathione levels in response to simulated microgravity using hind limb unloading in the brainstem and frontal cortex of mice [19]. Lipid peroxidation was observed over a wide range of areas in the brain, including the brainstem, cerebellum, frontal cortex, hippocampus, and striatum. In erythrocytes, increased lipid peroxidation was observed after spaceflight [13].

Bed rest in a $6°$-head-down tilt posture is often used to simulate the effect of microgravity using human subjects. The unloading condition derived from bedrest induces pro-oxidative conditions, namely decreased expression of the genes related to antioxidation, such as cytochrome c, nicotinamide nucleotide transhydrogenase, and glutathione S-transferase $\kappa 1$ [57]. Using a hindlimb unloading (HLU) rodent model is another experimental method frequently used to simulate the effect of microgravity. Increased ROS generation along with a decrease in the anti-oxidative protein SOD was observed in rat hippocampus in response to HLU [58].

Cardiovascular ROS Generation in Response to Microgravity

Another study demonstrated that 3-week HLU caused an increase in superoxide anion levels in the basilar and carotid arteries of rats via the local renin–angiotensin system [59]. In this study, upregulated expression of eNOS was observed in the carotid artery. The effect of microgravity on the cardiovascular system seems to be different depending on the region. Although 4-week hindlimb unloading led to an increase in superoxide levels along with an elevation in the levels of the pro-oxidative enzymes NOX2 and NOX4 and a decrease in the levels of the anti-oxidative enzymes Mn-SOD and GPx-1 in cerebral arteries, this effect was not observed in mesenteric arteries [60,61]. In human umbilical vein endothelial cells, the expression of pro-oxidative thioredoxin-interacting protein was increased in response to 10-day spaceflight [15].

5. Combination of Radiation and Microgravity

The effect of a combination of radiation and microgravity on ROS generation seems to be synergistic. Mao et al. studied the effect of low-dose radiation (LDR) and microgravity on oxidative damage in mouse brain, using HLU to simulate microgravity [18]. Surprisingly, exposure to a combination of LDR and HLU—but not LDR or HLU alone—for 7 days caused lipid oxidation in the brain cortex. After 9 months of exposure, lipid peroxidation was observed in the LDR and HLU conditions alone, but was more evident in the LDR + HLU condition. A stronger effect from the combination than from LDR alone was similarly observed in the hippocampus. Therefore, radiation and microgravity have been suggested to have a synergistic effect on lipid oxidation. A synergistic effect was implied in NOX2 expression as well. In contrast, Mao et al. showed reduced SOD activity in simulated microgravity [18]. This effect seems to be specific to the microgravity condition.

ROS production in mouse embryonic stem cells in the presence of H_2O_2 that mimics the effect of radiation exposure increased in response to simulated microgravity [62]. The production of ROS from any of the ROS sources (e.g., mitochondria, XO, endothelial NOS, and NOX) can facilitate ROS production from the other sources [63]. This may be the underlying mechanism behind the synergistic effect of radiation and microgravity on the ROS production described above.

6. Conclusions

The space environment is predominantly characterized by space radiation and microgravity. These two factors are prominently involved in ROS generation in biological systems. As we discussed here, ROS production is facilitated in specific organs and tissues, including neuronal and cardiovascular systems. However, Stein et al. reported that oxidative stress decreases during spaceflight and then increases after returning to Earth, based on the measurement of a biomarker of lipid oxidation—8-iso-prostaglandin F2α—in urine [64,65]. The discrepancy between organ/tissue/cellular level increase and individual level decrease of oxidative stress during space flight may be attributable to several factors. Firstly, food intake decreases [66,67] and anabolic response is impaired [68] in spaceflight, which may decrease the production of ROS in mitochondria. Secondly, the extent of oxidative stress can be different among tissue types. While several tissues described above show pro-oxidative conditions, some tissue types such as fibroblasts [13] and macrophages [69] show anti-oxidative conditions. Thirdly, the detection sensitivity for oxidative stress from samples such as blood or urine can vary depending on the method. A combination of biomarkers is suggested to provide a more accurate assessment of oxidative stress [70].

In this review, we mainly discussed the alteration of the expression level of ROS-related proteins. However, response to gravity is a mechanobiological process. While the role of the hippo pathway—which is involved in mechanosensitive control of organ size [71]—in gravity sensing has been reported [72,73], much remains to be elucidated to understand the mechanotransduction of gravity.

Men age faster in space. Physiological changes in microgravity and the aging process share common features, such as muscle and bone atrophy, balance and coordination problems (after returning to 1 *g* environment), decreased functional capacity of the cardiovascular system, mild hypothyroidism, increased stress hormones, decreased sex steroids, impaired anabolic response to food intake, and systemic inflammatory response [68,74]. According to NASA researchers Vernikos and Schneider, the processes of muscle and bone atrophy and loss of functional capacity of the cardiovascular system occur about ten times faster in space than on Earth [74]. In contrast, oxidative stress—which we discussed here in the context of a process derived from space radiation and microgravity—is thought to be crucial for the aging process [75]. Understanding oxidative stress in the space environment may help us understand the aging process and contribute to solving the problem of aging. Future research in this field will open the door to a safe passage to Mars and beyond.

Acknowledgments: This study was supported by Grant-in-Aid for Scientific Research on Innovative Areas, No. 15H05936.

Author Contributions: Ken Takahashi, Hiroki Okumura, and Rui Guo drafted the paper. Keiji Naruse designed the project.

Conflicts of Interest: The authors declare no conflict of interest.

Abbreviations

LEO	Low Earth Orbit
NASA	National Aeronautics and Space Administration
ROS	Reactive Oxygen Species
NADP$^+$	Nicotinamide Adenine Dinucleotide Phosphate
NOX	Nicotinamide Adenine Dinucleotide Phosphate Oxidase
XO	Xanthine Oxidase
NOS	Nitric Oxide Synthase
SOD	Superoxide Dismutase
GPx	Glutathione Peroxidase
CAT	Catalase
HZE	High-Charge and High-Energy
LDR	Low-Dose Radiation
HLU	Hindlimb Unloading
eNOS	Endothelial Isoform of Nitric Oxide Synthase

References

1. Witze, A. NASA rethinks approach to Mars exploration. *Nature* **2016**, *538*, 149–150. [CrossRef] [PubMed]
2. Cucinotta, F.A. Review of NASA approach to space radiation risk assessments for Mars exploration. *Health Phys.* **2015**, *108*, 131–142. [CrossRef] [PubMed]
3. Messina, P.; Vennemann, D. The European space exploration programme: Current status of ESA's plans for Moon and Mars exploration. *Acta Astronaut.* **2005**, *57*, 156–160. [CrossRef] [PubMed]
4. Grimm, D.; Grosse, J.; Wehland, M.; Mann, V.; Reseland, J.E.; Sundaresan, A.; Corydon, T.J. The impact of microgravity on bone in humans. *Bone* **2016**, *87*, 44–56. [CrossRef] [PubMed]
5. Stein, T.P. Weight, muscle and bone loss during space flight: Another perspective. *Eur. J. Appl. Physiol.* **2013**, *113*, 2171–2181. [CrossRef] [PubMed]
6. Bullard, R.W. Physiological problems of space travel. *Annu. Rev. Physiol.* **1972**, *34*, 205–234. [CrossRef] [PubMed]
7. Narici, M.V.; de Boer, M.D. Disuse of the musculo-skeletal system in space and on earth. *Eur. J. Appl. Physiol.* **2011**, *111*, 403–420. [CrossRef] [PubMed]
8. Zhu, H.; Wang, H.; Liu, Z. Effects of real and simulated weightlessness on the cardiac and peripheral vascular functions of humans: A review. *Int. J. Occup. Med. Environ. Health* **2015**, *28*, 793–802. [CrossRef] [PubMed]
9. Graveline, D.E. Cardiovascular deconditioning: Role of blood volume and sympathetic neurohormones. *Life Sci. Space Res.* **1964**, *2*, 287–298. [PubMed]
10. Coupe, M.; Fortrat, J.O.; Larina, I.; Gauquelin-Koch, G.; Gharib, C.; Custaud, M.A. Cardiovascular deconditioning: From autonomic nervous system to microvascular dysfunctions. *Respir. Physiol. Neurobiol.* **2009**, *169*, S10–S12. [CrossRef] [PubMed]
11. Delp, M.D.; Charvat, J.M.; Limoli, C.L.; Globus, R.K.; Ghosh, P. Apollo lunar astronauts show higher cardiovascular disease mortality: Possible deep space radiation effects on the vascular endothelium. *Sci. Rep.* **2016**, *6*, 29901. [CrossRef] [PubMed]
12. Cucinotta, F.A.; Hamada, N.; Little, M.P. No evidence for an increase in circulatory disease mortality in astronauts following space radiation exposures. *Life Sci. Space Res.* **2016**, *10*, 53–56. [CrossRef] [PubMed]
13. Rizzo, A.M.; Corsetto, P.A.; Montorfano, G.; Milani, S.; Zava, S.; Tavella, S.; Cancedda, R.; Berra, B. Effects of long-term space flight on erythrocytes and oxidative stress of rodents. *PLoS ONE* **2012**, *7*, e32361. [CrossRef] [PubMed]
14. Guan, J.; Wan, X.S.; Zhou, Z.; Ware, J.; Donahue, J.J.; Biaglow, J.E.; Kennedy, A.R. Effects of dietary supplements on space radiation-induced oxidative stress in Sprague-Dawley rats. *Radiat. Res.* **2004**, *162*, 572–579. [CrossRef] [PubMed]
15. Versari, S.; Longinotti, G.; Barenghi, L.; Maier, J.A.; Bradamante, S. The challenging environment on board the International Space Station affects endothelial cell function by triggering oxidative stress through thioredoxin interacting protein overexpression: The ESA-SPHINX experiment. *FASEB J.* **2013**, *27*, 4466–4475. [CrossRef] [PubMed]
16. Mao, X.W.; Pecaut, M.J.; Stodieck, L.S.; Ferguson, V.L.; Bateman, T.A.; Bouxsein, M.; Jones, T.A.; Moldovan, M.; Cunningham, C.E.; Chieu, J.; et al. Spaceflight environment induces mitochondrial oxidative damage in ocular tissue. *Radiat. Res.* **2013**, *180*, 340–350. [CrossRef] [PubMed]
17. Mao, X.W.; Pecaut, M.J.; Stodieck, L.S.; Ferguson, V.L.; Bateman, T.A.; Bouxsein, M.L.; Gridley, D.S. Biological and metabolic response in STS-135 space-flown mouse skin. *Free Radic. Res.* **2014**, *48*, 890–897. [CrossRef] [PubMed]
18. Mao, X.W.; Nishiyama, N.C.; Pecaut, M.J.; Campbell-Beachler, M.; Gifford, P.; Haynes, K.E.; Becronis, C.; Gridley, D.S. Simulated microgravity and low-dose/low-dose-rate radiation induces oxidative damage in the mouse brain. *Radiat. Res.* **2016**, *185*, 647–657. [CrossRef] [PubMed]
19. Wise, K.C.; Manna, S.K.; Yamauchi, K.; Ramesh, V.; Wilson, B.L.; Thomas, R.L.; Sarkar, S.; Kulkarni, A.D.; Pellis, N.R.; Ramesh, G.T. Activation of nuclear transcription factor-κB in mouse brain induced by a simulated microgravity environment. *In Vitro Cell Dev. Biol. Anim.* **2005**, *41*, 118–123. [CrossRef] [PubMed]
20. Wang, J.; Zhang, J.; Bai, S.; Wang, G.; Mu, L.; Sun, B.; Wang, D.; Kong, Q.; Liu, Y.; Yao, X.; et al. Simulated microgravity promotes cellular senescence via oxidant stress in rat PC12 cells. *Neurochem. Int.* **2009**, *55*, 710–716. [CrossRef] [PubMed]

21. Hollander, J.; Gore, M.; Fiebig, R.; Mazzeo, R.; Ohishi, S.; Ohno, H.; Ji, L.L. Spaceflight downregulates antioxidant defense systems in rat liver. *Free Radic. Biol. Med.* **1998**, *24*, 385–390. [CrossRef]
22. Baqai, F.P.; Gridley, D.S.; Slater, J.M.; Luo-Owen, X.; Stodieck, L.S.; Ferguson, V.; Chapes, S.K.; Pecaut, M.J. Effects of spaceflight on innate immune function and antioxidant gene expression. *J. Appl. Physiol.* **2009**, *106*, 1935–1942. [CrossRef] [PubMed]
23. Ikemoto, M.; Nikawa, T.; Kano, M.; Hirasaka, K.; Kitano, T.; Watanabe, C.; Tanaka, R.; Yamamoto, T.; Kamada, M.; Kishi, K. Cysteine supplementation prevents unweighting-induced ubiquitination in association with redox regulation in rat skeletal muscle. *Biol. Chem.* **2002**, *383*, 715–721. [CrossRef] [PubMed]
24. Lawler, J.M.; Song, W.; Demaree, S.R. Hindlimb unloading increases oxidative stress and disrupts antioxidant capacity in skeletal muscle. *Free Radic. Biol. Med.* **2003**, *35*, 9–16. [CrossRef]
25. Finkel, T.; Holbrook, N.J. Oxidants, oxidative stress and the biology of ageing. *Nature* **2000**, *408*, 239–247. [CrossRef] [PubMed]
26. Hayyan, M.; Hashim, M.A.; AlNashef, I.M. Superoxide ion: Generation and chemical implications. *Chem. Rev.* **2016**, *116*, 3029–3085. [CrossRef] [PubMed]
27. Glasauer, A.; Chandel, N.S. ROS. *Curr. Biol.* **2013**, *23*, R100–R102. [CrossRef] [PubMed]
28. Valko, M.; Leibfritz, D.; Moncol, J.; Cronin, M.T.; Mazur, M.; Telser, J. Free radicals and antioxidants in normal physiological functions and human disease. *Int. J. Biochem. Cell. Biol.* **2007**, *39*, 44–84. [CrossRef] [PubMed]
29. Mates, J.M.; Segura, J.A.; Alonso, F.J.; Marquez, J. Intracellular redox status and oxidative stress: Implications for cell proliferation, apoptosis, and carcinogenesis. *Arch. Toxicol.* **2008**, *82*, 273–299. [CrossRef] [PubMed]
30. Kamata, H.; Hirata, H. Redox regulation of cellular signalling. *Cell. Signal.* **1999**, *11*, 1–14. [CrossRef]
31. Sugamura, K.; Keaney, J.F., Jr. Reactive oxygen species in cardiovascular disease. *Free Radic. Biol. Med.* **2011**, *51*, 978–992. [CrossRef] [PubMed]
32. Lambeth, J.D. Nox enzymes, ROS, and chronic disease: An example of antagonistic pleiotropy. *Free Radic. Biol. Med.* **2007**, *43*, 332–347. [CrossRef] [PubMed]
33. Prieto-Bermejo, R.; Hernández-Hernández, A. The Importance of NADPH oxidases and redox signaling in angiogenesis. *Antioxidants* **2017**, *6*, 32. [CrossRef] [PubMed]
34. Gavazzi, G.; Banfi, B.; Deffert, C.; Fiette, L.; Schappi, M.; Herrmann, F.; Krause, K.H. Decreased blood pressure in NOX1-deficient mice. *FEBS Lett.* **2006**, *580*, 497–504. [CrossRef] [PubMed]
35. Yetik-Anacak, G.; Catravas, J.D. Nitric oxide and the endothelium: History and impact on cardiovascular disease. *Vasc. Pharmacol.* **2006**, *45*, 268–276. [CrossRef] [PubMed]
36. Li, Q.; Youn, J.Y.; Cai, H. Mechanisms and consequences of endothelial nitric oxide synthase dysfunction in hypertension. *J. Hypertens.* **2015**, *33*, 1128–1136. [CrossRef] [PubMed]
37. Rochette, L.; Lorin, J.; Zeller, M.; Guilland, J.C.; Lorgis, L.; Cottin, Y.; Vergely, C. Nitric oxide synthase inhibition and oxidative stress in cardiovascular diseases: Possible therapeutic targets? *Pharmacol. Ther.* **2013**, *140*, 239–257. [CrossRef] [PubMed]
38. Sihver, L.; Ploc, O.; Puchalska, M.; Ambrozova, I.; Kubancak, J.; Kyselova, D.; Shurshakov, V. Radiation environment at aviation altitudes and in space. *Radiat. Prot. Dosim.* **2015**, *164*, 477–483. [CrossRef] [PubMed]
39. Cucinotta, F.A.; Durante, M. Cancer risk from exposure to galactic cosmic rays: Implications for space exploration by human beings. *Lancet Oncol.* **2006**, *7*, 431–435. [CrossRef]
40. Kryston, T.B.; Georgiev, A.B.; Pissis, P.; Georgakilas, A.G. Role of oxidative stress and DNA damage in human carcinogenesis. *Mutat. Res.* **2011**, *711*, 193–201. [CrossRef] [PubMed]
41. Wang, Y.; Boerma, M.; Zhou, D. Ionizing radiation-induced endothelial cell senescence and cardiovascular diseases. *Radiat. Res.* **2016**, *186*, 153–161. [CrossRef] [PubMed]
42. LaVerne, J.A. Track effects of heavy ions in liquid water. *Radiat. Res.* **2000**, *153*, 487–496. [CrossRef]
43. Gonon, G.; Groetz, J.E.; de Toledo, S.M.; Howell, R.W.; Fromm, M.; Azzam, E.I. Nontargeted stressful effects in normal human fibroblast cultures exposed to low fluences of high charge, high energy (HZE) particles: Kinetics of biologic responses and significance of secondary radiations. *Radiat. Res.* **2013**, *179*, 444–457. [CrossRef] [PubMed]
44. Li, M.; Gonon, G.; Buonanno, M.; Autsavapromporn, N.; de Toledo, S.M.; Pain, D.; Azzam, E.I. Health risks of space exploration: Targeted and nontargeted oxidative injury by high-charge and high-energy particles. *Antioxid. Redox Signal.* **2014**, *20*, 1501–1523. [CrossRef] [PubMed]

45. Newberg, A.B. Changes in the central nervous system and their clinical correlates during long-term spaceflight. *Aviat. Space Environ. Med.* **1994**, *65*, 562–572. [PubMed]

46. DeFelipe, J.; Arellano, J.I.; Merchan-Perez, A.; Gonzalez-Albo, M.C.; Walton, K.; Llinas, R. Spaceflight induces changes in the synaptic circuitry of the postnatal developing neocortex. *Cereb. Cortex* **2002**, *12*, 883–891. [CrossRef] [PubMed]

47. Van Ombergen, A.; Demertzi, A.; Tomilovskaya, E.; Jeurissen, B.; Sijbers, J.; Kozlovskaya, I.B.; Parizel, P.M.; van de Heyning, P.H.; Sunaert, S.; Laureys, S.; et al. The effect of spaceflight and microgravity on the human brain. *J. Neurol.* **2017**, 1–5. [CrossRef] [PubMed]

48. Kim, J.S.; Yang, M.; Kim, S.H.; Shin, T.; Moon, C. Neurobiological toxicity of radiation in hippocampal cells. *Histol. Histopathol.* **2013**, *28*, 301–310. [PubMed]

49. Gauger, G.E.; Tobias, C.A.; Yang, T.; Whitney, M. The effect of space radiation of the nervous system. *Adv. Space Res.* **1986**, *6*, 243–249. [CrossRef]

50. Darby, S.C.; Ewertz, M.; McGale, P.; Bennet, A.M.; Blom-Goldman, U.; Bronnum, D.; Correa, C.; Cutter, D.; Gagliardi, G.; Gigante, B.; et al. Risk of ischemic heart disease in women after radiotherapy for breast cancer. *N. Engl. J. Med.* **2013**, *368*, 987–998. [CrossRef] [PubMed]

51. Lipshultz, S.E.; Cochran, T.R.; Franco, V.I.; Miller, T.L. Treatment-related cardiotoxicity in survivors of childhood cancer. *Nat. Rev. Clin. Oncol.* **2013**, *10*, 697–710. [CrossRef] [PubMed]

52. Boerma, M.; Nelson, G.A.; Sridharan, V.; Mao, X.W.; Koturbash, I.; Hauer-Jensen, M. Space radiation and cardiovascular disease risk. *World J. Cardiol.* **2015**, *7*, 882–888. [CrossRef] [PubMed]

53. Soucy, K.G.; Lim, H.K.; Kim, J.H.; Oh, Y.; Attarzadeh, D.O.; Sevinc, B.; Kuo, M.M.; Shoukas, A.A.; Vazquez, M.E.; Berkowitz, D.E. HZE ^{56}Fe-ion irradiation induces endothelial dysfunction in rat aorta: Role of xanthine oxidase. *Radiat. Res.* **2011**, *176*, 474–485. [CrossRef] [PubMed]

54. Datta, K.; Suman, S.; Kallakury, B.V.; Fornace, A.J., Jr. Exposure to heavy ion radiation induces persistent oxidative stress in mouse intestine. *PLoS ONE* **2012**, *7*, e42224. [CrossRef] [PubMed]

55. Qu, L.; Chen, H.; Liu, X.; Bi, L.; Xiong, J.; Mao, Z.; Li, Y. Protective effects of flavonoids against oxidative stress induced by simulated microgravity in SH-SY5Y cells. *Neurochem. Res.* **2010**, *35*, 1445–1454. [CrossRef] [PubMed]

56. Gambara, G.; Salanova, M.; Ciciliot, S.; Furlan, S.; Gutsmann, M.; Schiffl, G.; Ungethuem, U.; Volpe, P.; Gunga, H.C.; Blottner, D. Gene expression profiling in slow-type calf soleus muscle of 30 days space-flown mice. *PLoS ONE* **2017**, *12*, e0169314. [CrossRef] [PubMed]

57. Salanova, M.; Gambara, G.; Moriggi, M.; Vasso, M.; Ungethuem, U.; Belavy, D.L.; Felsenberg, D.; Cerretelli, P.; Gelfi, C.; Blottner, D. Vibration mechanosignals superimposed to resistive exercise result in baseline skeletal muscle transcriptome profiles following chronic disuse in bed rest. *Sci. Rep.* **2015**, *5*, 17027. [CrossRef] [PubMed]

58. Wang, Y.; Javed, I.; Liu, Y.; Lu, S.; Peng, G.; Zhang, Y.; Qing, H.; Deng, Y. Effect of prolonged simulated microgravity on metabolic proteins in rat hippocampus: Steps toward safe space travel. *J. Proteome Res.* **2016**, *15*, 29–37. [CrossRef] [PubMed]

59. Zhang, R.; Bai, Y.G.; Lin, L.J.; Bao, J.X.; Zhang, Y.Y.; Tang, H.; Cheng, J.H.; Jia, G.L.; Ren, X.L.; Ma, J. Blockade of AT1 receptor partially restores vasoreactivity, NOS expression, and superoxide levels in cerebral and carotid arteries of hindlimb unweighting rats. *J. Appl. Physiol.* **2009**, *106*, 251–258. [CrossRef] [PubMed]

60. Zhang, R.; Ran, H.H.; Peng, L.; Xu, F.; Sun, J.F.; Zhang, L.N.; Fan, Y.Y.; Peng, L.; Cui, G. Mitochondrial regulation of NADPH oxidase in hindlimb unweighting rat cerebral arteries. *PLoS ONE* **2014**, *9*, e95916. [CrossRef] [PubMed]

61. Peng, L.; Ran, H.H.; Zhang, Y.; Zhao, Y.; Fan, Y.Y.; Peng, L.; Zhang, R.; Cao, F. NADPH Oxidase Accounts for Changes in Cerebrovascular Redox Status in Hindlimb Unweighting Rats. *Biomed. Environ. Sci.* **2015**, *28*, 799–807. [CrossRef]

62. Ran, F.; An, L.; Fan, Y.; Hang, H.; Wang, S. Simulated microgravity potentiates generation of reactive oxygen species in cells. *Biophys. Rep.* **2016**, *2*, 100–105. [CrossRef] [PubMed]

63. Dikalov, S. Cross talk between mitochondria and NADPH oxidases. *Free Radic. Biol. Med.* **2011**, *51*, 1289–1301. [CrossRef] [PubMed]

64. Stein, T.P. Space flight and oxidative stress. *Nutrition* **2002**, *18*, 867–871. [CrossRef]

65. Stein, T.P.; Leskiw, M.J. Oxidant damage during and after spaceflight. *Am. J. Physiol. Endocrinol. Metab.* **2000**, *278*, E375–E382. [PubMed]

66. Da Silva, M.S.; Zimmerman, P.M.; Meguid, M.M.; Nandi, J.; Ohinata, K.; Xu, Y.; Chen, C.; Tada, T.; Inui, A. Anorexia in space and possible etiologies: An overview. *Nutrition* **2002**, *18*, 805–813. [CrossRef]

67. Heer, M.; Boerger, A.; Kamps, N.; Mika, C.; Korr, C.; Drummer, C. Nutrient supply during recent European missions. *Pflügers Arch. Eur. J. Physiol.* **2000**, *441*, R8–R14. [CrossRef]

68. Biolo, G.; Heer, M.; Narici, M.; Strollo, F. Microgravity as a model of ageing. *Curr. Opin. Clin. Nutr. Metab. Care* **2003**, *6*, 31–40. [CrossRef] [PubMed]

69. Adrian, A.; Schoppmann, K.; Sromicki, J.; Brungs, S.; von der Wiesche, M.; Hock, B.; Kolanus, W.; Hemmersbach, R.; Ullrich, O. The oxidative burst reaction in mammalian cells depends on gravity. *Cell. Commun. Signal.* **2013**, *11*, 98. [CrossRef] [PubMed]

70. Veglia, F.; Cighetti, G.; de Franceschi, M.; Zingaro, L.; Boccotti, L.; Tremoli, E.; Cavalca, V. Age- and gender-related oxidative status determined in healthy subjects by means of OXY-SCORE, a potential new comprehensive index. *Biomarkers* **2006**, *11*, 562–573. [CrossRef] [PubMed]

71. Zhou, Q.; Li, L.; Zhao, B.; Guan, K.L. The hippo pathway in heart development, regeneration, and diseases. *Circ. Res.* **2015**, *116*, 1431–1447. [CrossRef] [PubMed]

72. Porazinski, S.; Wang, H.; Asaoka, Y.; Behrndt, M.; Miyamoto, T.; Morita, H.; Hata, S.; Sasaki, T.; Krens, S.F.; Osada, Y.; et al. YAP is essential for tissue tension to ensure vertebrate 3D body shape. *Nature* **2015**, *521*, 217–221. [CrossRef] [PubMed]

73. Asaoka, Y.; Nishina, H.; Furutani-Seiki, M. YAP is essential for 3D organogenesis withstanding gravity. *Dev. Growth Differ.* **2017**, *59*, 52–58. [CrossRef] [PubMed]

74. Vernikos, J.; Schneider, V.S. Space, gravity and the physiology of aging: Parallel or convergent disciplines? A mini-review. *Gerontology* **2010**, *56*, 157–166. [CrossRef] [PubMed]

75. Hohn, A.; Weber, D.; Jung, T.; Ott, C.; Hugo, M.; Kochlik, B.; Kehm, R.; Konig, J.; Grune, T.; Castro, J.P. Happily (n)ever after: Aging in the context of oxidative stress, proteostasis loss and cellular senescence. *Redox Biol.* **2017**, *11*, 482–501. [CrossRef] [PubMed]

International Journal of
Molecular Sciences

MDPI

Article

Spaceflight Activates Autophagy Programs and the Proteasome in Mouse Liver

Elizabeth A. Blaber [1,2], Michael J. Pecaut [3] and Karen R. Jonscher [4,*]

1   Universities Space Research Association, Mountain View, CA 94040, USA; e.blaber@nasa.gov
2   NASA Ames Research Center, Moffett Field, CA 94035, USA
3   Department of Basic Sciences, Division of Radiation Research, Loma Linda University School of Medicine, Loma Linda, CA 92350, USA; mpecaut@llu.edu
4   Department of Anesthesiology, University of Colorado Anschutz Medical Campus, Aurora, CO 80045, USA
*   Correspondence: karen.jonscher@ucdenver.edu; Tel.: +1-(303)-724-3979

Received: 31 August 2017; Accepted: 13 September 2017; Published: 27 September 2017

Abstract: Increased oxidative stress is an unavoidable consequence of exposure to the space environment. Our previous studies showed that mice exposed to space for 13.5 days had decreased glutathione levels, suggesting impairments in oxidative defense. Here we performed unbiased, unsupervised and integrated multi-'omic analyses of metabolomic and transcriptomic datasets from mice flown aboard the Space Shuttle Atlantis. Enrichment analyses of metabolite and gene sets showed significant changes in osmolyte concentrations and pathways related to glycerophospholipid and sphingolipid metabolism, likely consequences of relative dehydration of the spaceflight mice. However, we also found increased enrichment of aminoacyl-tRNA biosynthesis and purine metabolic pathways, concomitant with enrichment of genes associated with autophagy and the ubiquitin-proteasome. When taken together with a downregulation in nuclear factor (erythroid-derived 2)-like 2-mediated signaling, our analyses suggest that decreased hepatic oxidative defense may lead to aberrant tRNA post-translational processing, induction of degradation programs and senescence-associated mitochondrial dysfunction in response to the spaceflight environment.

Keywords: spaceflight; autophagy; proteasome; metabolomics; tRNA biosynthesis; senescence

1. Introduction

Long-duration spaceflight is associated with significant risks including prolonged exposure to microgravity, continuous exposure to low-dose/low-dose rate radiation, psychological and environmental stress and contact with potentially dangerous levels of microbial contamination [1–5]. Radiation is known to induce single (SSB) and double strand (DSB) DNA breaks, permanently damaging nuclear and mitochondrial DNA and leading to early apoptosis or necrosis [6–8]. At low-dose/low-dose rates, radiation immediately triggers oxidative stress via a spike in the level of reactive oxygen species (ROS). Acute psychological stress increases hepatic lipid peroxidation as well as levels of ROS [9], however chronically stressed animals compensate [9,10]. Therefore, exposure to the space environment, characterized by changes in physiological and psychological stress, as well as exposure to low-dose/low-dose rate radiation, may systemically alter ROS levels in a complex fashion [11–13].

Excessive ROS production, without a corresponding upregulation in antioxidant or ROS scavenger pathways, can cause damage to cellular components including DNA, proteins and lipids, inducing pro-inflammatory cytokines and the nuclear factor κ-light-chain-enhancer of activated B cells (NF-κB) pathway [14]. This can lead to cell cycle arrest, activation of senescence or apoptosis and upregulation of inflammatory signaling molecules, causing widespread organelle, cell and tissue damage [15]. Under conditions of chronic exposure to oxidative stress (such as in spaceflight), an imbalance occurs between ROS production and antioxidant quenching resulting in increased cellular and tissue damage [16–18]. Mitochondria are particularly vulnerable to damage by excess ROS and we hypothesized that the liver, a mitochondria-rich metabolic organ, may be a target of spaceflight-induced deficits. In support of this hypothesis, astronauts were shown to exhibit a mild diabetogenic phenotype following spaceflight, the severity of which was linked with flight duration [19,20].

We previously demonstrated that mice flown aboard the Space Shuttle Atlantis (Space Transportation System (STS)-135) for 13.5 days exhibited a significantly impaired response to oxidative stress evidenced by decreased hepatic levels of the antioxidant glutathione (GSH, reduced), with concomitant increased ophthalmate, a biomarker for depletion of glutathione, and increased ratio of glutathione disulfide (GSSG):GSH [21]. Other spaceflight-induced changes in hepatic genes linked to oxidative defense have also been observed [13]. Our targeted analyses of metabolomics and transcriptomics datasets obtained from livers of spaceflight mice showed dysregulation of pathways involved in both lipid metabolism and the immune response, with signs of retinoid export and activation of peroxisome proliferator-activated receptor (PPAR) pathways suggestive of nascent hepatic fat accretion and collagen deposition. Proteomics data acquired from mice exposed to space for 30 days exhibited similar patterns when compared with mice re-acclimated to ground conditions [22].

Our previous integrated data analysis utilized a limited set of genes and metabolites that were elevated in abundance in spaceflight mice as compared with ground controls and were tightly correlated with histological evidence of increased hepatic lipid accumulation. The goal of the present study was to perform unbiased, integrated analyses using the entire metabolomics and transcriptomics datasets to determine whether additional insights into the effects of exposure to the space environment on liver metabolism and cellular function could be gleaned. Utilizing several enrichment algorithms, we determined that degradation and senescence programs were altered in spaceflight mice in concert with attenuation of oxidative defense networks.

2. Results and Discussion

2.1. Short Duration Exposure to the Space Environment Significantly Alters Hepatic Metabolite Profiles

Previously, we selected specific genes and metabolites to interrogate from our large-scale-'omics datasets to address defined hypotheses. Here, we performed unsupervised Partial Least Squares Discriminant Analysis (PLS-DA) using the entire metabolomics dataset to determine whether 13.5 days of exposure to the space environment was sufficient to induce significant changes in metabolism in livers of mice flown in space (FLT) as compared with matched ground controls (AEM). The scores plot of the PLS-DA analysis demonstrates clear separation between the FLT mice (green) and AEM controls (red) (Figure 1A). Volcano plot analysis (Figure 1B) shows that 10 biochemicals increased in FLT mice with a fold change (FC) greater than 2 and a *p (corr)*-value less than 0.05, while only 3 biochemicals decreased in the FLT mice as compared to the AEM controls (Table 1) from a total of 14 upregulated (FC > 2) and 11 downregulated (FC < 0.5) biochemicals (Figure S1A and Table S1). Variable Influence on Projection (VIP) analysis of the top 15 most important features contributing to the clustering is plotted in Figure S1B and the most significant features in the volcano plot also appear within the top 15 VIP features. Hierarchical clustering with organization of features by VIP score (Figure 1C) also shows clear separation of metabolite features between the two groups, with relatively fewer metabolites decreased in abundance in the FLT mice as compared with the AEM controls.

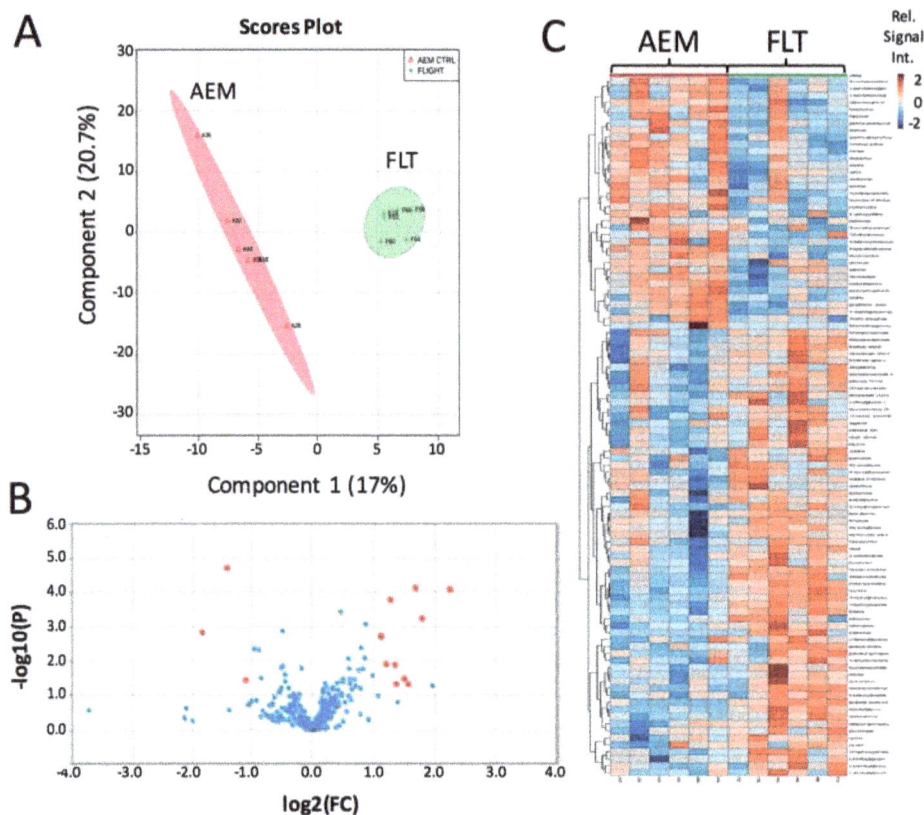

Figure 1. Brief exposure to the space environment results in significant metabolite changes in mouse liver. (**A**) PLS-DA analysis was performed on normalized metabolomics data that was subsequently log-transformed and auto-scaled. The first two components are plotted; (**B**) volcano plot comparing flown in space (FLT) vs. matched ground controls (AEM) considering unequal variance and using a fold change (FC) threshold of 2 and a *p (corr)*-value threshold of 0.05. Data points in red indicate significant named biochemical features. Data points in blue are not significant; (**C**) heat map and hierarchical clustering performed using a Pearson score for the distance measure, with features organized by VIP score from the PLS-DA analysis. Red—AEM, green—FLT; red indicates compounds with high signal abundance and blue those with low signal abundance. Color intensity correlates with relative signal abundance. *n* = 6/group.

Table 1. Significantly changing biochemicals in spaceflight identified in volcano plot analysis.

Biochemicals	FC	log2 (FC)	p	$-\log10\ (p)$
Reduced				
4-Guanidinobutanoate [1],*	0.377	−1.409	1.98×10^{-5}	4.703
Glycerophosphorylcholine *	0.282	−1.826	1.52×10^{-3}	2.817
3-Ureidopropionate	0.466	−1.102	3.83×10^{-2}	1.417
Increased				
3-hydroxybutyrate *	3.229	1.691	7.54×10^{-5}	4.123
Glutarate pentanedioate *	4.766	2.253	8.40×10^{-5}	4.076

Table 1. *Cont.*

Biochemicals	FC	log2 (FC)	p	$-\log10\ (p)$
Propionylcarnitine *	2.440	1.287	1.68×10^{-4}	3.776
3-methylglutarylcarnitine *	3.478	1.798	5.90×10^{-4}	3.229
Dimethylglycine *	2.195	1.134	1.87×10^{-3}	2.728
Hexadecanedioate *	2.189	1.131	2.07×10^{-3}	2.684
Ophthalmate	2.326	1.218	1.30×10^{-2}	1.887
Hydroxyisovaleroyl carnitine	2.567	1.360	1.37×10^{-2}	1.862
Putrescine	2.857	1.514	3.48×10^{-2}	1.459
Cholate	2.602	1.380	4.98×10^{-2}	1.302
Taurodeoxycholate	2.978	1.574	5.00×10^{-2}	1.301

[1], biochemical with the most significant change in volcano plot and FC analysis. *, indicates biochemical is also one of the top 15 VIP features contributing to the PLS-DA analysis. FC–fold change comparing FLT to AEM controls. n = 6/group.

2.2. Altered Betaine and Glutathione Metabolism Are Central Defects in Spaceflight Mouse Livers

Metabolite set enrichment was performed in MetaboAnalyst using pathway-associated metabolite sets based on "normal metabolic pathways" (Figure 2). Enrichment was performed using either all metabolites (Figure 2A) or a more limited subset consisting only of those significantly changing in abundance (Figure 2B). Significance was determined by two-tailed Student's *t*-test. A fold-change cutoff was not applied, therefore some of the significant metabolites differ from those identified in the volcano plot. When enrichment analysis was performed using all metabolites, relatively few *p (corr)*-values for enriched metabolite sets reached significance ($p < 0.05$). However, the top several sets (*methionine metabolism, branched chain fatty acid oxidation, betaine metabolism* and *glycine, serine and threonine metabolism*) were retained when the limited subset of metabolites was used for analysis (Figure 2B). Furthermore, enrichment of *glutathione* and *glutamate metabolism*, important components of the response to oxidative stress, rose to significance in the analysis of the limited dataset, suggesting the utility of performing the same analysis multiple times on different subsets of the data.

We next identified specific metabolites that were contributing to the enrichment scores. The *glycine, serine and threonine metabolism* enrichment score was based on decrease of cystathionine and increase of betaine and dimethylglycine. Eight other metabolites contributed to the enrichment, however their abundance changes were not significant between groups. The *glutathione metabolism* enrichment score was based on decreased abundance of glutathione and cysteinylglycine in FLT mice as compared with AEM controls, with additional contributions from four non-significantly changing metabolites. *Fatty acid oxidation* enrichment scores (both *branched chain* and *very long chain*) were based on abundance of propionylcarnitine and carnitine, with coenzyme A and acetyl carnitine contributing to the score as well. It should be noted that MetaboAnalyst does not recognize many lipids, therefore lipid metabolic pathways are likely under-represented. Contributors to enrichment of *methionine metabolism* were similar to those of *betaine metabolism* and consisted primarily of betaine, dimethylglycine and choline, with additional contributions from *S*-adenosylhomocysteine and 5-methyltetrahydrofolic acid, suggesting activation of the *S*-adenosyl methionine (SAM) cycle in the FLT mice. Abundances of these metabolites in FLT mouse livers and AEM controls are summarized in Table S2. For validation, metabolite pathway analysis was performed and similar pathways were found to have a high impact score based on enrichment and topology analysis (Figure S2).

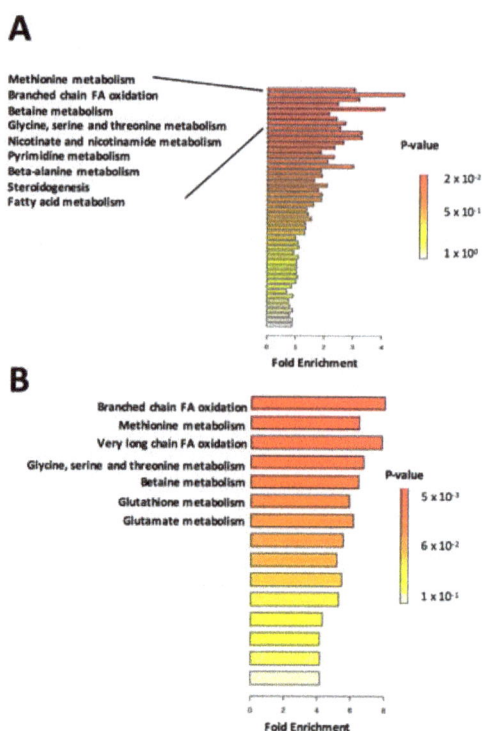

Figure 2. Metabolite set enrichment analysis reveals metabolic pathways enriched in livers of FLT mice. Enrichment analysis was performed in MetaboAnalyst using (**A**) all metabolites or (**B**) only metabolites with significantly changing abundances between groups ($p < 0.05$). $n = 5$/group. Eighty-eight metabolite sets based on "normal metabolic pathways" were used for the analysis. For clarity of presentation, only the most significantly enriched metabolite sets are annotated.

We previously reported a significant increase in abundance of betaine in FLT livers [4,5]. Betaine is metabolized from choline, which decreased in FLT mice [4,5]. Betaine is also a methyl donor and provides the methyl group for metabolism of homocysteine to methionine, generating dimethylglycine as well. Furthermore, the transmethylation cycle provides substrates used for the synthesis of cystathionine and GSH via transsulferation [23]; labile methyl groups are required for these processes as well as support of folate metabolism and synthesis of methylated compounds. Since methionine is lost in the transmethylation pathway, it is possible that choline is supplied to preserve pathway function [24]. However, the observed increase in abundance of betaine with a concomitant decrease in cystathionine and GSH suggests increased choline metabolism in spaceflight is not linked to augmented activation of one carbon metabolic pathways, since abundance levels of relevant metabolites such as *S*-adenosylhomocysteine, serine, sarcosine and glycine are unchanged. Therefore, other mechanisms are likely involved.

Dietary choline and methionine induce lipotrophic effects through upregulation of very low-density lipoprotein export and fatty acid oxidation, and choline deficiency has been associated with oxidative stress, inflammation, and steatosis. Shown in Figure 3, histological assessment confirms that spaceflight indeed results in increased presence of inflammatory cells as well as augmented steatosis. Although we previously measured upregulation of mRNA expression levels of PPAR-α, a transcriptional regulator of fatty acid oxidation, it is likely that PPAR activation is mediated through increased retinol abundance instead of elevated choline [5]. Choline may be replenished by recycling from phosphatidylcholine, which may impair integrity of lipid membranes. Metabolomics analysis of choline-containing lysolipids revealed an average 40% decrease in abundance of these lipids in FLT mice as compared to AEM controls, although differences between groups were not significant for individual features (Figure S3). These results suggest that augmented metabolism of betaine, in excess of what is needed for methionine metabolism, may lead to injurious choline deficits in spaceflight mice. Whether this is causally related to decreased glutathione and the ability to respond to oxidative stress remains an open question.

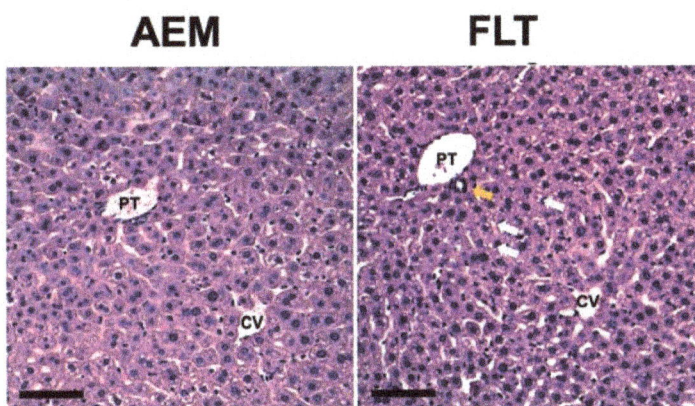

Figure 3. Increased steatosis and infiltration of inflammatory cells in livers from mice exposed to spaceflight. H&E staining was performed on fixed liver sections from n = 4–5 mice/group to investigate liver histology. Inspection of the H&E stained sections revealed that the AEM ground control mice had small cytoplasmic lipid droplets predominantly located in zone 2 whereas the FLT mice had an increase in slightly larger droplets distributed in a panlobular pattern. Multiple lipid droplets are indicated using white arrows. Furthermore, FLT mice showed increased accumulation of mononuclear inflammatory cells, particularly near portal ducts (yellow arrow). Representative images are shown from each group. PT = portal triad, CV = central vein. Scale bar = 100 μm.

Notably, betaine serves as an organic osmolyte, protecting cells from effects of dehydration. Water intake in FLT mice was decreased by ~20% as compared to AEM controls, although food intake was unchanged (Table 2), therefore a likely cause for the upregulation of betaine is dehydration. We also measured upregulation of taurine, another osmolyte. Abundance of 4-guanidinobutanoate was strikingly decreased (Figure S1A) and was the most important feature contributing to the clustering of groups in the PLS-DA analysis (Figure S1B). Guanidino compounds are metabolized from arginine, and dehydration modifies abundance of these compounds in the kidney [25]. Kidney injury not only results in reduced arginine synthesis but change in levels of guanidino compounds and their metabolism in muscles and liver [26]. The marked decrease in abundance of 4-guanidinobutanoate may therefore be associated with dehydration in the FLT mice. Many of the most striking changes in metabolite and transcript abundances appeared to be related to increased dehydration of the FLT mice as compared to AEM controls; although this is a possible limitation of the study, a recent proteomics study on livers from male mice in the "Bion-M1" study flown in space for 30 days and re-acclimated showed results similar to those from our previously published data [22]. Therefore, either the Bion-M1 mice were also dehydrated, or the enriched pathways that we measured indeed have functional significance, potentially related to a shift in metabolic requirements due to unloading. Taken together, the data suggest that dehydration, coupled with oxidative stress, combine to deplete choline stores, leading to impaired lipid membrane metabolism and contributing to increased steatosis in livers from mice exposed to spaceflight.

Table 2. Average food and water intake measurements for STS-135 mice and AEM controls.

Intake	AEM [a]	FLT	FLT/AEM	*p*-Value
Food Intake (g) [b]	4.08 ± 0.10	4.09 ± 0.18	1.00	0.865
Water Intake (g)	3.38 ± 0.22	2.73 ± 0.01	0.81	0.038

[a], Food and water consumption were measured over the 13.5-day flight. Values represent mean $\pm$ SEM; [b], Intake values are means calculated for 3 cages of 5 mice per group. Table adapted from [5].

2.3. Spaceflight Causes Broad Alterations in Transcriptome Profiles in the Liver

To understand the role of transcriptional regulation in the observed alterations to metabolites, we performed an unbiased analysis of transcriptome datasets obtained with Affymetrix Genechip 1.0 ST arrays using GeneSpring software. We found significant alterations (p (*corr*) < 0.05) in 3005 out of 28,852 genes, or approximately 10% of identifiable probesets (Figure 4A,C). Of these, 601 genes were found to have biological significance; expression levels of 449 genes were upregulated (FC > 1.5) and 152, downregulated (FC < −1.5) (Figure 4B).

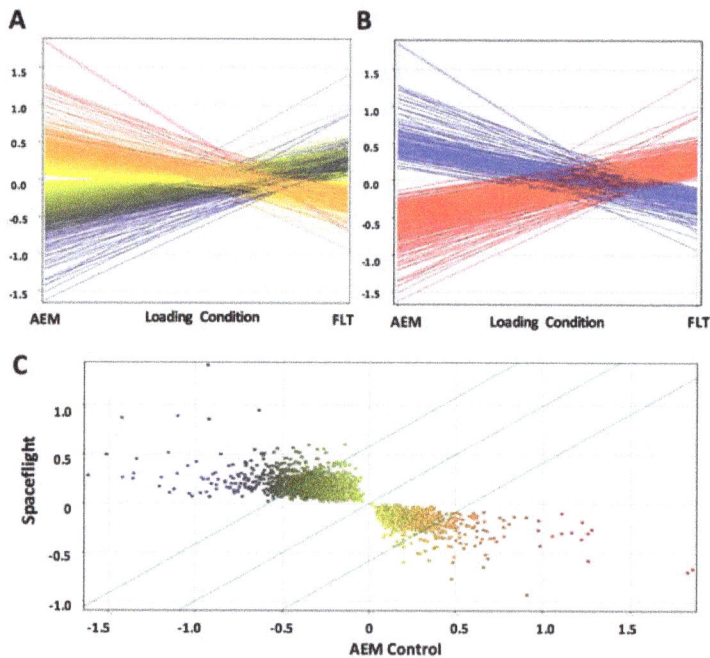

Figure 4. Hepatic gene expression is significantly altered by exposure to the space environment. Profile (**A**) and scatter (**C**) plots of all significantly regulated genes (p (*corr*) < 0.05), and (**B**) profile plots of biologically significant genes with differential regulation FC +/− 1.5. Red lines (**B**) and points (**C**) indicate significantly upregulated genes in AEM mice, whilst blue indicate significantly upregulated genes in FLT. Yellow indicates genes that are statistically but not biologically significant.

Analysis of Gene Ontology (GO) biological functions (Figure 5) revealed that most upregulated genes were involved in metabolism or basic cellular processes, including transcription, translation, and DNA repair. Of note, several autophagy-related genes were altered as were genes involved in oxidative stress and regulation of peroxisomes, in particular fatty acid synthesis and degradation (Figure 5A). Our previous studies found significant alterations in both mRNA and metabolites associated with activation of PPAR-mediated pathways, as well as alterations in fatty acid oxidation in response to spaceflight [4,5]. Furthermore, as peroxisomes catalyze redox reactions and are potential regulators of oxidative stress-mediating signaling pathways, it is possible that peroxisomes and mitochondria may cooperate to determine cell fate decisions [27]. Specifically, peroxisomes house several enzymes that can produce or degrade ROS and reactive nitrogen species (RNS) and therefore may act as modulators of oxidative balance [28–30]. Recent studies have shown that disturbances in peroxisomal metabolism play a role in the accumulation of cellular damage due to oxidative stress and therefore, cellular aging. There is also evidence that peroxisomes can act as upstream initiators of mitochondrial ROS signaling pathways [31]. However, the precise mechanisms by which this occurs are yet to be fully elucidated.

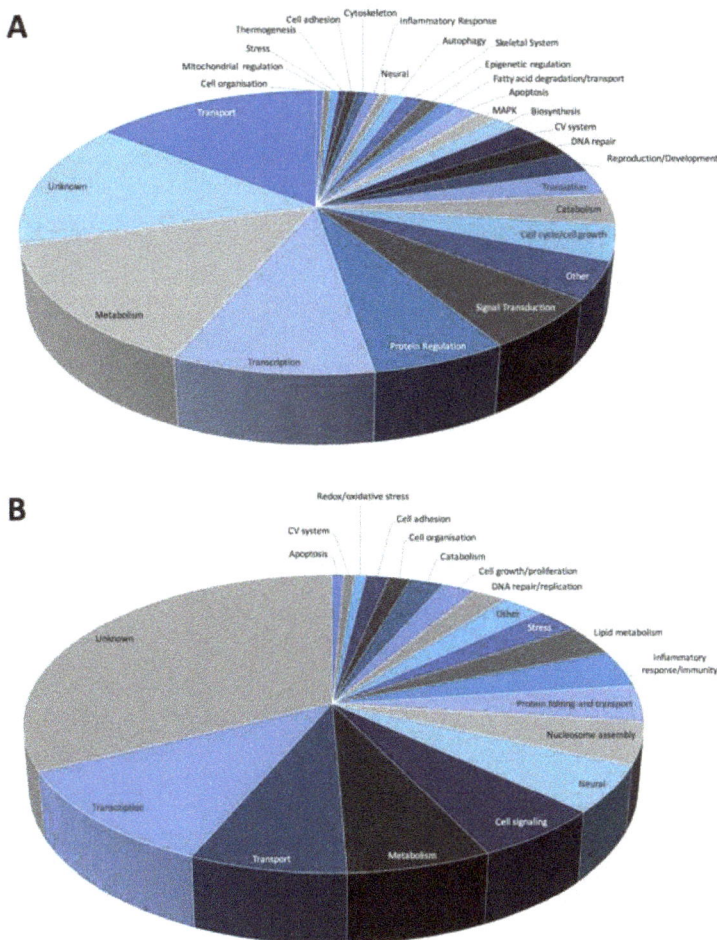

Figure 5. GO biological functions associated with (**A**) up- and (**B**) downregulated datasets. Change in regulation was determined as the ratio of average expression of FLT to AEM values for each transcript, *n* = 6/group.

GO analysis of downregulated datasets denoted alterations in the regulation of transcription, lipid metabolism and cell signaling (Figure 5B). Several processes related to activation and regulation of the inflammatory response/immunity were also downregulated. Previous studies have shown significant deficits in immunity in response to spaceflight, including suppression of proliferation and differentiation in hematopoietic stem cell lineages [32,33], as well as shifts in immune cell phenotypes characterized by increased numbers of bone marrow-derived T cells and decreased bone marrow-derived B cell populations [34]. Furthermore, studies using cluster of differentiation (CD) 34+ bone marrow progenitor cells revealed decreases in total cell number in microgravity samples, and additionally decreased erythropoiesis with concomitant increased macrophage differentiation [35].

2.4. Pathways Involved in Lipid Membrane Metabolism and Protein Biosynthesis Are Enriched in Multi-'Omics Datasets from Livers of Spaceflight Mice

To further validate our metabolomics results, we performed an integrated analysis of transcriptomics and metabolomics datasets using the MetaboAnalyst Integrated Analysis function. This algorithm performs an enrichment analysis to determine whether the observed genes and metabolites in a given pathway appear more often than expected by random chance within the dataset. An additional topology analysis evaluates whether a given gene or metabolite plays an important role in a biological response based on its position within a pathway. An over-representation analysis based on hypergeometric testing using 17,403 genes and 247 metabolites was used for the enrichment analysis and topology was assessed with "Betweenness Centrality", which measures the number of shortest paths from all nodes to all others passing through a given node within the pathway. Integrated analysis of both genes and metabolites (Figure 6A) confirmed the impact of spaceflight on lipid membrane metabolism. Eight of the top 20 enriched pathways relate to lipid membrane metabolism (including *glycosphingolipid biosynthesis*, *glycerophospholipid metabolism*, *arachidonic acid metabolism*, *inositol phosphate metabolism*, *glycophosphatidylinositol-anchor biosynthesis* and *sphingolipid metabolism*). Previously, we reported that Ingenuity Pathway Analysis revealed *endocytosis* as an enriched gene pathway [4], which was also the most highly enriched pathway in the gene-centric analysis (Figure 6B). Enrichment of inflammatory pathways was also evident by the presence of multiple pathways related to cancer. Supporting our other metabolite set enrichment analysis, *glycine, serine and threonine metabolism* had high enrichment and topology scores in the metabolite-centric analytical workflow (Figure 6C).

Interestingly, several studies have shown that peroxisomes may alter lipid production and concentration in response to changes in metabolism, mediating cellular signaling through sphingolipids. In our analyses, we observed enrichment of *sphingolipid biosynthesis* and *sphingolipid metabolism*, as well as alterations in peroxisome gene expression levels, suggesting that spaceflight may alter peroxisome-related signaling pathways, including sphingolipids, to regulate cellular processes [36]. Sphingolipids, specifically, are important messengers for signaling events resulting in activation of cellular proliferation, differentiation or senescence [27,36]. Furthermore, sphingolipids have been linked to insulin resistance, oxidative stress and lipid peroxidation in hepatocytes, suggesting a potential role of sphingolipids in the progression of nonalcoholic fatty liver disease [37]. These results therefore indicate a potential connection between peroxisome redox metabolism and mitochondrial oxidative stress mediated by sphingolipid signaling pathways and resulting in upregulation of inflammatory/stress-related signaling, such as NF-κB.

Of note, *aminoacyl-tRNA biosynthesis* emerged as the top enriched metabolite pathway in the metabolite-centric analysis which, when taken together with enrichment of *purine metabolism* in the gene-metabolite centric analysis, suggested that spaceflight may increase biosynthesis or even post-transcriptional modification of tRNAs, the building blocks of mRNA decoding and protein translation. Post-transcriptional modification of tRNAs critically influences tRNA functions such as folding, stability and decoding [38,39]. Defects in tRNA modifications and modification enzymes are associated with oxidative stress [40] and human diseases including cancer, diabetes and cardiomyopathy [38]. Indeed, results from a recent study in *Caenorhabditis elegans* associating defects in post-transcriptional modification of mitochondrial tRNAs with dysfunctional oxidative phosphorylation suggest that the cell's maladaptive response to hypomodified mitochondrial tRNAs may be a mechanism underlying disease development [41]. Although speculative, the idea that exposure to the space environment may lead to aberrant tRNA post-transcriptional modifications is provocative and warrants further investigation.

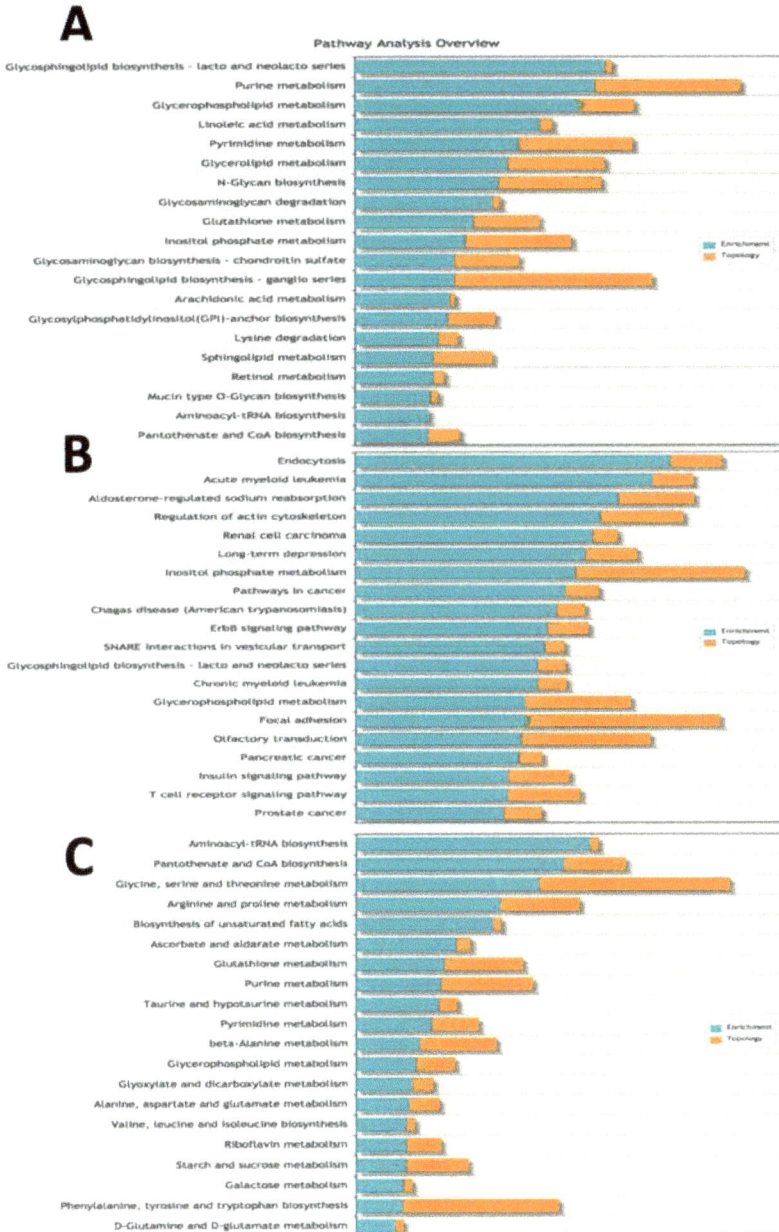

Figure 6. Integrated enrichment analysis using multi-'omics datasets from livers of spaceflight mice compared with AEM ground controls. Data were submitted to the MetaboAnalyst Integrated Pathway Analysis module. (**A**) gene-metabolite; (**B**) gene and (**C**) metabolite centric workflows were compared for *n* = 6 mice per group.

Since MetaboAnalyst preferentially utilizes metabolic pathways for enrichment analyses, we further interrogated the transcriptomic dataset using EGAN (Exploratory Gene Association Networks) to confirm importance of enriched pathways [42]. We performed an association analysis using all transcripts significantly changing between groups (Figure 7). Three main clusters emerged. The first cluster focused on nucleic acid metabolism and included *purine metabolism, pyrimidine metabolism* and *nicotinate and nicotinamide metabolism*. *Purine metabolism* enrichment was dominated by strong upregulation of phosphodiesterase 4D (*Pde4d*, Table 3), an enzyme with 3′,5′-cyclic-adenosine monophosphate (AMP) phosphodiesterase activity that degrades cAMP, an important second messenger mediating signaling in multiple pathways. Adenosine monophosphate deaminase 2 (*Ampd2*, Table 3), an enzyme that converts AMP to inosine monophosphate (IMP) and maintains cellular guanine nucleotide pools [43], was downregulated, potentially attenuating protein synthesis. This protein also mediates gluconeogenesis in the rodent liver [44] and, together with *Pde4d* upregulation (Table 3), suggests pathways impacting cellular quiescence may be altered by spaceflight.

Previous studies in spaceflight have noted alterations in gene expression related to quiescence and senescence pathways in multiple tissues. Specifically, exposure of bone marrow-derived mesenchymal stem cells to spaceflight following addition of osteogenic differentiation factors resulted in increased expression of genes related to neural development, neural morphogenesis and transmission of nerve impulses and synapses in studies conducted on the International Space Station [45]. This same study found increased expression of cell cycle arrest molecules indicating either increased differentiation of cells in space or activation of cellular quiescence or senescence [45]. Our previous studies have also found significant alterations in the proliferation and differentiation potential of both embryonic and bone marrow stem cells during 13–15 days of spaceflight, with upregulation of the cell cycle arrest and senescence marker cyclin dependent kinase inhibitor (CDKN)1a/p21 [46–49]. Notably, analysis of our liver transcriptome dataset also found upregulation of CDKN1a/p21 (3.15 fold, *p* (*corr*) < 0.05, Table 3), as well as upregulation of INK4C/p18 with concomitant attenuation of cyclin gene expression (Figure S4). Similarly, GeneSet Enrichment Analysis (GSEA) using our whole transcriptome profile showed alterations in several pathways, including quiescent cell activation, cell cycle regulation, and activation of oxidative phosphorylation. When taken in total, these results suggest potential systemic induction of cellular senescence due to short-term exposure to the space environment.

The second major cluster that emerged from our EGAN analysis was *aminoacyl-tRNA biosynthesis*, supporting the MetaboAnalyst enrichment results, showing upregulation of expression of all genes within the cluster. Modifications regulate the turnover of RNAs, and improperly modified tRNAs are targeted for degradation [50], therefore it is possible that upregulation of aminoacyl-tRNA biosynthesis actually targets cells for attenuation of protein translation due to hypomodifications or alterations of the epitranscriptome. Finally, we observed a third cluster of genes corresponding to enrichment of *ubiquitin-mediated proteolysis*. We performed Ingenuity Pathway Analysis as shown in Figure S5 and determined that the components of the 20S proteasome were significantly upregulated, supporting the observed pattern of molecular catabolism. Analysis of significantly upregulated datasets using GO analysis also indicated activation of catabolism, primarily in ATP-dependent and glycolytic pathways.

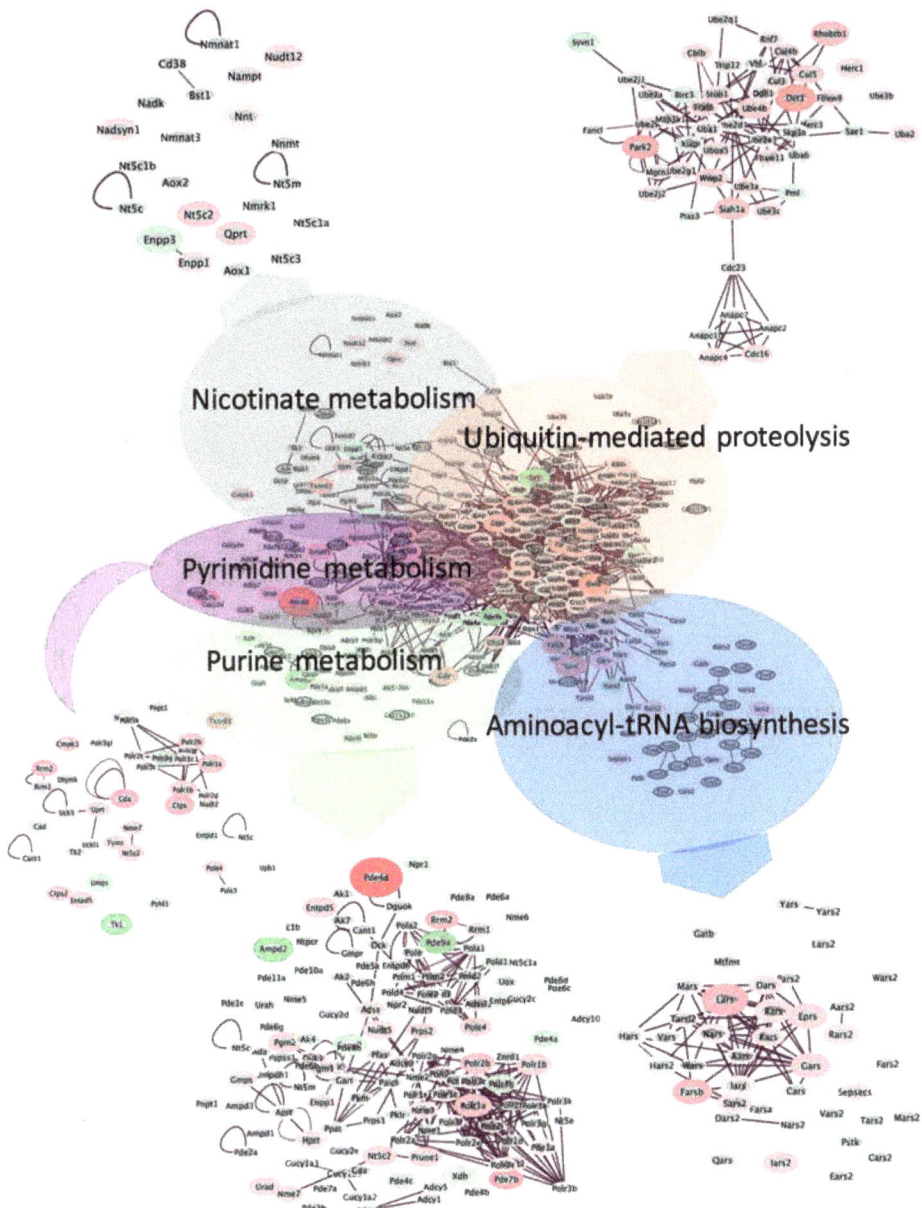

Figure 7. Cluster analysis using EGAN (Exploratory Gene Association Networks) software was performed using all significantly changing transcripts ($p < 0.05$). Genes associated with the top enriched pathways were clustered using a radial force-driven display. Insets are zoomed out views of each cluster. Only significantly changing genes are included in insets for clarity of presentation. Intensity of color (red = upregulated, green = downregulated, grey = not detected) is associated with degree of fold change and width of the bounding circle is inversely related to *p (corr)*-value. $n = 6$ animals per group were used for the analysis.

Table 3. Differentially regulated genes and biological processes in livers of FLT mice.

Gene Name	Gene ID	p *(corr)* Value	Fold Change	GO Biological Process
Agpat9	1-Acylglycerol-3-phosphate-o-acyltransferase 9	8.10×10^{-4}	3.740	Lipid metabolic process
Cdkn1a	Cyclin-dependent kinase inhibitor1A (P21)	4.24×10^{-4}	3.153	Regulation of cyclin-dependent protein serine
Elovl3	Elongation of very long chain fatty acids-like3	3.76×10^{-3}	3.055	Lipid metabolic process
Pnpla2	Patatin-like phospholipase domain containing 2	1.17×10^{-3}	2.645	Lipid metabolic process
Pde4d	Phosphodiesterase 4D, cAMP specific	7.68×10^{-4}	2.161	cAMP catabolic process
Pex11a	Peroxisomal biogenesis factor 11 alpha	1.34×10^{-3}	2.066	Peroxisome organization
Pex3	Peroxisomal biogenesis factor 3	8.82×10^{-3}	2.037	Peroxisome organization
Pex19	Peroxisomal biogenesis factor 19	3.23×10^{-3}	1.953	Protein targeting to peroxisome
Cirbp	Cold inducible RNA binding protein	1.15×10^{-2}	1.895	Response to stress
Pex16	Peroxisomal biogenesis factor 16	4.73×10^{-4}	1.887	Protein targeting to peroxisome
Acot8	Acyl-Coa-thioesterase 8	4.73×10^{-4}	1.776	Peroxisome organization
Atg2a	Autophagy related 2A	2.59×10^{-3}	1.707	Autophagy
Vwa8	Von willebrand factor A domain containing 8	1.94×10^{-3}	1.651	ATP catabolic process
Abcd3	ATP-binding cassette, sub-familyD (ALD), member 3	1.94×10^{-3}	1.651	ATP catabolic process
Abcg8	ATP-binding cassette, sub-family G (WHITE), member 8	2.08×10^{-2}	1.628	ATP catabolic process
Atp10d	ATPase, class V, type 10D	7.48×10^{-3}	1.614	ATP catabolic process
Map1lc3b	Microtubule-associated protein 1 light chain 3 β	2.50×10^{-2}	1.553	Autophagy
Abcg5	ATP binding cassette subfamily G member 5	4.11×10^{-2}	1.550	ATP catabolic process
Ppara	peroxisome proliferator activated receptor α	2.10×10^{-3}	1.550	Negative regulation of transcription
Mtor	Mechanistic target of rapamycin	5.06×10^{-3}	1.541	Positive regulation of protein phosphorylation
Wipi1	WD repeat domain, phosphoinositide interacting 1	3.81×10^{-2}	1.537	Autophagic vacuole assembly
Ppargc1b	PPARG coactivator 1 β	1.24×10^{-2}	1.535	Transcription from mitochondrial promoter
Atg14	Autophagy related 14	2.54×10^{-2}	1.505	Autophagic vacuole assembly
Pex1	Peroxisomal biogenesis factor 1	8.32×10^{-3}	1.417	Protein targeting to peroxisome
Pex11b	Peroxisomal biogenesis factor	2.49×10^{-2}	1.365	Peroxisome organization
Pex10	Peroxisomal biogenesis factor 10	3.85×10^{-2}	1.357	Peroxisome organization
Wipi2	WD repeat domain, phosphoinositide interacting 2	1.67×10^{-2}	1.277	Autophagic vacuole assembly
Map1lc3a	Microtubule associated protein 1 light chain 3 α	1.30×10^{-2}	1.256	Autophagic vacuole assembly
Ampd2	Adenosine monophosphate deaminase 2	1.01×10^{-2}	−1.608	AMP biosynthetic process
Nfe2l2	Nuclear factor, erythroid 2 like 2	5.80×10^{-3}	−1.643	Transcription, DNA-dependent
Cyp26a1	Cytochrome P450 family 26 subfamily A member 1	4.97×10^{-2}	−2.015	Central nervous system development
Hsp90aa1	Heat shock protein 90 α family class A member 1	7.68×10^{-4}	−2.653	ATP catabolic process
Hspb1	Heat shock protein family B (small) member 1	1.03×10^{-3}	−5.755	Response to stress

Red indicates significantly upregulated genes in FLT compared to AEM controls, whilst blue indicates significantly downregulated genes in FLT compared to AEM controls.

The ubiquitin–proteasome system (UPS) and autophagy are the two main intracellular degradation pathways [51]. Autophagy primarily mediates the degradation of long-lived proteins and organelles, maintaining intracellular homeostasis. Since activation of the proteasome is associated with increased autophagy, we used BIOMART [52] to obtain a list of genes associated with autophagy GO terms and screened that list for expression changes using EGAN (Figure 8). We found significant upregulation in expression of a number of genes, including *Atg2a*, important for autophagosome formation as well as regulation of lipid droplet morphology and dispersion; microtubule-associated genes (*Map1lc3a, Map1lc3b*); *Wipi1* and *2*, involved in pre-autophagosome formation and the autophagy response to starvation; and *Mtor*, a central mediator of cellular response to stressors such as DNA damage and oxidative stress, supporting our GSEA results (Table 3). Alterations in the rate of autophagy have been shown to regulate ROS formation and redox balance under specific circumstances [53]. However, proteins damaged by ROS/RNS form protein aggregates that are degraded through the 20S proteasome in order to maintain cellular homeostasis [53]. Exposure to chronic or sustained oxidative stress can lead to inactivation of the proteasome, resulting in accumulation of protein conjugates. Heavily oxidized protein aggregates are also not suitable for degradation by the proteasome [54]. These aggregates may be specific targets of autophagy-related pathways mediated by heat shock protein chaperones, which were also altered in our analyses (Table 3). These results suggest that spaceflight leads to upregulation of multiple autophagy-related pathways and are consistent with activation of the proteasome and attenuation of protein synthesis in livers from FLT mice as compared with AEM controls.

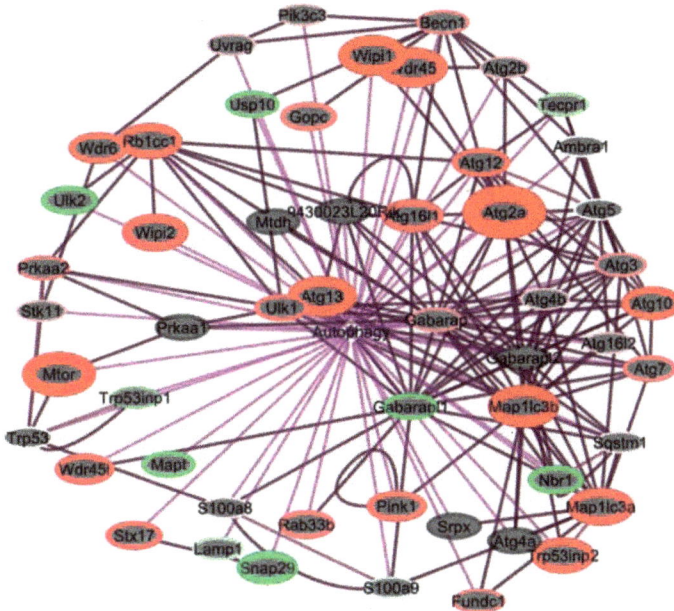

Figure 8. Autophagy programs are upregulated in livers from spaceflight mice. Cluster analysis using EGAN software was performed using genes associated with autophagy programs generated by BIOMART. Genes were clustered using a radial force-driven display. Intensity of color (red = upregulated, green = downregulated, grey = not detected) is associated with degree of fold change and width of the bounding circle is inversely related to *p (corr)*-value. *n* = 6 animals per group were used for the analysis.

Finally, we sought to determine whether decreased response to oxidative stress was associated with induction of autophagy and we performed a pathway analysis of the nuclear factor (erythroid-derived 2)-like 2 (NFE2L2/NRF2)-mediated response to oxidative stress (Figure 9).

We found significant downregulation of *Nrf2* expression levels in FLT mice as compared to AEM controls (Table 3), as well as diminished expression of downstream pathway members, suggesting that exposure to the space environment leads to attenuation of oxidative defense. Notably, spaceflight studies using mice flown on STS-131 for 15 days also revealed downregulation of this oxidative stress mitigator in bone marrow tissues [46,47]. Unfortunately, we did not have sufficient sample to directly probe for changes in protein oxidation, although oxidative damage is a likely cause for the observed increase in autophagy programs and upregulation of the proteasome in the FLT mice.

Autophagy-related programs are essential for liver regeneration [55] and repair and induction of autophagy may occur in response to exposure to the space environment. Several studies have characterized the molecular mechanisms involved in regeneration of the liver in response to a variety of stress conditions and in response to partial hepatectomy (PHx), whereby a significant portion of the rodent liver is removed and the remaining portion regenerates and restores the liver to its original size. This process is highly regulated and includes several distinct stages, including withdrawal of hepatocytes from quiescence, cell cycle entry and progression, cessation of cell division and return of hepatocytes to quiescence [56–59]. Although hepatocytes are the first cells to replicate, they are followed sequentially by other cell types within the liver including stellate cells and sinusoidal endothelial cells [59].

A recent study demonstrated that autophagy is critical in the prevention of hepatocyte senescence during the early stages of liver regeneration, and inhibition of autophagy-related genes results in delayed liver regeneration, aggregation of unfolded proteins, and activation of senescence in hepatocytes with a corresponding increase in senescence-associated secretory phenotype (SASP)-related molecules [58]. This also coincides with considerable damage to the mitochondria, reduced β-oxidation and reduced intrahepatic ATP generation, leading to dysregulation of hepatocellular lipid stores [58]. As we found mild upregulation of autophagy-related genes as well as upregulation of senescence, it is possible that autophagy pathways were initially activated to degrade oxidized proteins. However, as chronic oxidative stress has been shown to induce hepatocyte senescence, it is likely that senescence signaling pathways were activated in response to accumulation of oxidized proteins and failure of autophagy mechanisms to clear these proteins. Indeed, other studies have also demonstrated links between increased oxidative stress and increased hepatocyte senescence, resulting in steatosis due to mitochondrial dysfunction [60–62]. This senescence-associated mitochondrial dysfunction is a regulated process driven by signaling through p21 and through p38 mitogen-activated protein kinase (MAPK), both of which we found to be upregulated in the current study (Table 3). Such changes are very similar to those that occur during ageing and obesity-related pathologies, such as insulin resistance, and have been associated with impaired energy generation and increased production of ROS. Studies conducted by us and others have shown spaceflight leads to increased oxidative stress, increased insulin resistance, and systemic induction of aging-related pathologies. It is possible that in the liver, increased oxidative stress and altered autophagy pathways may cause hepatocyte senescence through activation of p21 and mitochondrial dysfunction, resulting in hepatic steatosis (as we previously reported [5]) and impaired regenerative capacity. These molecular changes may have important implications for the onset of obesity-related diseases and regenerative health in the course of long-duration space exploration.

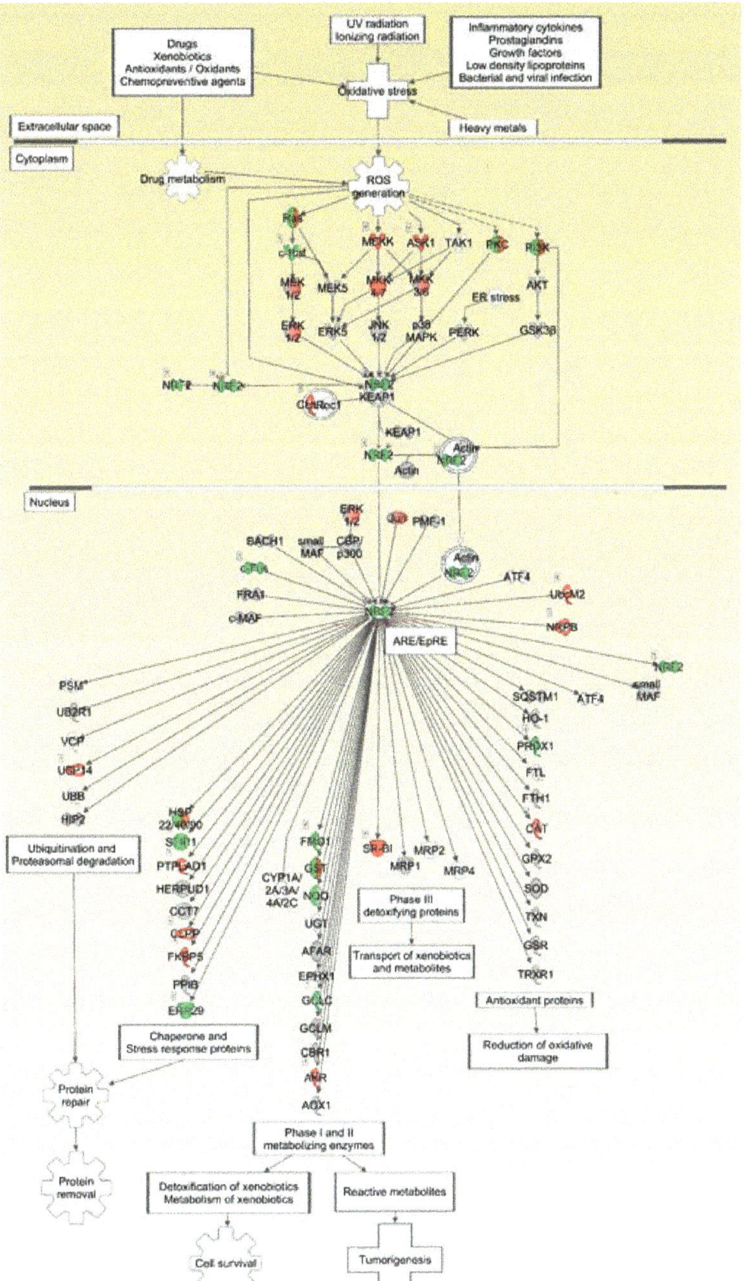

Figure 9. NFE2L2/NRF2-mediated pathways are downregulated in spaceflight mouse livers. Ingenuity Pathway Analysis was used for analysis of mRNA transcript levels in livers from FLT and AEM control mice. Grey = unchanged, green = downregulated, red = upregulated. Intensity of color correlates with degree of fold-change. *n* = 6/group.

3. Materials and Methods

Animal studies were reviewed and approved by multiple Animal Care and Use Committee (ACUC) boards, including the NASA Ames Research Center ACUC (NAS-11-002-Y1; 31 May 2011), the NASA Kennedy Space Center (KSC) ACUC (FLT-11-078; 23 May 2011) and the University of Colorado at Boulder Institutional ACUC (1104.11; 10 May 2011). No protocol was required for assays performed at the University of Colorado Anschutz Medical Center or Loma Linda University since only tissues obtained after euthanasia (no live animals) were analyzed at our sites. All NASA studies involving vertebrate animals were carried out in strict accordance with the recommendations in the Guide for the Care and Use of Laboratory Animals of the National Institutes of Health.

3.1. Animals and Sample Collection

Nine-week old weight-matched female C57BL/6J mice were selected for this study because they produce fewer odor annoyance issues and were the only gender approved for flight (n = 15/group). Additionally, all historical data (e.g., STS-118) were obtained on female mice so the use of female mice permitted comparisons across missions, particularly for the musculoskeletal studies that were primary. Mice were housed in Animal Enclosure Modules (AEMs, 10 mice per AEM habitat; 5 per side separated by a wire mesh) and either flown on the Space Shuttle Atlantis (STS-135) for 13.5 days (FLT) or housed at the Space Life Science Laboratory (SLSL) at KSC (Ground AEM controls; AEM).

Following two days of acclimation after receipt, mice were provided a NASA NuRFB foodbar (TD 04197; 47% carbohydrate, 17.9% protein, 3.9% lipids, 2.8% fiber, 2.80 kcal/g). All mice were placed into AEM housing one day before flight. Environmental parameters for ground control mice were matched as closely as possible with flight conditions using 48 h delayed telemetry data. Conditions were controlled for temperature, humidity, and a 12/12 h light/dark cycle; foodbars and water were provided *ad libitum*.

Tissues were harvested at the SLSL within 3–5 h after return of the Space Shuttle Atlantis and were distributed amongst a team of investigations through NASA's Biospecimen Sharing Program; we received one half lobe of liver from six mice per group. Mice were euthanized using 4% isoflurane followed by cardiac puncture and exsanguination. Liver lobes were extracted and dissected. A portion of the liver was prepared in 4% paraformaldehyde and the rest snap frozen in liquid nitrogen then shipped to either Loma Linda University or University of Colorado Anschutz Medical Campus and stored appropriately prior to use.

3.2. Transcriptomics

Samples were prepared and analyzed as previously described [4,5]. Briefly, RNA was isolated using an RNeasy kit (Qiagen, Germantown, MD, USA) and the Ambion WT expression kit (Thermo Fisher Scientific, Waltham, MA, USA) was employed to prepare mRNA for whole transcriptome microarray analysis using an Affymetrix GeneChip 1.0 ST array (Thermo Fisher Scientific). Arrays were scanned using a GeneChip Scanner 3000 7G (Thermo Fisher Scientific) and Command Console Software v. 3.2.3 (Thermo Fisher Scientific) to produce. CEL intensity files which were processed with CARMAweb (Comprehensive R-based Microarray Analysis web service). GeneSpring software (Agilent, Santa Clara, CA, USA) was used to perform statistical analysis. Specifically, samples were first filtered on signal intensity values in order to remove background noise, statistical analysis was performed on normalized and filtered samples using a moderated t-test with Benjamini Hochberg FDR correction factor. Samples were then filtered based on fold change, whereby p (corr) < 0.05 and fold change $+/-$ 1.5 or more was considered significant. CARMAweb files were then imported into Ingenuity Pathway Analysis (Qiagen) software for subsequent analysis.

3.3. Metabolomics

Frozen liver pieces ($n = 6$/group) were shipped to Metabolon, Inc. (Morrisville, NC, USA) and stored at $-80\ °C$ before use. Samples were prepared for the appropriate instrument, either Liquid Chromatography/Mass Spectrometry (LC/MS) or Gas Chromatography/Mass Spectrometry (GC/MS), as described previously [4,5]. Briefly, automated sample preparation was conducted to extract metabolites for analysis by LC and GC. Extracts were placed briefly on a TurboVap® (Zymark, Clackamas, OR, USA) to remove organic solvent. Each extract was then frozen and dried under vacuum.

3.4. Liquid Chromatography/Mass Spectrometry (LC/MS)

Extracts were split into two aliquots, dried, then reconstituted in acidic or basic LC-compatible solvents, each of which contained 11 or more injection standards at fixed concentrations. Metabolite features were measured using a Waters ACQUITY UPLC (Waters, Milford, MA, USA) and a Thermo-Finnigan LTQ-FT MS (Thermo Fisher Scientific) in two independent injections using separate dedicated columns as described previously [4,5]. One aliquot was analyzed in positive ion mode under acidic conditions and gradient eluted from a Waters BEH C_{18} 2.1 mm $\times$ 100 mm column, containing 1.7 μm resin, using water and methanol as mobile phases with both containing 0.1% formic acid. The other aliquot was analyzed in negative ion mode using basic conditions, which also employed water/methanol, and contained 6.5 mM ammonium bicarbonate for ion pairing. Mass spectrometric analysis alternated between MS and data-dependent MS^2 scans using dynamic exclusion, scanning from 80–1000 m/z. Accurate mass measurements were made on precursor ions with greater than 2 million counts; typical mass error was less than 5 ppm.

3.5. Gas Chromatography/Mass Spectrometry (GC/MS)

Extracts destined for GC/MS were vacuum desiccated for 24 h then derivatized under dried nitrogen using bis(trimethylsilyl)-trifluoroacetamide (BSTFA). Volatile metabolites were separated using 5% phenyl/95% dimethyl polysiloxane fused silica (20 m $\times$ 0.18 mm ID; 0.18 μm film thickness) and a temperature ramp from 40 to 300 °C within 16 min; helium was used as the carrier gas. Compounds were analyzed using a Thermo-Finnigan Trace DSQ fast-scanning single-quadrupole MS (Thermo Fisher Scientific) equipped with electron impact ionization set to scan from 50–750 m/z at unit mass resolving power, as we previously described [4,5].

3.6. Compound Identification

Mass spectral data were loaded into a relational database and peaks were identified using Metabolon's peak integration software [63]. Compounds were identified by comparison to library entries of purified standards based on the combination of retention time and mass spectra. Data were normalized to correct for variation resulting from instrument inter-day tuning differences. Each compound was corrected by registering the medians to equal 1.00 in run-day blocks and normalizing each data point proportionately. Missing values, assumed to be below the limit of detection of the instrument, were imputed with the observed minimum after normalization.

3.7. Data Availability

Transcriptomics and metabolomics data are publicly accessible via the NASA GeneLAB data base. Transcriptomics data are found at the website (Available online: https://genelab-data.ndc.nasa.gov/genelab/accession/GLDS-25/). Metabolomics data are available at the website (Available online: https://genelab-data.ndc.nasa.gov/genelab/accession/GLDS-108/).

3.8. Integrated Data Analysis

MetaboAnalyst software (Available online: www.metaboanalyst.ca) was used for pathway and enrichment analysis of metabolomics data as well as integrated analysis of transcriptomics and metabolomics datasets. Data were log-transformed and auto-scaled then subjected to various analytical modules within the software [64]. Ingenuity Pathway Analysis (IPA) and Exploratory Gene Association Networks (EGAN) [42] (Available online: akt.ucsf.edu/EGAN) were used to assess pathway activation based on changes in mRNA expression levels. Human Metabolite Database (HMDB) accession numbers were queried and names of genes associated with metabolites exhibiting significant changes in abundance were extracted manually. These genes were used to construct a subset list of genes that may be functionally significant, based on metabolite abundance changes, and this subset was also subjected to pathway and enrichment analysis.

3.9. Histology

Fixed tissue sections were processed for hematoxylin and eosin (H&E) staining as described previously [65]. Histologic images were captured on an Olympus BX51 microscope equipped with a 4mp Macrofire digital camera (Optronics, Tokyo, Japan) using the PictureFrame Application 2.3 (Optronics). All images were cropped and assembled using Photoshop CS2 (Adobe Systems, Inc., San Jose, CA, USA).

3.10. Statistical Analysis

Data were analyzed with GraphPad Prism V6.0 (GraphPad Software, La Jolla, CA, USA) using a Mann-Whitney U-test or unpaired t-test with Welch's correction for comparison between groups. Means $\pm$ SEM were reported. The ROUT method with $Q = 1\%$ was used to identify outliers for exclusion from analysis. p *(corr)*-values less than 0.05 were selected to indicate significance.

4. Conclusions

The unbiased analyses presented here both support our previous results and extend them to show that exposure to the space environment for only 13.5 days results in increased oxidative stress due to elevated ROS and impaired oxidative defense (by way of attenuation of NRF2-related pathways) in the mouse liver. Furthermore, our multi-'omics studies suggest that accumulation of oxidized proteins coupled with mitochondrial dysfunction may lead to activation of hepatocyte senescence, resulting in hepatocyte lipid accumulation and steatosis. Further investigation into the potential for liver damage in the course of long-duration space exploration is needed, as are development of effective countermeasures to protect astronaut health.

Supplementary Materials: Supplementary materials can be found at www.mdpi.com/1422-0067/18/10/2062/s1.

Acknowledgments: The authors thank Amgen and their support team for the opportunity to participate in this study, and Virginia Ferguson, Mary Bouxsein, Ted Bateman, Louis Stodieck and BioServe Space Technologies at the University of Colorado at Boulder for organizing the tissue sharing project. We also appreciate Naomi Clayman, Paula Dumars and Vera Vizir of the University of Colorado Denver and NASA, as well as support staff at the Kennedy Space Center who assisted with and organized tissue collection. We are grateful to David Orlicky at the University of Colorado Anschutz Medical Campus for histological expertise. Michael J. Pecaut appreciates funding from NASA (NNX13AN34G).

Author Contributions: Michael J. Pecaut conceived and designed the metabolomics and transcriptomics experiments. Michael J. Pecaut, Elizabeth A. Blaber and Karen R. Jonscher analyzed the data. Elizabeth A. Blaber and Karen R. Jonscher wrote the paper.

Conflicts of Interest: The authors declare no conflict of interest.

References

1. Meehan, R.; Whitson, P.; Sams, C. The role of psychoneuroendocrine factors on spaceflight-induced immunological alterations. *J. Leukoc. Biol.* **1993**, *54*, 236–244. [PubMed]

2. Alwood, J.S.; Kumar, A.; Tran, L.H.; Wang, A.; Limoli, C.L.; Globus, R.K. Low-dose, ionizing radiation and age-related changes in skeletal microarchitecture. *J. Aging Res.* **2011**, *2012*. [CrossRef] [PubMed]

3. Cucinotta, F.A.; Schimmerling, W.; Wilson, J.W.; Peterson, L.E.; Badhwar, G.D.; Saganti, P.B.; Dicello, J.F. Space radiation cancer risks and uncertainties for mars missions. *Radiat. Res.* **2001**, *156*, 682–688. [CrossRef]

4. Pecaut, M.J.; Mao, X.W.; Bellinger, D.L.; Jonscher, K.R.; Stodieck, L.S.; Ferguson, V.L.; Bateman, T.A.; Mohney, R.P.; Gridley, D.S. Is spaceflight-induced immune dysfunction linked to systemic changes in metabolism? *PLoS ONE* **2017**, *12*, e0174174. [CrossRef] [PubMed]

5. Jonscher, K.R.; Alfonso-Garcia, A.; Suhalim, J.L.; Orlicky, D.J.; Potma, E.O.; Ferguson, V.L.; Bouxsein, M.L.; Bateman, T.A.; Stodieck, L.S.; Levi, M.; et al. Spaceflight activates lipotoxic pathways in mouse liver. *PLoS ONE* **2016**, *11*, e0152877.

6. Gobbel, G.T.; Bellinzona, M.; Vogt, A.R.; Gupta, N.; Fike, J.R.; Chan, P.H. Response of postmitotic neurons to X-irradiation: Implications for the role of DNA damage in neuronal apoptosis. *J. Neurosci.* **1998**, *18*, 147–155. [PubMed]

7. Shinohara, C.; Gobbel, G.T.; Lamborn, K.R.; Tada, E.; Fike, J.R. Apoptosis in the subependyma of young adult rats after single and fractionated doses of x-rays. *Cancer Res.* **1997**, *57*, 2694–2702. [PubMed]

8. Bellinzona, M.; Gobbel, G.T.; Shinohara, C.; Fike, J.R. Apoptosis is induced in the subependyma of young adult rats by ionizing irradiation. *Neurosci. Lett.* **1996**, *208*, 163–166. [CrossRef]

9. Jafari, M.; Salehi, M.; Zardooz, H.; Rostamkhani, F. Response of liver antioxidant defense system to acute and chronic physical and psychological stresses in male rats. *EXCLI J.* **2014**, *13*, 161–171. [PubMed]

10. Duda, W.; Curzytek, K.; Kubera, M.; Iciek, M.; Kowalczyk-Pachel, D.; Bilska-Wilkosz, A.; Lorenc-Koci, E.; Leskiewicz, M.; Basta-Kaim, A.; Budziszewska, B.; et al. The effect of chronic mild stress and imipramine on the markers of oxidative stress and antioxidant system in rat liver. *Neurotox. Res.* **2016**, *30*, 173–184. [CrossRef] [PubMed]

11. Mao, X.W.; Pecaut, M.J.; Stodieck, L.S.; Ferguson, V.L.; Bateman, T.A.; Bouxsein, M.L.; Gridley, D.S. Biological and metabolic response in STS-135 space-flown mouse skin. *Free Radic. Res.* **2014**, *48*, 890–897. [CrossRef] [PubMed]

12. Mao, X.W.; Pecaut, M.J.; Stodieck, L.S.; Ferguson, V.L.; Bateman, T.A.; Bouxsein, M.; Jones, T.A.; Moldovan, M.; Cunningham, C.E.; Chieu, J.; et al. Spaceflight environment induces mitochondrial oxidative damage in ocular tissue. *Radiat. Res.* **2013**, *180*, 340–350. [CrossRef] [PubMed]

13. Baqai, F.P.; Gridley, D.S.; Slater, J.M.; Luo-Owen, X.; Stodieck, L.S.; Ferguson, V.; Chapes, S.K.; Pecaut, M.J. Effects of spaceflight on innate immune function and antioxidant gene expression. *J. Appl. Physiol.* **2009**, *106*, 1935–1942. [CrossRef] [PubMed]

14. Lee, M.Y.; Wang, Y.; Vanhoutte, P.M. Senescence of cultured porcine coronary arterial endothelial cells is associated with accelerated oxidative stress and activation of NFKB. *J. Vasc. Res.* **2010**, *47*, 287–298. [CrossRef] [PubMed]

15. Kondo, H.; Limoli, C.; Searby, N.D.; Almeida, E.A.; Loftus, D.J.; Vercoutere, W.; Morey-Holton, E.; Giedzinski, E.; Mojarrab, R.; Hilton, D.; et al. Shared oxidative pathways in response to gravity-dependent loading and γ-irradiation of bone marrow-derived skeletal cell progenitors. *Radiats. Biol. Radioecol.* **2007**, *47*, 281–285. [PubMed]

16. Evans, J.L.; Goldfine, I.D.; Maddux, B.A.; Grodsky, G.M. Oxidative stress and stress-activated signaling pathways: A unifying hypothesis of type 2 diabetes. *Endocr. Rev.* **2002**, *23*, 599–622. [CrossRef] [PubMed]

17. Madamanchi, N.R.; Vendrov, A.; Runge, M.S. Oxidative stress and vascular disease. *Arterioscler Arterioscler. Thromb. Vasc. Biol.* **2005**, *25*, 29–38. [CrossRef] [PubMed]

18. Csanyi, G.; Miller, F.J., Jr. Oxidative stress in cardiovascular disease. *Int. J. Mol. Sci.* **2014**, *15*, 6002–6008. [CrossRef] [PubMed]

19. Da Silva, M.S.; Zimmerman, P.M.; Meguid, M.M.; Nandi, J.; Ohinata, K.; Xu, Y.; Chen, C.; Tada, T.; Inui, A. Anorexia in space and possible etiologies: An overview. *Nutrition* **2002**, *18*, 805–813. [CrossRef]

20. Tobin, B.W.; Uchakin, P.N.; Leeper-Woodford, S.K. Insulin secretion and sensitivity in space flight: Diabetogenic effects. *Nutrition* **2002**, *18*, 842–848. [CrossRef]

21. Dello, S.A.; Neis, E.P.; de Jong, M.C.; van Eijk, H.M.; Kicken, C.H.; Olde, D.S.W.; Dejong, C.H. Systematic review of ophthalmate as a novel biomarker of hepatic glutathione depletion. *Clin. Nutr.* **2013**, *32*, 325–330. [CrossRef] [PubMed]

22. Anselm, V.; Novikova, S.; Zgoda, V. Re-adaption on earth after spaceflights affects the mouse liver proteome. *Int. J. Mol. Sci.* **2017**, *18*. [CrossRef] [PubMed]
23. Chandler, T.L.; White, H.M. Choline and methionine differentially alter methyl carbon metabolism in bovine neonatal hepatocytes. *PLoS ONE* **2017**, *12*, e0171080. [CrossRef] [PubMed]
24. Lobley, G.E.; Connell, A.; Revell, D. The importance of transmethylation reactions to methionine metabolism in sheep: Effects of supplementation with creatine and choline. *Br. J. Nutr.* **1996**, *75*, 47–56. [CrossRef] [PubMed]
25. Levillain, O.; Marescau, B.; Possemiers, I.; de Deyn, P. Dehydration modifies guanidino compound concentrations in the different zones of the rat kidney. *Pflugers Arch.* **2002**, *444*, 143–152. [CrossRef] [PubMed]
26. Levillain, O.; Marescau, B.; Possemiers, I.; al Banchaabouchi, M.; de Deyn, P.P. Influence of 72% injury in one kidney on several organs involved in guanidino compound metabolism: A time course study. *Pflugers Arch.* **2001**, *442*, 558–569. [CrossRef] [PubMed]
27. Nordgren, M.; Fransen, M. Peroxisomal metabolism and oxidative stress. *Biochimie* **2014**, *98*, 56–62. [CrossRef] [PubMed]
28. Fransen, M.; Nordgren, M.; Wang, B.; Apanasets, O. Role of peroxisomes in ROS/RNS-metabolism: Implications for human disease. *Biochim. Biophys. Acta* **2012**, *1822*, 1363–1373. [CrossRef] [PubMed]
29. Bonekamp, N.A.; Volkl, A.; Fahimi, H.D.; Schrader, M. Reactive oxygen species and peroxisomes: Struggling for balance. *Biofactors* **2009**, *35*, 346–355. [CrossRef] [PubMed]
30. Antonenkov, V.D.; Grunau, S.; Ohlmeier, S.; Hiltunen, J.K. Peroxisomes are oxidative organelles. *Antioxid. Redox Signal.* **2010**, *13*, 525–537. [CrossRef] [PubMed]
31. Manivannan, S.; Scheckhuber, C.Q.; Veenhuis, M.; van der Klei, I.J. The impact of peroxisomes on cellular aging and death. *Front. Oncol.* **2012**, *2*, 50. [CrossRef] [PubMed]
32. Zayzafoon, M.; Meyers, V.E.; McDonald, J.M. Microgravity: The immune response and bone. *Immunol. Rev.* **2005**, *208*, 267–280. [CrossRef] [PubMed]
33. Borchers, A.T.; Keen, C.L.; Gershwin, M.E. Microgravity and immune responsiveness: Implications for space travel. *Nutrition* **2002**, *18*, 889–898. [CrossRef]
34. Pecaut, M.J.; Nelson, G.A.; Peters, L.L.; Kostenuik, P.J.; Bateman, T.A.; Morony, S.; Stodieck, L.S.; Lacey, D.L.; Simske, S.J.; Gridley, D.S. Genetic models in applied physiology: Selected contribution: Effects of spaceflight on immunity in the C57BL/6 mouse. I. Immune population distributions. *J. Appl. Physiol.* **2003**, *94*, 2085–2094. [CrossRef] [PubMed]
35. Davis, T.A.; Wiesmann, W.; Kidwell, W.; Cannon, T.; Kerns, L.; Serke, C.; Delaplaine, T.; Pranger, A.; Lee, K.P. Effect of spaceflight on human stem cell hematopoiesis: Suppression of erythropoiesis and myelopoiesis. *J. Leukoc. Biol.* **1996**, *60*, 69–76. [PubMed]
36. Bikman, B.T.; Summers, S.A. Ceramides as modulators of cellular and whole-body metabolism. *J. Clin. Investig.* **2011**, *121*, 4222–4230. [CrossRef] [PubMed]
37. Kasumov, T.; Li, L.; Li, M.; Gulshan, K.; Kirwan, J.P.; Liu, X.; Previs, S.; Willard, B.; Smith, J.D.; McCullough, A. Ceramide as a mediator of non-alcoholic fatty liver disease and associated atherosclerosis. *PLoS ONE* **2015**, *10*, e0126910. [CrossRef] [PubMed]
38. Bohnsack, M.T.; Sloan, K.E. The mitochondrial epitranscriptome: The roles of RNA modifications in mitochondrial translation and human disease. *Cell Mol. Life Sci.* **2017**. [CrossRef] [PubMed]
39. Vare, V.Y.; Eruysal, E.R.; Narendran, A.; Sarachan, K.L.; Agris, P.F. Chemical and conformational diversity of modified nucleosides affects tRNA structure and function. *Biomolecules* **2017**, *7*. [CrossRef] [PubMed]
40. Dewe, J.M.; Fuller, B.L.; Lentini, J.M.; Kellner, S.M.; Fu, D. TRMT1-catalyzed tRNA modifications are required for redox homeostasis to ensure proper cellular proliferation and oxidative stress survival. *Mol. Cell Biol.* **2017**. [CrossRef] [PubMed]
41. Navarro-Gonzalez, C.; Moukadiri, I.; Villarroya, M.; Lopez-Pascual, E.; Tuck, S.; Armengod, M.E. Mutations in the caenorhabditis elegans orthologs of human genes required for mitochondrial tRNA modification cause similar electron transport chain defects but different nuclear responses. *PLoS Genet* **2017**, *13*, e1006921. [CrossRef] [PubMed]
42. Paquette, J.; Tokuyasu, T. Egan: Exploratory gene association networks. *Bioinformatics* **2010**, *26*, 285–286. [CrossRef] [PubMed]

43. Akizu, N.; Cantagrel, V.; Schroth, J.; Cai, N.; Vaux, K.; McCloskey, D.; Naviaux, R.K.; van Vleet, J.; Fenstermaker, A.G.; Silhavy, J.L.; et al. AMPD2 regulates GTP synthesis and is mutated in a potentially treatable neurodegenerative brainstem disorder. *Cell* **2013**, *154*, 505–517. [CrossRef] [PubMed]

44. Hudoyo, A.W.; Hirase, T.; Tandelillin, A.; Honda, M.; Shirai, M.; Cheng, J.; Morisaki, H.; Morisaki, T. Role of AMPD2 in impaired glucose tolerance induced by high fructose diet. *Mol. Genet. Metab. Rep.* **2017**, *13*, 23–29. [CrossRef] [PubMed]

45. Monticone, M.; Liu, Y.; Pujic, N.; Cancedda, R. Activation of nervous system development genes in bone marrow derived mesenchymal stem cells following spaceflight exposure. *J. Cell. Biochem.* **2010**, *111*, 442–452. [CrossRef] [PubMed]

46. Blaber, E.A.; Dvorochkin, N.; Torres, M.L.; Yousuf, R.; Burns, B.P.; Globus, R.K.; Almeida, E.A. Mechanical unloading of bone in microgravity reduces mesenchymal and hematopoietic stem cell-mediated tissue regeneration. *Stem Cell Res.* **2014**, *13*, 181–201. [CrossRef] [PubMed]

47. Blaber, E.A.; Dvorochkin, N.; Lee, C.; Alwood, J.S.; Yousuf, R.; Pianetta, P.; Globus, R.K.; Burns, B.P.; Almeida, E.A. Microgravity induces pelvic bone loss through osteoclastic activity, osteocytic osteolysis, and osteoblastic cell cycle inhibition by CDKN1A/P21. *PLoS ONE* **2013**, *8*, e61372. [CrossRef] [PubMed]

48. Blaber, E.; Sato, K.; Almeida, E.A. Stem cell health and tissue regeneration in microgravity. *Stem Cells Dev.* **2014**, *23*, 73–78. [CrossRef] [PubMed]

49. Blaber, E.A.; Finkelstein, H.; Dvorochkin, N.; Sato, K.Y.; Yousuf, R.; Burns, B.P.; Globus, R.K.; Almeida, E.A. Microgravity reduces the differentiation and regenerative potential of embryonic stem cells. *Stem Cells Dev.* **2015**, *24*, 2605–2621. [CrossRef] [PubMed]

50. Chanfreau, G.F. Impact of RNA modifications and RNA-modifying enzymes on eukaryotic ribonucleases. *Enzymes* **2017**, *41*, 299–329. [PubMed]

51. Kraft, C.; Peter, M.; Hofmann, K. Selective autophagy: Ubiquitin-mediated recognition and beyond. *Nat. Cell Biol.* **2010**, *12*, 836–841. [CrossRef] [PubMed]

52. Smedley, D.; Haider, S.; Durinck, S.; Pandini, L.; Provero, P.; Allen, J.; Arnaiz, O.; Awedh, M.H.; Baldock, R.; Barbiera, G.; et al. The biomart community portal: An innovative alternative to large, centralized data repositories. *Nucleic Acids Res.* **2015**, *43*, W589–W598. [CrossRef] [PubMed]

53. Navarro-Yepes, J.; Burns, M.; Anandhan, A.; Khalimonchuk, O.; del Razo, L.M.; Quintanilla-Vega, B.; Pappa, A.; Panayiotidis, M.I.; Franco, R. Oxidative stress, redox signaling, and autophagy: Cell death versus survival. *Antioxid. Redox Signal.* **2014**, *21*, 66–85. [CrossRef] [PubMed]

54. Kriegenburg, F.; Poulsen, E.G.; Koch, A.; Kruger, E.; Hartmann-Petersen, R. Redox control of the ubiquitin-proteasome system: From molecular mechanisms to functional significance. *Antioxid. Redox Signal.* **2011**, *15*, 2265–2299. [CrossRef] [PubMed]

55. Wang, K. Autophagy and apoptosis in liver injury. *Cell Cycle* **2015**, *14*, 1631–1642. [CrossRef] [PubMed]

56. Michalopoulos, G.K. Liver regeneration. *J. Cell Physiol.* **2007**, *213*, 286–300. [CrossRef] [PubMed]

57. Michalopoulos, G.K. Advances in liver regeneration. *Expert. Rev. Gastroenterol. Hepatol.* **2014**, *8*, 897–907. [CrossRef] [PubMed]

58. Toshima, T.; Shirabe, K.; Fukuhara, T.; Ikegami, T.; Yoshizumi, T.; Soejima, Y.; Ikeda, T.; Okano, S.; Maehara, Y. Suppression of autophagy during liver regeneration impairs energy charge and hepatocyte senescence in mice. *Hepatology* **2014**, *60*, 290–300. [CrossRef] [PubMed]

59. Diehl, A.M.; Chute, J. Underlying potential: Cellular and molecular determinants of adult liver repair. *J. Clin. Investig.* **2013**, *123*, 1858–1860. [CrossRef] [PubMed]

60. Ogrodnik, M.; Miwa, S.; Tchkonia, T.; Tiniakos, D.; Wilson, C.L.; Lahat, A.; Day, C.P.; Burt, A.; Palmer, A.; Anstee, Q.M.; et al. Cellular senescence drives age-dependent hepatic steatosis. *Nat. Commun.* **2017**, *8*, 15691. [CrossRef] [PubMed]

61. Apostolopoulou, M.; Gordillo, R.; Koliaki, C.; Gancheva, S.; Jelenik, T.; Herder, C.; Markgraf, D.; Scherer, P.E.; Roden, M. Serum and hepatic sphingolipids relate to insulin resistance, hepatic mitochondrial capacity and oxidative stress in non-alcoholic fatty liver disease. In *Diabetologie und Stoffwechsel*; Thieme: New York, NY, USA, 2017; pp. S1–S84.

62. Meikle, P.J.; Summers, S.A. Sphingolipids and phospholipids in insulin resistance and related metabolic disorders. *Nat. Rev. Endocrinol.* **2017**, *13*, 79–91. [CrossRef] [PubMed]

63. Dehaven, C.D.; Evans, A.M.; Dai, H.; Lawton, K.A. Organization of GC/MS and LC/MS metabolomics data into chemical libraries. *J. Cheminform.* **2010**, *2*, 9. [CrossRef] [PubMed]

64. Xia, J.; Wishart, D.S. Using metaboanalyst 3.0 for comprehensive metabolomics data analysis. *Curr. Protoc. Bioinform.* **2016**, *55*. [CrossRef]
65. Russell, T.D.; Palmer, C.A.; Orlicky, D.J.; Fischer, A.; Rudolph, M.C.; Neville, M.C.; McManaman, J.L. Cytoplasmic lipid droplet accumulation in developing mammary epithelial cells: Roles of adipophilin and lipid metabolism. *J. Lipid. Res.* **2007**, *48*, 1463–1475. [CrossRef] [PubMed]

International Journal of
Molecular Sciences

MDPI

Article

Re-Adaption on Earth after Spaceflights Affects the Mouse Liver Proteome

Viktoria Anselm [1,†], Svetlana Novikova [2,†] and Victor Zgoda [2,*]

1   Interfaculty Institute of Biochemistry (IFIB), Hoppe-Seyler-Straße 4, Tuebingen 72076, Germany;
    mail@v-anselm.de
2   Orekhovich Institute of Biomedical Chemistry of Russian Academy of Medical Sciences, Pogodinskaya 10,
    Moscow 119121, Russia; novikova.s.e3101@gmail.com
*   Correspondence: vic@ibmh.msk.su; Tel.: +7-(499)-246-8465
†   These authors contributed equally to this work.

Received: 3 July 2017; Accepted: 8 August 2017; Published: 12 August 2017

Abstract: Harsh environmental conditions including microgravity and radiation during prolonged spaceflights are known to alter hepatic metabolism. Our studies have focused on the analysis of possible changes in metabolic pathways in the livers of mice from spaceflight project "Bion-M 1". Mice experienced 30 days of spaceflight with and without an additional re-adaption period of seven days compared to control mice on Earth. To investigate mice livers we have performed proteomic profiling utilizing shotgun mass spectrometry followed by label-free quantification. Proteomic data analysis provided 12,206 unique peptides and 1086 identified proteins. Label-free quantification using MaxQuant software followed by multiple sample statistical testing (ANOVA) revealed 218 up-regulated and 224 down-regulated proteins in the post-flight compared to the other groups. Proteins related to amino acid metabolism showed higher levels after re-adaption, which may indicate higher rates of gluconeogenesis. Members of the peroxisome proliferator-activated receptor pathway reconstitute their level after seven days based on a decreased level in comparison with the flight group, which indicates diminished liver lipotoxicity. Moreover, bile acid secretion may regenerate on Earth due to reconstitution of related transmembrane proteins and CYP superfamily proteins elevated levels seven days after the spaceflight. Thus, our study demonstrates reconstitution of pharmacological response and decreased liver lipotoxicity within seven days, whereas glucose uptake should be monitored due to alterations in gluconeogenesis.

Keywords: spaceflight; mouse; liver; proteome; metabolism; cytochrome P450

1. Introduction

Spaceflights provide a unique opportunity to study microgravity-related changes in organs on a biochemical level. In 2013, the Institute of Biomedical Problems of the Russian Academy of Sciences (IBMP RAS) targeted this research field with the project "Bion-M1" by sending mice in bio-satellites to the Earth orbit in a 30-days flight [1]. As a metabolically highly active organ, the liver of these mice is particularly interesting for us to investigate the effects of spaceflights. Open questions on the proteome dynamics during prolonged spaceflights and their effect on this vital organ involved in metabolism is of great clinical interest in terms of therapy and medical monitoring. Liver function involves a number of enzymatic systems. Among them, members of the cytochrome P450 superfamily crucial for hepatic drug metabolism are the well characterized [2]. The majority of proteins belonging to the protein families CYP1, CYP2, and CYP3 are involved in metabolism of drugs and xenobiotics with substrate specificity intrinsic to the each particular subfamily [3]. As the level of these CYP enzymes determines the overall drug dosage and therapeutic success, their level and activity should be examined to provide proper medication during prolonged spaceflight and re-adaptation on Earth. Furthermore, the systems controlling carbohydrates, proteins, and lipid metabolism need to be studied in the context of spaceflights in order to develop complete nutrition. Effects of 13-day spaceflights on liver function were previously observed including loss of retinol and activation of peroxisome proliferator-activated receptor (PPAR) pathways indicating early signs of non-alcoholic fatty liver disease (NAFLD) [4]. Quantitative mass-spectrometry using stable isotope labeling or label-free approach represents thereby a powerful method to reveal differentially expressed proteins [5]. Thus, using mass-spectrometry, it was shown that PPAR pathways are crucial for lipid metabolism in the liver [6].

To extend our knowledge concerning liver proteins affected by microgravity, we tried to give insight into their expression levels during re-adaption on Earth. Here we present an approach for proteome profiling of mice livers using shotgun mass spectrometry coupled with label-free quantification as a powerful method to reveal a broad range of differentially expressed proteins. Moreover, functional analysis along clusters of differentially expressed proteins provides a comprehensive overview of the metabolic response of a mammalian system to environmental changes related to spaceflights.

2. Results

We studied the effects of microgravity on mice livers by comparing the proteome of mice livers from a 30-day spaceflight, seven day re-adapted mice, and control mice (Figure 1A). Control group animals were housed in the same environmental conditions (e.g., temperature, humidity, gas composition) as in the spaceflight [1]. Three biological replicates were provided for the re-adaption (post-flight) group as well as for the control group, and four replicates were provided for the spaceflight group (flight). The sample homogenates were analyzed by LC-MS/MS using a nano-flow HPLC system and Q Exactive HF mass spectrometer followed by data processing (Figure 1B). Raw data analysis with MaxQuant and statistical data analysis with Perseus provided 12,206 unique peptides and 1086 identified proteins (Table S1), therefrom 1046 proteins were used for relative protein quantification (Table S2) based on at least three unique peptides.

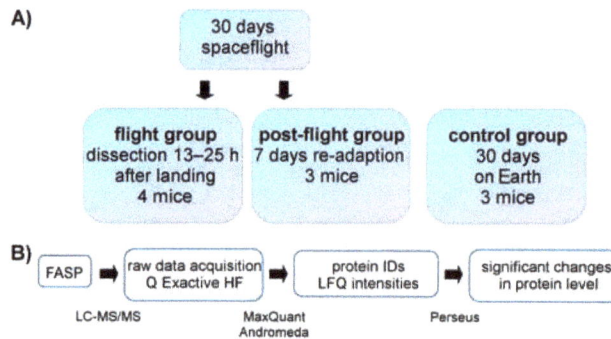

Figure 1. Sample overview and methodological workflow. (**a**) Sample overview for all three groups investigated. The control group on Earth was held in housing and climate conditions corresponding to the conditions of flight group; (**b**) Workflow of sample preparation, processing, and analysis. IDs—identifications; FASP—filter-aided sample preparation; LFQ—label-free quantification.

The initially observed label-free quantification (LFQ) intensities were used to examine significant changes in the protein levels between flight, post-flight and control condition by applying an ANOVA test. Quantified proteins of the control group were compared to their levels in the flight group as well as in the post-flight group using unsupervised hierarchical clustering (Figure 2).

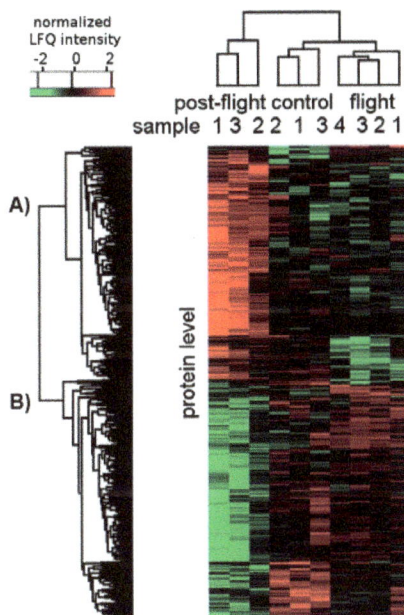

Figure 2. Hierarchical clustering of all proteins with a significantly changed expression profile between at least two groups. Significance was calculated using multiple-sample test (ANOVA model, FDR = 0.05, S0 = 0.1). Samples are clustered in columns and proteins are clustered in rows. Red marked proteins are significantly up-regulated, black marked proteins show no abundance change, and proteins presented in green are down-regulated. Clusters of proteins, which are significantly (**A**) more abundant or (**B**) less abundant in the post-flight group than in the flight or control group are shown. LFQ—Label-free quantification; FDR—false discovery rate

Multiple sample testing (ANOVA test, false discovery rate (FDR) = 0.05) resulted in 442 proteins with significantly changed abundances (Table S3). Proteins with significant expression differences are grouped mainly in two large clusters. These clusters comprise 218 (cluster A) and 224 (cluster B) proteins that are up- and down-regulated in the post-flight compared to the other groups, respectively. Furthermore, we observed clustering of three separate sample groups between flight, post-flight, and control condition.

We analyzed all affected proteins regarding their actual function against GeneOntology and Kyoto Encyclopedia of Genes and Genomes (KEGG) databases. Possible clusters of the significant proteins were surveyed by unsupervised hierarchical clustering considering their function, biological process, or cellular component based on ANOVA multiple-sample testing (FDR < 0.05; Figure 2, Table S4). Top 10 annotation terms are represented in Table 1.

In particular, various terms related to ribonucleoproteins or proteins involved in nitrogen compound metabolic processes are enriched in "A" cluster among terms with the highest FDRs (Table 1, Table S4). Additionally, numerous proteins related to amino acid metabolism, especially associated with amine catabolic processes, are part of the list with enriched proteins. The observed protein catabolism activation indicates enhanced gluconeogenesis since degradation protein products serve as carbon substrates during gluconeogenesis [7]. Interestingly, the second protein cluster (Figure 2B) comprising proteins down-regulated in the post-flight group (*n* = 224), is represented by mostly membrane components (Table 1, Table S4). Proteins assigned with terms related to membrane compartments are represented among terms with the highest FDRs in this cluster.

Table 1. Top 10 annotation terms with the highest FDR in enrichment analysis of protein clusters A and B from unsupervised hierarchical clustering. (GOCC = GeneOntology cell compartment; KEGG = Kyoto Encyclopedia of Genes and Genomes; GOBP = GeneOntology biological process; GOMF = GeneOntology molecular function.)

Type	Name	p-Value	Enrichment	Total	In Cluster	Cluster Size	Benj. Hoch. FDR
	Cluster A: Significantly Higher Protein Levels in Post-flight Group						
GOCC name	Ribonucleoprotein complex	1.62×10^{-14}	1.92	55	52	218	1.90×10^{-12}
KEGG name	Ribosome	1.70×10^{-11}	1.97	38	37	218	3.13×10^{-9}
GOBP name	Cellular amino acid metabolic process	2.23×10^{-12}	1.82	59	53	218	7.79×10^{-9}
GOBP name	Cellular nitrogen compound metabolic process	1.07×10^{-11}	1.54	116	88	218	1.87×10^{-8}
GOMF name	Structural constituent of ribosome	3.68×10^{-11}	1.97	37	36	218	3.37×10^{-8}
GOBP name	Nitrogen compound metabolic process	3.29×10^{-11}	1.52	119	89	218	3.83×10^{-8}
GOCC name	Mitochondrial matrix	1.69×10^{-9}	1.91	36	34	218	9.86×10^{-8}
GOBP name	Cellular amine metabolic process	6.56×10^{-10}	1.71	63	53	218	5.73×10^{-7}
GOBP name	Translation	1.69×10^{-9}	1.91	36	34	218	9.81×10^{-7}
GOBP name	Macromolecule biosynthetic process	1.41×10^{-9}	1.76	53	46	218	9.84×10^{-7}
	Cluster B: Significantly Lower Protein Levels in Post-flight Group						
GOCC name	Membrane part	1.04×10^{-30}	1.65	173	145	224	4.85×10^{-28}
GOCC name	Intrinsic to membrane	6.62×10^{-23}	1.80	102	93	224	1.03×10^{-20}
GOCC name	Integral to membrane	6.62×10^{-23}	1.80	102	93	224	1.55×10^{-20}
GOCC name	Organelle membrane	3.49×10^{-14}	1.47	157	117	224	3.26×10^{-12}
GOCC name	Endoplasmic reticulum part	7.21×10^{-12}	1.72	69	60	224	5.63×10^{-10}
GOCC name	Endoplasmic reticulum	5.52×10^{-11}	1.84	46	43	224	3.69×10^{-9}
GOCC name	Endoplasmic reticulum membrane	2.47×10^{-9}	1.72	54	47	224	1.28×10^{-7}
GOCC name	Plasma membrane part	5.85×10^{-8}	1.67	52	44	224	2.74×10^{-6}
GOCC name	Membrane	4.95×10^{-7}	1.19	260	157	224	2.10×10^{-5}
GOMF name	Transporter activity	1.46×10^{-6}	1.78	31	28	224	2.67×10^{-4}

Volcano plots visualizing significance versus fold-change were obtained for control/flight, control/post-flight and flight/post-flight pairs (Figure S1). Applying permutation-based FDR calculation (FDR = 0.05, S0 = 0.1) [8], we determined differentially expressed proteins (Figure S1). Further investigation of significantly differentially expressed proteins (ANOVA test, FDR = 0.05) against KEGG database revealed protein groups of bile secretion, PPAR signaling pathway, retinol metabolism, and cytochromes superfamily (Table S4, Table S5, Table 2).

The proteins associated with bile secretion are less represented with a FDR of 0.019 (Table S4). The level of 10 quantified and bile secretion related proteins (Table 2) is significantly lower in post-flight than in the flight or control group (ANOVA test, FDR = 0.05). The majority of these proteins have ion or cation transporter activity to regulate the bile flow from liver to gall bladder by building an osmotic gradient [9]. A decreased abundance of these transporters in the post-flight group may indicate less bile secretion leading to an accumulation of harmful bile acids in hepatocytes [10]. A major part of our observed bile secretion related proteins is only significantly down-regulated in the post-flight group compared to the flight group (Table 2) which indicates considerable reconstitution of bile secretion within seven days. Notably, there was no difference in the abundance of these transmembrane proteins between the control and flight group, which could be owed to less sensitive fold change detection between flight and control. This could be caused by the high significance threshold observed with multiple testing correction using permutation based FDR (Figure S1, A). Comparisons between control and post-flight or flight and post-flight demonstrate considerable lower significance thresholds.

Table 2. Fold-change of proteins annotated against KEGG database.

Protein Name	Uniprot ID	Fold Change: Control/Post-flight	Fold Change: Flight/Post-flight	Student's t-Test q-Value	ANOVA q-Value
Bile Secretion[1]					
Solute carrier organic anion transporter family member 1B2	Q9JL3	-	1.7	0.0187	0.0143
ATP-binding cassette sub-family G member 2	Q7TMS5	-	2.7	0.0106	0.0131
Scavenger receptor class B member 1	Q4FK30	*		>0.05	0.0198
Solute carrier family 22 member 1	O08966	*		>0.05	0.0197
Sodium/potassium-transporting ATPase subunit β-1	Q545P0	-	1.6	0.0127	0.0132
Sodium/potassium-transporting ATPase subunit beta-3	Q544Q7	-	2.4	0.0171	0.0116
Aquaporin-1	Q02013	*		>0.05	0.0211
Solute carrier organic anion transporter family member 1A1	Q53ZW9	-	4.0	0.0293	0.0291
Sodium/potassium-transporting ATPase subunit alpha-1	Q8VDN2	1.9	1.6	0.0165	0.0122
Epoxide hydrolase 1	Q9D379	-	1.9	0.0077	0.0278
PPAR Signaling Pathway[1]					
Long-chain specific acyl-CoA dehydrogenase, mitochondrial	P51174	-	0.6	0.0340	0.0374
Long-chain-fatty-acid-CoA ligase 1	D3Z041	-	1.9	0.0052	0.0075
Very long-chain acyl-CoA synthetase	O35488	-	1.5	0.0247	0.0341
Cytochrome P450 4A10	O88833	*		>0.05	0.0154
Fatty acid-binding protein, liver	Q3V2F7	-	3.9	0.0080	0.0187
Acyl-CoA-binding protein	Q548W7	*		>0.05	0.0374
Medium-chain specific acyl-CoA dehydrogenase, mitochondrial	P45952	0.5	0.6	0.0457	0.0132
Carnitine O-palmitoyltransferase 2, mitochondrial	P52825	-	1.7	0.0110	0.0115
Carnitine O-palmitoyltransferase 1, liver isoform	Q7TQD5	*		>0.05	0.0153
CD36 antigen, isoform CRA_a	Q3UAI3	-	2.1	0.0044	0.0255
Cytochrome P450 4A12A	Q91WL5	-	5.0	0.0113	0.0128
Peroxisomal bifunctional enzyme	Q9DBM2	-	0.6	0.0141	0.0116
Phosphoenol pyruvate carboxy kinase, cytosolic [GTP]	Q9Z2V4	*		>0.05	0.0271

Int. J. Mol. Sci. **2017**, *18*, 1763

Table 2. *Cont.*

Protein Name	Uniprot ID	Fold Change: Control/Post-flight	Fold Change: Flight/Post-flight	Student's t-Test q-Value	ANOVA q-Value
Retinol Metabolism [1]					
Cytochrome P450 1A2	B6VGH4	-	2.7	0.0117	0.0125
Cytochrome P450 2B10	Q9WUD0	*		>0.05	0.0120
Cytochrome P450 4A10	O88833	*		>0.05	0.0154
Cytochrome P450 3A11	Q3UEN8	-	0.7	0.0198	0.0166
UDP-glucuronosyltransferase	Q3UEP4	-	2.1	0.0167	0.0132
Cytochrome P450 3A13	Q3UW87	-	0.4	0.0121	0.0279
UDP-glucuronosyltransferase 1-1	Q63886	-	1.2	0.0420	0.0185
UDP-glucuronosyltransferase 1-6	Q64435	-	2.0	0.0179	0.0410
Cytochrome P450 2C54	Q6XVG2	*		>0.05	0.0476
UDP-glucuronosyltransferase 2A3	Q8BWQ1	-	1.9	0.0109	0.0132
UDP-glucuronosyltransferase	Q8K154	-	1.9	0.0172	0.0126
UDP-glucuronosyltransferase	Q8R084	-	3.5	0.0118	0.0119
Cytochrome P450 2C70	Q91W64	*		>0.05	0.0340
Cytochrome P450 4A12A	Q91WL5	-	5.0	0.0113	0.0128
Dehydrogenase/reductase SDR family member 4	Q99LB2	-	1.7	0.0108	0.0131
Cytochromes [2]					
Cytochrome P450 2D26	Q8CIM7	1.5	-	0.0474	0.0120
Cytochrome P450 4V3	B2RSR0	-	2.3	0.0175	0.0176
Cytochrome P450 1A2	B6VGH4	-	2.7	0.0117	0.0125
Cytochrome P450 2B10	Q9WUD0	*		>0.05	0.0120
Cytochrome P450 4A10	O88833	*		>0.05	0.0154
Cytochrome P450 2D9	P11714	-	6.1	0.0117	0.0106
Cytochrome P450 2D10	P24456	-	2.1	0.0134	0.0152
Cytochrome P450 2F2	P33267	-	1.6	0.0264	0.0198
Cytochrome P450 3A11	Q3UEN8	-	0.7	0.0198	0.0166
Cytochrome P450 3A13	Q3UW87	-	0.4	0.0121	0.0279
Cytochrome P450 2C54	Q6XVG2	*		>0.05	0.0476
Cytochrome P450 2C70	Q91W64	*		>0.05	0.0340
Cytochrome P450 4A12A	Q91WL5	-	5.0	0.0113	0.0128

* Proteins with ANOVA-significant but two-sample t-test (Student's t-test) insignificant q-value; [1] Proteins derived from partially same KEGG names (Kyoto Encyclopedia of Genes and Genomes); [2] Members of the Cytochrome P450 superfamily with significant protein abundance changes.

The PPAR signaling pathway members are mostly up-regulated. The most prominent level increased (fold change >2) in flight group comparing with post-flight group was found for cytochrome P450 4A12A (Cyp4a12) involved in fatty acid ω-oxidation, fatty acid-binding protein (Fabp1) used for fatty acid transport through blood stream, and CD36 antigen (Cd36) acting as fatty acid translocase with lipoprotein binding function [11]. In contrast, long-chain specific acyl-CoA dehydrogenase (Acadl), medium-chain specific acyl-CoA dehydrogenase, (Acadm) and peroxisomal bifunctional enzyme (Ehhadh) essential for fatty acid β-oxidation in mitochondria and peroxisomes are down-regulated.

Regarding spaceflights induced lipotoxicity [4], our examinations give further insights into the reconstitution of liver metabolism. Increased expression levels of the liver residing fatty acid binding protein (Fabp1, Q3V2F7) indicate an active PPAR pathway [12] which, in turn, initiates adipose cell differentiation ultimately resulting in NAFLD [13]. Fabp1, as well as other lipid binding protein levels are significantly increased in the flight group compared to the post-flight group (Table 2) suggesting an activation of PPAR. The observed significant change diminishes comparing control mice and re-adapted mice.

Differentially expressed proteins are enriched for proteins related toretinol metabolism. Different UDP-glucuronosyl transferases, Cyp1a2, and dehydrogenase/reductase SDR family member 4 (Dhrs4) metabolizing retinoids are up-regulated, whereas Cyp3a13 level is decreased in flight group compared with re-adapted mice (Table 2). Retinoic acids are mobilized in hepatic stellate cells as part of retinol metabolism [14] and act as upstream PPAR activators [15,16]. UDP-glucuronosyl transferases perform glucuronidation of retinoic acids for their solubilization [17,18] followed by secretion. After secretion, retinoic acids activate PPARs in hepatocytes.

Apart from retinol metabolism and PPAR pathway, various members of the cytochrome P450 superfamily were detected to have a significantly changed abundance in the post-flight group. We observed up-regulation of CYP2D subfamily members (Cyp2d9, Cyp2d10, and Cyp2d26), as well as increased level of Cyp4v3, Cyp1a2, Cyp2f2, and Cyp4a12a. Members of CYP3A subfamily (Cyp3a11 and Cyp3a13) were down-regulated. Cyp4a12a showed a five-fold (p-value = 0.0113) increased level in the flight group compared to the post-flight group. CYP4A subfamily members are involved in fatty acid ω-oxidation in microsomes [19]. Reconstitution of Cytochrome P450 1A2 was confirmed since there is no significant change between control and post-flight along with detected fold change between flight and post-flight (Table 2). The subfamily CYP2D appears 6.1-(p-value = 0.0117) and 2.1-fold (p-value = 0.0134) elevated in the flight group compared to post-flight group (Table 2). Repeatedly, no significant difference in cytochromes profile between control and post-flight shows that re-adaption to the usual drug metabolizing CYP levels occurs within seven days after landing.

3. Discussion

Manned spaceflights are promising ways to shed light on basic questions regarding origins of life and the human future. To enable this, questions about the impact of environmental conditions during spaceflights can be addressed testing model organisms such as mice. One of these questions is how long-term adaption to spaceflight environment and re-adaption to Earth is reflected in the liver proteome. Our study design was established to exclude housing effects, which are based on the same environmental conditions concerning temperature, humidity, feeding and gas compositions for all animals studied. We performed label-free proteomic profiling and revealed no significant differences between the proteomes of flight and control group due to insufficient fold change detection combined with high variances within groups. Probably, small sample size contributes to the low significance.

We have revealed significant differential protein expression between flight and post-flight groups. Interestingly, up- and down-regulated proteins are linked with fatty acid oxidation. Remarkably, we revealed up-regulated proteins involved in fatty acid transport (Fabp1), cell uptake (Cd36), translocation into mitochondria (carnitine O-palmitoyltransferase 2, mitochondrial, Cpt), and activation for β-oxidation (long-chain-fatty-acid-CoA ligase 1 (Acsl1)). Moreover, Cyp4a12 related to fatty acid ω-oxidation is up-regulated whereas enzymes essential for β-oxidation both in mitochondria

(Acadl and Acadm) and peroxisomes (Ehhadh) are down-regulated. These results could indicate impaired fatty acid β-oxidation, which could lead to fatty acid accumulation, in turn, possibly causing lipotoxicity that underlies NAFLD pathogenesis. Yet, we should interpret these results with caution since a small sample size and label-free approach applied. Notably, the small sample size is linked with the fact that only 16 out of 45 mice (36%) survived the spaceflight and samples were distributed among large number of scientific teams [1]. Nevertheless, project "Bion-M1" provided unique samples to investigate spaceflight effects on living beings.

Functional annotation against KEGG database revealed PPAR and retinol metabolism signaling pathways, which are enriched by down-regulated proteins in the post-flight group compared with the flight group. This result indicates re-adaption occurs partially within seven days associated with diminishing liver lipotoxicity, which is consistent with the data of Jonscher et al. [4]. Jonscher et al. linked matabolome and transcriptome profile including modulated PPARα pathways with lipotoxicity and NAFLD-like phenotype. Analysis of PPAR α, δ, and γ mRNA level of the samples performed in our laboratory did not show significant level change between the samples (data not shown). However, it has been shown previously that activity of PPARs and its pathway members is regulated by posttranslational modifications including phosphorylation [20]. Further validation of PPAR signaling pathway by the targeted proteomics or PTM analysis will help to obtain direct evidences for the PPAR pathway involvement in re-adaptation after space flight.

Moreover, we observed down-regulated fatty acid binding proteins and UDP-glucuronosyl transferases that can contribute to NAFLD recovery. Since NAFLD leads to severe complications such as increased risk of cirrhosis, type 2 diabetes, and cardiovascular complications [21] it is highly important to maintain a hypocaloric diet both during spaceflight and re-adaption to Earth.

Strikingly, enhanced expression of the CYP4A subfamily, that is responsible for ω-oxidation of fatty acids [4], is induced by the PPAR pathway, which coincides with the observed activation of the PPAR pathway. Additionally, our observation can be supported by a previously observed increase in ω-oxidation products in mice livers which experienced a 13-day spaceflight [4]. However, we detected increased expression of Cyp4a12a in flight compared to post-flight, which is consistent with the previously observed increased amount of ω-oxidation products in the flight group [4]. The cytochrome P450 family member Cyp4a12a should be investigated further regarding its amount to confirm an increased level in the flight group and to be measured during different time periods for re-adaption.

In addition, recovery of other elevated CYP levels was shown, which indicates reconstitution of the hepatic pharmacological response. Our results of reconstituted levels of Cyp1a2 underline previously detected elevated levels of cytochrome P450 1A2 during flight compared to post-flight [22], showing the importance to monitor liver performance during flights. Numerous drugs are turned over in the liver by CYP1A2 [23] illustrating its immense clinical importance. However, we should consider that the observed reconstitution of protein levels could be a consequence of stress during biosatellite re-entry and the landing procedure. To address this concern, mice could pass 30 days on a manned space station with following euthanization on the space station, thus, avoiding stressful landing conditions. This approach could also elucidate direct effects of microgravity or radiation on the liver by excluding stressful landing as a source for proteomic changes, which cannot be excluded from our study.

With relation to energy metabolism, up-regulation in proteins involved in amino acid metabolic processes in the re-adaptation group may indicate an increased level of gluconeogenesis. Initiation of gluconeogenesis is started during lack of cellular glucose caused by short-term fasting [7]. Therefore, prolonged spaceflights may result in reduced glucose uptake during assimilation to Earth conditions. We also assume restoration of bile acid secretion seven days after landing due to restored levels of transmembrane proteins related to bile acid secretion.

In this study, we did not perform quantitative validation using a targeted mass-spectrometry technique. However, Cyp1a1 and Cyp2d9 up-regulation, as well as Cyp3a11 down-regulation is consistent with our previous data on CYP450 absolute abundance obtained by selected reaction monitoring (SRM) with isotopically labeled peptide standards on the same animals (mice from flight, post-flight, and ground control group) [22]. Label-free quantification is a promising approach due to low-cost and using of high-resolution mass-spectrometers, which yields high precise measurements of several peptides per protein (at least three unique peptides in our study). Shao et al. successfully used labeled-free platform combines shotgun mass-spectrometry, targeted SRM, and computational method ("Standard curve slope") to identify sex-dependent CYP450 expression patterns. Relative expression of Cyp1a1, Cyp2d10, and Cyp2d26 was analyzed in male and female rat liver microsomes [24].

4. Materials and Methods

4.1. Samples

This study was approved by IACUC of MSU Institute of Mitoengineering (Protocol No-35, 1 November 2012) and of Biomedical Ethics Commission of IBMP (protocol No-319, 4 April 2013) and performed in accordance to the European Convention for the Protection of Vertebrate Animals used for Experimental and Other Scientific Purposes. Male C57/BL6 mice (four–five months old) were selected for spaceflight and ground control experiments. At the time of spaceflight and at the start of the related ground control experiments, the mice had average weight 22 ± 2 g. All mice were pathogen-free. To avoid violent behavior mice were trained before the flight and ground control experiment to form the stable groups of three mice each. Animals were adapted to paste diet. Stable cohorts of three mice were housed in each habitat during spaceflight and ground control experiment. During spaceflight the mice were fed with paste-like food with 76–78% water content developed at IBMP RAS. The mice were kept in a natural light-dark cycle (12 h light and 12 h dark). For more details about experimental animals see Andreev-Andrievsky et al. [1]. In our experiment, we compared liver proteomes of the mice which passed 30 days (from 19 April 2013 to 19 May 2013) in a biosatellite in space and sampled 13–25 h after landing (flight group) with mice of the same spaceflight sampled seven days after landing (post-flight group). Mice held in a biosatellite on Earth in the similar cage from 26 July 2013 to 26 August 2013 under the corresponding housing and climate conditions (temperature, humidity, and atmosphere gas composition) served as control group. The environmental conditions as well as food delivery were continuously recorded during spaceflight. All mice were euthanized by cervical dislocation; livers were immediately sampled and stored at $-80\ ^\circ\text{C}$ until further procedures.

4.2. Sample Preparation

Twenty milligrams of each mouse liver were washed with PBS, homogenized in 200 μL lysis buffer (4% SDS in 0.1 M Tris-HCl pH 8.5) and centrifuged at $3000\times$ g at $4\ ^\circ\text{C}$ for 5 min. Total protein content was measured according to the BCA method [25]. A total protein amount of 100 μg for each sample was used for tryptic digestion according to the common FASP protocol [26]. Briefly, detergents in the samples were exchanged with 100 mM Tris-HCl (pH 8.5) using Microcon filters (10 kDa cut off, Millipore, Bedford, MA, USA). Protein disulfide bridges were reduced with 100 mM 1,4-dithiothreitol in 100 mM Tris-HCl (pH 8.5), alkylation of thiols was performed with 55 mM iodacetamide in 8 M urea/100 mM Tris-HCl (pH 8.5). Tryptic digestion with a trypsin (Sequencing Grade Modified, Promega, Madison, WI, USA) to protein ratio of 1:100 was carried out overnight at $37\ ^\circ\text{C}$ in a 50 mM tetraethylammonium bicarbonate (pH 8.5) followed by an additional digestion step under the same conditions with 2 h duration.

4.3. LC-MS/MS

We separated 2 μg peptides for each sample with high-performance liquid chromatography (HPLC, Ultimate 3000 Nano LC System, Thermo Scientific, Rockwell, IL, USA) in a 15-cm long C18 column with an inner diameter of 75 μm (Acclaim® PepMap™ RSLC, Thermo Fisher Scientific, Rockwell, IL, USA). The peptides were eluted with a gradient from 5–35% buffer B (80% acetonitrile, 0.1% formic acid) over 115 min at a flow rate of $0.3\ \mu\text{L}^{-1}$ min. Total run time including 90 min to reach 99% buffer B, flushing 10 min with 99% buffer B and 15 min re-equilibration to buffer A (0.1% formic acid) amounted 155 min. Further analysis was performed with a Q Exactive HF mass spectrometer (Q ExactiveTM HF Hybrid Quadrupole-OrbitrapTM Mass spectrometer, Thermo Fisher Scientific, Rockwell, IL, USA). Mass spectra were acquired at a resolution of 60,000 (MS) and 15,000 (MS/MS) in a m/z range of 400–1500 (MS) and 200–2000 (MS/MS). An isolation threshold of 100,000 counts was determined for precursor's selection and up to top 25 precursors were chosen for fragmentation with high-energy collisional dissociation (HCD) at 25 NCE and 100 ms accumulation time. Precursors with a charged state of +1 were rejected and all measured precursors were excluded from measurement for 20 s. At least three technical runs were measured for each sample. The mass spectrometry proteomics data have been deposited to the ProteomeXchange Consortium (http://proteomecentral.proteomexchange.org) via the PRIDE partner repository [27] with the dataset identifier PXD005102.

4.4. Data Processing

The obtained raw data were processed using the MaxQuant software [28] (version 1.5.5.1, Jürgen Cox, Max Planck Institute of Biochemistry, Martinsried, Germany) with the built-in search engine Andromeda [29]. Protein sequences of the complete mouse proteome provided by Uniprot (August 2016) was used for protein identification with Andromeda. Carbamidomethylation of cysteines was set as fixed modification and protein N-terminal acetylation as well as oxidation of methionines was set as variable modification for the peptide search. A maximum mass deviation of 4.5 ppm was allowed for precursor's identification and 20 ppm were set as match tolerance for fragment identification (acquisition in Orbitrap). Up to two missed cleavages were allowed for trypsin digestion. The software option "Match between runs" was enabled and features within a time window of 0.7 min were used to match between runs. The false discovery rates (FDR) for peptide and protein identifications were set to 1%. Only unique peptides were used for label-free quantification (LFQ) according to the method described by Cox et al. [30].

4.5. Statistical Analysis

The obtained LFQ intensities were filtered and statistically analyzed using the Perseus environment. Protein groups identified only by peptides with modified sites, contaminant matches and matches to the reverse database were removed. Proteins identified by at least two unique peptides were determined as identified, whereas only proteins identified with three unique peptides were used for quantification. Solely protein groups that were at least once quantified in each group were included in our observations. Normalization was performed by built-in MaxQuant algorithm [30]. The LFQ intensities were transformed by log2(x) and missing LFQ intensity values (NaN) were replaced using low LFQ intensity values from the normal distribution (width = 0.3, down shift = 1.8). Unpaired Student's t-tests were used to compare two groups, whereas the ANOVA test was used for multiple-sample testing. To introduce an artificial variance for small variance values, the constant S0 was set to 0.1 for all statistical tests. A permutation-based FDR of 5% was used for truncation of all tests results. Additionally, only proteins with a probability for significant protein abundance changes with a p-value <0.05 were used for fold change visualization in the presented tables.

5. Conclusions

Studying the alterations of protein level induced by spaceflights in model organisms such as mice revealed proteome changes suggesting impaired fatty acid oxidation leading to lipotoxicity, as well as altered glucose up-take and bile secretion. These results offer the opportunity to adjust the drug treatment and nutrition for future long-termed spaceflights of humans and the duration of medical monitoring after spaceflights.

Supplementary Materials: Supplementary materials can be found at www.mdpi.com/1422-0067/18/8/1763/s1.

Int. J. Mol. Sci. **2017**, *18*, 1763

Acknowledgments: This work was supported by the Russian Scientific Foundation [grant number 16-44-03007]. The authors are grateful to the "Human Proteome" Core Facility of the Orekhovich Institute of Biomedical Chemistry (Russia) for the culturing of HL-60 cell line and for the opportunity to use flow cytometry and mass-spectrometry equipment. We thank Barbara J. Petzuch for thoroughly proof reading this article and Arthur T. Kopylov for supportive idea discussion.

Author Contributions: Victor Zgoda conceived and designed the experiments; Svetlana Novikova and Viktoria Anselm performed the experiments; Viktoria Anselm analyzed the data; All authors contributed to paper writing.

Conflicts of Interest: The authors declare no conflict of interest.

Abbreviations

PPAR	Peroxisome proliferator-activated receptor
FDR	False discovery rates
CYP	Cytochrome P450
FASP	Filter aided sample preparation
LFQ	Label-free quantification
NAFLD	Non-alcoholic fatty liver disease
FABP	Fatty acid binding protein
ID	Identification number
GOCC	GeneOntology cell compartment
GOMF	GeneOntology molecular function
GOBP	GeneOntology biological process
KEGG	Kyoto Encyclopedia of Genes and Genomes

References

1. Andreev-Andrievskiy, A.; Popova, A.; Boyle, R.; Alberts, J.; Shenkman, B.; Vinogradova, O.; Dolgov, O.; Anokhin, K.; Tsvirkun, D.; Soldatov, P.; et al. Mice in Bion-M 1 Space Mission: Training and Selection. *PLoS ONE* **2014**, *9*, e104830. [CrossRef] [PubMed]
2. McDonnell, A.M.; Dang, C.H. Basic review of the cytochrome p450 system. *J. Adv. Pract. Oncol.* **2013**, *4*, 263–268. [PubMed]
3. Aebersold, R.; Mann, M. Mass spectrometry-based proteomics. *Nature* **2003**, *422*, 198–207. [CrossRef] [PubMed]
4. Jonscher, K.R.; Alfonso-Garcia, A.; Suhalim, J.L.; Orlicky, D.J.; Potma, E.O.; Ferguson, V.L.; Bouxsein, M.L.; Bateman, T.A.; Stodieck, L.S.; Levi, M.; et al. Spaceflight Activates Lipotoxic Pathways in Mouse Liver. *PLoS ONE* **2016**, *11*, e0152877. [CrossRef]
5. Gillet, L.C.; Leitner, A.; Aebersold, R. Mass Spectrometry Applied to Bottom-Up Proteomics: Entering the High-Throughput Era for Hypothesis Testing. *Annu. Rev. Anal. Chem.* **2016**, *9*, 449–472. [CrossRef] [PubMed]
6. Pawlak, M.; Lefebvre, P.; Staels, B. Molecular mechanism of PPARα action and its impact on lipid metabolism, inflammation and fibrosis in non-alcoholic fatty liver disease. *J. Hepatol.* **2015**, *62*, 720–733. [CrossRef] [PubMed]
7. Rui, L. Energy Metabolism in the Liver. In *Comprehensive Physiology*; John Wiley & Sons, Inc.: Hoboken, NJ, USA, 2014; Vol. 4, pp. 177–197.
8. Tyanova, S.; Temu, T.; Sinitcyn, P.; Carlson, A.; Hein, M.Y.; Geiger, T.; Mann, M.; Cox, J. The Perseus computational platform for comprehensive analysis of (prote)omics data. *Nat. Methods* **2016**, *13*, 731–740. [CrossRef] [PubMed]
9. Boyer, J.L. Bile Formation and Secretion. In *Comprehensive Physiology*; John Wiley & Sons, Inc.: Hoboken, NJ, USA, 2013; Vol. 3, pp. 1035–1078.
10. Oswald, M.; Kullak-Ublick, G.A.; Paumgartner, G.; Beuers, U. Expression of hepatic transporters OATP-C and MRP2 in primary sclerosing cholangitis. *Liver* **2001**, *21*, 247–253. [CrossRef] [PubMed]
11. Frayn, K.N.; Arner, P.; Yki-Järvinen, H. Fatty acid metabolism in adipose tissue, muscle and liver in health and disease. *Essays Biochem.* **2006**, *42*, 89–103. [CrossRef] [PubMed]

12. Vildhede, A.; Wiśniewski, J.R.; Norén, A.; Karlgren, M.; Artursson, P. Comparative Proteomic Analysis of Human Liver Tissue and Isolated Hepatocytes with a Focus on Proteins Determining Drug Exposure. *J. Proteome Res.* **2015**, *14*, 3305–3314. [CrossRef] [PubMed]

13. Merrick, B.A.; Madenspacher, J.H. Complementary gene and protein expression studies and integrative approaches in toxicogenomics. *Toxicol. Appl. Pharmacol.* **2005**, *207*, 189–194. [CrossRef] [PubMed]

14. Shirakami, Y.; Lee, S.-A.; Clugston, R.D.; Blaner, W.S. Hepatic metabolism of retinoids and disease associations. *Biochim. Biophys. Acta-Mol. Cell Biol. Lipids* **2012**, *1821*, 124–136. [CrossRef] [PubMed]

15. Shaw, N.; Elholm, M.; Noy, N. Retinoic acid is a high affinity selective ligand for the peroxisome proliferator-activated receptor beta/delta. *J. Biol. Chem.* **2003**, *278*, 41589–41592. [CrossRef] [PubMed]

16. Ziouzenkova, O.; Plutzky, J. Retinoid metabolism and nuclear receptor responses: New insights into coordinated regulation of the PPAR-RXR complex. *FEBS Lett.* **2008**, *582*, 32–38. [CrossRef] [PubMed]

17. Radominska, A.; Little, J.M.; Lehman, P.A.; Samokyszyn, V.; Rios, G.R.; King, C.D.; Green, M.D.; Tephly, T.R. Glucuronidation of retinoids by rat recombinant UDP: Glucuronosyltransferase 1.1 (bilirubin UGT). *Drug Metab. Dispos.* **1997**, *25*, 889–892. [PubMed]

18. Eun, J.W.; Bae, H.J.; Shen, Q.; Park, S.J.; Kim, H.S.; Shin, W.C.; Yang, H.D.; Jin, C.Y.; You, J.S.; Kang, H.J.; et al. Characteristic molecular and proteomic signatures of drug-induced liver injury in a rat model. *J. Appl. Toxicol.* **2015**, *35*, 152–164. [CrossRef] [PubMed]

19. Wanders, R.J. A.; Komen, J.; Kemp, S. Fatty acid omega-oxidation as a rescue pathway for fatty acid oxidation disorders in humans. *FEBS J.* **2011**, *278*, 182–194. [CrossRef] [PubMed]

20. Burns, K.A.; Vanden Heuvel, J.P. Modulation of PPAR activity via phosphorylation. *Biochim. Biophys. Acta* **2007**, *1771*, 952–960. [CrossRef] [PubMed]

21. Fan, J.-G.; Cao, H.-X. Role of diet and nutritional management in non-alcoholic fatty liver disease. *J. Gastroenterol. Hepatol.* **2013**, *28*, 81–87. [CrossRef] [PubMed]

22. Moskaleva, N.; Moysa, A.; Novikova, S.; Tikhonova, O.; Zgoda, V.; Archakov, A. Spaceflight Effects on Cytochrome P450 Content in Mouse Liver. *PLoS ONE* **2015**, *10*, e0142374. [CrossRef] [PubMed]

23. Wang, B.; Zhou, S.F. Synthetic and natural compounds that interact with human cytochrome P450 1A2 and implications in drug development. *Curr. Med. Chem.* **2009**, *16*, 4066–4218. [CrossRef] [PubMed]

24. Shao, Y.; Yin, X.; Kang, D.; Shen, B.; Zhu, Z.; Li, X.; Li, H.; Xie, L.; Wang, G.; Liang, Y. An integrated strategy for the quantitative analysis of endogenous proteins: A case of gender-dependent expression of P450 enzymes in rat liver microsome. *Talanta* **2017**, *170*, 514–522. [CrossRef] [PubMed]

25. Walker, J.M. The Bicinchoninic Acid (BCA) Assay for Protein Quantitation. In *Basic Protein and Peptide Protocols*; Humana Press: Totowa, New Jersey, USA, 1994; Vol. 32, pp. 5–8.

26. Wiśniewski, J.R.; Zougman, A.; Nagaraj, N.; Mann, M. Universal sample preparation method for proteome analysis. *Nat. Methods* **2009**, *6*, 359–362. [CrossRef] [PubMed]

27. Vizcaíno, J.A.; Côté, R.G.; Csordas, A.; Dianes, J.A.; Fabregat, A.; Foster, J.M.; Griss, J.; Alpi, E.; Birim, M.; Contell, J.; et al. The PRoteomics IDEntifications (PRIDE) database and associated tools: Status in 2013. *Nucleic Acids Res.* **2013**, *41*, D1063–D1069. [CrossRef] [PubMed]

28. Cox, J.; Mann, M. MaxQuant enables high peptide identification rates, individualized p.p.b.-range mass accuracies and proteome-wide protein quantification. *Nat. Biotechnol.* **2008**, *26*, 1367–1372. [CrossRef] [PubMed]

29. Cox, J.; Neuhauser, N.; Michalski, A.; Scheltema, R.A.; Olsen, J.V.; Mann, M. Andromeda: A Peptide Search Engine Integrated into the MaxQuant Environment. *J. Proteome Res.* **2011**, *10*, 1794–1805. [CrossRef] [PubMed]

30. Cox, J.; Hein, M.Y.; Luber, C.A.; Paron, I.; Nagaraj, N.; Mann, M. Accurate proteome-wide label-free quantification by delayed normalization and maximal peptide ratio extraction, termed MaxLFQ. *Mol. Cell Proteom.* **2014**, *13*, 2513–2526. [CrossRef] [PubMed]

International Journal of
Molecular Sciences

MDPI

Review

Metabolic Pathways of the Warburg Effect in Health and Disease: Perspectives of Choice, Chain or Chance

Jorge S. Burns [1,2,]* and Gina Manda [3]

[1] Advanced Polymer Materials Group, University Politehnica of Bucharest, Gh Polizu 1-7, 011061 Bucharest, Romania
[2] Department of Medical and Surgical Sciences for Children & Adults, University Hospital of Modena and Reggio Emilia, 41121 Modena, Italy
[3] "Victor Babes", National Institute of Pathology, 050096 Bucharest, Romania; gina.manda@gmail.com
* Correspondence: jsburns@unimore.it

Received: 17 October 2017; Accepted: 13 December 2017; Published: 19 December 2017

Abstract: Focus on the Warburg effect, initially descriptive of increased glycolysis in cancer cells, has served to illuminate mitochondrial function in many other pathologies. This review explores our current understanding of the Warburg effect's role in cancer, diabetes and ageing. We highlight how it can be regulated through a chain of oncogenic events, as a chosen response to impaired glucose metabolism or by chance acquisition of genetic changes associated with ageing. Such chain, choice or chance perspectives can be extended to help understand neurodegeneration, such as Alzheimer's disease, providing clues with scope for therapeutic intervention. It is anticipated that exploration of Warburg effect pathways in extreme conditions, such as deep space, will provide further insights crucial for comprehending complex metabolic diseases, a frontier for medicine that remains equally significant for humanity in space and on earth.

Keywords: Warburg effect; mitochondria; radiation; metabolism; neurodegenerative diseases; space biology

1. Introduction: What Is the Warburg Effect?

With the successful delivery of Mars Exploration Rovers (MER); Sojourner, Spirit (MER-A), Opportunity (MER-B) and Curiosity, plans for sending humans to Mars are in progress [1]. Beyond the considerable engineering challenges, pioneering habitation in hostile environments will require full awareness of the biological impact of galactic cosmic radiation. Like narratives describing the fate of fictional invaders from Mars [2], the final outcome will also crucially depend on a comprehensive understanding and appreciation of the microbial world. Endo-symbiotically intertwined in the evolutionary tree of eukaryotic life, mitochondria determine our overall metabolism [3]. In mammalian cells, though not all species, these ubiquitous and diverse organelles use oxygen (O_2) as the terminal acceptor for anabolic processes, specializing our life to a dependency on oxic environments. Careful quantitative measurement of mitochondrial energy production mechanisms led to a hypothesis persisting for nearly a century as insightful for cellular stress responses to harsh environments and illness.

Dramatic changes in infant growth rate accompany the first breath of the newborn. The oxygen we breathe fuels the high energy-yielding oxyhydrogen gas reaction by which two molecules of hydrogen and one molecule of oxygen are converted to water, releasing -193 kJ/mol of energy under physiological conditions to fuel ATP synthesis from ADP and phosphate. An accompanying major source of cellular energy and new cell mass is glucose. Glycolysis breaks glucose down to pyruvate (Figure 1), that ultimately can be metabolized oxidatively to CO_2 by a network of enzymes known as the tricarboxylic acid (TCA) cycle. Linking the metabolic pathways of glycolysis and the TCA cycle, a pyruvate dehydrogenase complex (PDC) made of three enzymatic proteins in the mitochondrial matrix breaks down pyruvate to form acetyl-CoA, releasing CO_2 and NADH. Subsequently, during aerobic respiration, the TCA cycle coordinates catabolic breakdown of acetyl-CoA, releasing electron flow to the final O_2 acceptor through a respiratory chain concomitantly generating a transiently stored proton gradient across the inner mitochondrial membrane. This gradient of electric potential energetically fuels oxidative phosphorylation (OXPHOS) mediated by the ATP synthase complex, generating ATP from ADP. Energy from the proton gradient can also be dissipated as heat through passive proton leakage or via uncoupling proteins [4].

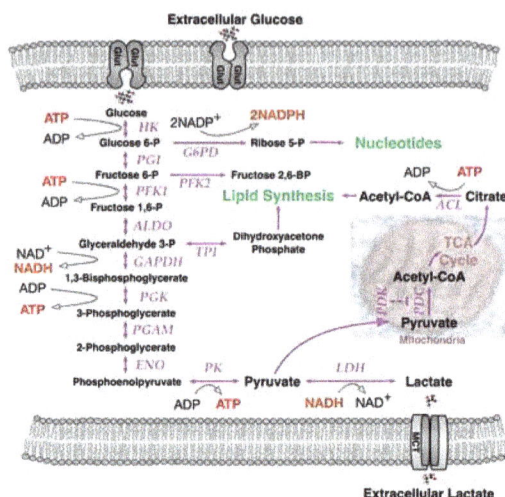

Figure 1. Glucose metabolism in cells. Extracellular glucose enters the cell via GLUT, a glucose transporter protein, that facilitates transport through the lipid bilayer plasma membrane. Subsequently, glucose is metabolized by pathway enzymes including HK, hexokinase; G6PD, glucose-6-phospahate dehydrogenase; PGI, phosphoglucose isomerase; PFK, phosphofructokinase; ALDO, aldolase; TPI, triose phosphate isomerase; GAPDH, glyceraldehyde 3-phosphate dehydrogenase; PGK, phosphoglycerate kinase; PGAM, phosphoglycerate mutase; ENO, enolase; PK, pyruvate kinase. Mitochondrial metabolism regulatory enzyme PDK, Pyruvate dehydrogenase kinase, phosphorylates and inactivates pyruvate dehydrogenase, the first component of PDC, the pyruvate dehydrogenase complex converting Pyruvate to Acetyl-CoA. Oxidation of Acetyl-CoA by TCA, Tricarboxylic acid cycle chemical reactions in the matrix of the mitochondria releases stored energy. The TCA metabolite citrate can be exported outside mitochondria to be broken down into oxaloacetate and acetyl-CoA by the enzyme ACL, ATP citrate lyase. Cytosolic acetyl-CoA serves as a central intermediate in lipid metabolism. When oxygen is in short supply, LDH, lactate dehydrogenase converts pyruvate, the final product of glycolysis, to lactate. MCT, monocarboxylate transporter proteins allow lactate to traverse cell membranes. Solid purple arrows, metabolite transition pathways; thin solid black arrow, coenzyme transition; Dotted T-bar, inhibition.

Appreciating the essential relationship between oxygen and growth at the cellular level, Otto Warburg and co-workers compared normal liver tissue to corresponding cancer cells and in 1924 described a surprising outcome, namely, the Warburg effect [5,6]. Measuring O_2 consumption in thin tissue slices metabolizing glucose, they observed that although the respiration of liver carcinoma tissue slices was 20% less than that of normal tissue, about ten-fold more glucose was metabolized than expected. Moreover, the amount of lactic acid, the glycolysis product, was two orders of magnitude higher in cancer cells than in the normal tissue. When the Warburg effect was discovered, an increased glycolysis in cancer cells under aerobic conditions was misinterpreted as evidence for respiration damage. However, we now understand that it reflects an altered regulation of glycolysis in relation to mitochondrial function. The Warburg effect actually comprises a complex collection of contributory changes in gene expression and respiratory function (Figure 2) that include: (i) high glucose transporter expression; (ii) high hexokinase expression; (iii) high pyruvate kinase muscle (PKM2) expression; (iv) expression of phosphoglycerate mutase (PGAM) that allows pyruvate production without ATP generation; (v) high pyruvate dehydrogenase kinase (PDK) levels; and (vi) high expression of specific transcription factors, principally MYC, HIF-1α, NF-κB and OCT1 that sustain the Warburg effect [7].

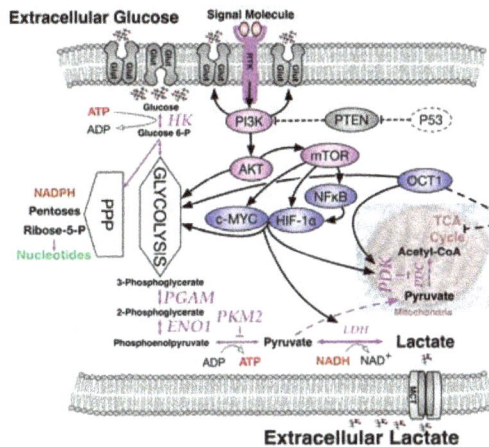

Figure 2. Molecular changes driving the Warburg effect. The shift to aerobic glycolysis in tumor cells reflects multiple oncogenic signaling pathways. Downstream from an active RTK, receptor tyrosine kinase, PI3K, Phosphatidylinositol 3-kinase activates AKT, protein kinase B/Akt strain transforming, stimulating glycolysis by directly regulating glycolytic enzymes. It also activates mechanistic target of rapamycin, mTOR, a protein kinase altering metabolism in various ways, including enhancement of three key transcription factors; the avian Myelocytomatosis viral oncogene proto-oncogene homologue (MYC); Nuclear factor kappa-light-chain-enhancer of activated B cells (NF-κB); and the hypoxia-inducible factor 1 alpha (HIF-1α), to promote a hypoxia-adaptive metabolism. NF-κB subunits bind and activate the *HIF-1α* gene. HIF-1α increases expression of glucose transporters, GLUT, glycolytic enzymes and pyruvate dehydrogenase kinase, PDK, that blocks pyruvate dehydrogenase complex, PDC driven entry of pyruvate into the tricarboxylic acid, TCA cycle. Transcription factor MYC cooperates with HIF-1α to activate several genes encoding glycolytic proteins, yet also stimulates mitochondrial biogenesis whilst inhibiting mitochondrial respiration, favoring substrates for macromolecular synthesis in dividing cells. Tumor suppressor p53 ordinarily opposes the glycolytic phenotype via Phosphatase and Tensin homolog, PTEN but loss of p53 function (dashed line) is frequent in tumor cells. Octamer binding protein 1 (OCT1) activates transcription of genes that drive glycolysis and suppress oxidative phosphorylation. Change to the pyruvate kinase M2, PKM2 isoform affects glycolysis by slowing the pyruvate kinase reaction, diverting substrates into an alternative biosynthetic and reduced nicotinamide adenine dinucleotide phosphate, (NADPH)-generating pentose phosphate pathway, PPP. Phosphoglycerate mutase, PGAM and α-enolase, ENO1 are commonly upregulated in cancer as is the pyruvate kinase M2 isozyme, PKM2 allowing a high rate of nucleic acid synthesis, especially in tumor cells. Lactate dehydrogenase, LDH enhances production of lactate, a signaling molecule that can stabilize HIF-1α and accumulate in the tumor microenvironment via Monocarboxylate Transporters, MCT, nourishing adjacent aerobic tumor cells that convert lactate to pyruvate for further metabolic processing. Solid black arrows, influenced targets; solid purple arrows, metabolite transition pathways; dotted purple arrows, reduced metabolite transition pathways; thin solid black arrow, coenzyme transition; dotted T-bar, inhibition.

Hexokinase catalyzes the first step in glycolysis by converting glucose to glucose-6-phosphate, making it available for metabolism via the pentose phosphate pathway, or glycolysis and the TCA cycle. Two predominant PKM isoforms are generated from the same gene by different splicing: the fetal form PKM2 uses exon 10, while the adult form PKM1 uses exon 9. The PKM2 protein, often aberrantly expressed in cancer cells, is subject to post-translational phosphorylation of a tyrosine residue that dramatically reduces its ability to convert phosphoenol pyruvate to pyruvate, thereby slowing the TCA cycle via precursor starvation [8]. Though slowed, the TCA cycle remains operational and pyruvate is still produced, but subsequent events conspire to its enhanced conversion to lactate. PGAM activity occurs only in the presence of PKM2 and governs an alternative pathway that converts phosphoenol pyruvate (PEP) to pyruvate without using pyruvate kinase and without producing ATP. In particular, proteomic analysis indicated that PEP, the cellular substrate for pyruvate kinase, contributed to PGAM His-11 phosphorylation to activate its catalytic site [9]. In addition, PDK phosphorylated the pyruvate dehydrogenase complex (PDC) to inactivate PDC, preventing the conversion of pyruvate to acetyl-CoA. Thus, enzyme kinetics for alternative pyruvate use, such as lactic acid production, is improved. Among several transcription factors maintaining the Warburg effect (mainly driven by MYC and HIF-1α with loss of regulatory p53 function) [10] those responding to low oxygen levels are highly significant, since low oxygen conditions (5% O_2 as opposed 18–20% ambient O_2) improve blastocyst stage embryo culture in a number of species [11].

Beyond describing how the Warburg effect favors anabolic metabolism, this review will focus on its establishment by the diverse mechanisms hinted at above; chosen adaptive responses that enforce oncogenic events, or stochastically acquired deterioration of mitochondrial function. Collectively, these perspectives provide important clues for understanding the significance of the Warburg effect within complex pathologies and extreme contexts such as space travel.

2. Paradoxes of Efficiency within Perpetual Pyruvate Pathways

A dividing cell would assumedly have high energy requirements, but a paradox of the Warburg effect was that its mechanism for providing free energy in the form of ATP was less efficient. However, this presumably provided a selective advantage to actively dividing cells, given an association between aerobic glycolysis and proliferation across species. Beyond homeostatic energetic requirements, growth and cell division require anabolic processes. The Warburg effect may be an evolutionary conserved mechanism for balancing ATP production with biomass production.

Focus on metabolic regulation in cancer provided a unifying theory for understanding interactions between prominent oncoproteins and tumor suppressors that deregulate glycolysis. The metabolic perspective extended tumorigenesis beyond a cumulative cascade of growth signal activation and tumor suppressor gene inactivation, introducing conceptually useful driving forces exploitable for cancer therapy.

Typically, the conventional paths for glucose metabolism and TCA cycle generate in adult tissues a large amount of ATP (up to 36 molecules of ATP per molecule of glucose). Pyruvate can also be reductively metabolized to organic acids or alcohols (e.g., lactic acid or ethanol) by anoxic glucose fermentation, a far less efficient pathway for generating ATP (only two molecules of ATP per molecule of glucose). Despite the much lower relative yield, ATP production rate remains high if the glucose supply is abundant. This alone does not explain the advantage to aerobic glycolysis, because most ATP is derived from other sources and is not limiting in proliferating cells. Consumption, rather than production of ATP is needed to overcome a high ATP/AMP ratio that would ordinarily inhibit key rate-limiting steps in glycolysis. Moreover, there is an associated greater need for NADPH to regulate glycolytic transcription factors and support fatty acid synthesis. Mechanisms that increase ATP consumption drive glycolytic conversion of glucose to lactate. This may again seem paradoxical if one imagines that generation of glycolytic intermediates for biosynthesis of macromolecules would be more helpful than excretion of carbon as lactate. Yet this may serve as a regulatory buffer for the process of biomass production, allowing cells to increase biosynthesis during cell proliferation only when precursor concentrations are appropriately abundant.

Pyruvate, the final product of glycolysis, has three main fates in mammalian cells: (1) conversion to lactic acid via lactate dehydrogenase (LDH); (2) conversion to alanine via alanine aminotransferase with concomitant conversion of glutamate to α-ketoglutarate; (3) conversion to acetyl-CoA in mitochondria via the pyruvate dehydrogenase (PDH) complex to enter the TCA cycle. The Warburg effect influences pyruvate fate by increasing its provision for anabolic processes. A high glycolytic flux in proliferating cells saturates the maximum PDH activity, leaving excess pyruvate for the action of LDH and alanine aminotransferase. Some cancer cells secrete conspicuously large amounts of alanine, that frequently parallel cellular lactate levels and may reflect a consequence of using glutamine to generate NADPH [12]. Conceptually, maximizing pathway flux maximizes ATP yield when TCA cycle enzyme production is limiting. Relatively high costs of enzyme synthesis ultimately decrease the ATP production rate. Accordingly, given choice between higher-flux, low-yield metabolic pathways (aerobic glycolysis) and lower-flux, high-yield pathways (invoking the TCA cycle), maximizing pathway flux surpasses reallocating proteins away from glycolytic processes towards the TCA cycle and respiratory chain enzymes [13]. Improved ATP economy, rather than *de novo* generation of biomaterial, may explain the maintenance of low-yield pathways such as the Warburg effect in tumor cells. Finding the Warburg effect also in non-replicating striated muscle tissues is consistent with this interpretation.

3. The Need for NADPH and Diversified Carbon Sources

Cells in metabolically stressed microenvironments may find themselves short of glucose as well as oxygen. Serum starved human fibroblasts can switch to anaerobic metabolic pathways that mimic the Warburg effect [14]. The AMP-activated protein kinase (AMPK), a known cell metabolism sensor, is activated by phosphorylation during situations of metabolic stress. AMPK regulates levels of NADPH, a key coenzyme that removes dangerous reactive oxygen species (ROS) in anabolic reactions, crucial to several metabolic processes for cell survival [15]. Ordinarily, the glucose-utilizing oxidative phase of the pentose phosphate pathway (PPP) generates most of the NADPH. When glucose is limiting, AMPK can inhibit activity of the NADPH-consuming metabolic enzyme acetyl-CoA carboxylase to indirectly maintain intracellular NADPH levels. NADPH is mostly a reducing agent in anabolic reactions, whilst reduced nicotinamide adenine dinucleotide (NADH), which differs by a phosphate group that allows distinction by a different set of enzymes, is principally involved in catabolic reactions. When oxygen is absent or in short supply, LDH catalyzes the conversion of pyruvate to lactate with concomitant interconversion of NADH to its oxidized form, (NAD^+). Regeneration of NAD^+ is necessary for continued flux through glycolysis and mediates conversion of glyceraldehyde-3-phosphate to 1,3-bisphosphoglycerate by glyceraldehyde 3-phosphate dehydrogenase (GAPDH). This is the sixth step of glycolysis that uses NAD^+ to produce NADH for maintaining the cellular redox state required for regulating gene expression. Both reduced forms of NAD^+ (NADH and NADPH), activate transcription factors such as the transcriptional regulator C-binding protein involved in cell growth, differentiation and transformation [16]. In a counterpoised manner, the oxidized forms (NAD^+ and $NADP^+$) can inhibit transcription factor binding to DNA [17]. Indicative of anabolic requirements for lipid biosynthesis, as many as 14 molecules of NADPH are required for generating each molecule of palmitoyl-CoA and 26 molecules of NADPH are required for cholesterol. Cytosolic isocitrate dehydrogenase (IDH1) generates NADPH by converting isocitrate to α-ketoglutarate. IDH1 may be a vulnerable requirement for particular cancer cells [18] although other cancers bear dominant inhibitor mutations of IDH1 and presumably have other means for NADPH production. In sum, lactate production and its by-product NAD^+ can enhance both glycolytic flux and incorporation of glucose metabolites into biomass, to allow faster cell growth. In situations such as embryogenesis, immune response and wound healing, when nutrients are abundant, rapid cell division provides a more significant selective advantage than efficient carbon utilization.

One additional route supplying high demand for NADPH in proliferating cells, involves NADPH synthesis from glutamine metabolism (Figure 3) via its oxidation to malate with subsequent activity of malate dehydrogenase generating pyruvate and NADPH [12]. Notably, glutamine uptake and metabolism proved to be regulated by the *MYC* oncogene [19]. *MYC*-transformed cells may become dependent on glutamine and exhibit elevated expression of glutamine transporters and glutamine catabolic enzymes. Alternatively, in less glutamine dependent MYC-deficient tumors, IDH1 activity may be the main source of NADPH, although further undefined sources remain likely. Glutamine is the most abundant amino acid in human plasma and a major contributor to the replenishment of TCA cycle intermediates (anaplerosis). Such replenishment sustains TCA cycle intermediate efflux (cataplerosis) maintaining precursors for many non-essential amino acids for biomass synthesis. Fatty acids, such as acetyl-CoA, along with nitrogen molecules in nucleotides, amino acids and amino sugars, can also be derived from glutamine, as well as glucose [20], through a process coined glutaminolysis that contributes to glyceroneogenesis (Figure 1). Increased levels of acetyl-CoA induced by MYC may not only enter the TCA cycle, but also serve as a donor for histone acetylation associated with gene activation [21]. MYC can activate the expression of glutaminase 1 (GLS1) that deamidates glutamine to produce glutamate [19]. Conversion of glutamine to glutamate fuels folate metabolism since addition of glutamates to folates increases the retention of folates within cells, promoting GLS1 activity. Under low oxygen conditions, reductive glutamine metabolism can contribute significantly to lipid biosynthesis. As a reciprocal carbon source alternative to glucose in diploid fibroblasts, glutamine could provid 30% of the ATP energy requirement when cells were cultured in standard 5.5 mM glucose-containing medium with increased glutamine utilization in lower concentrations of glucose [22]. In rapidly proliferating HeLa cells cultured in 10 mM glucose-containing medium, glutamine still provided over 50% of the ATP requirement, emphasizing its key role in supporting a proliferative metabolism.

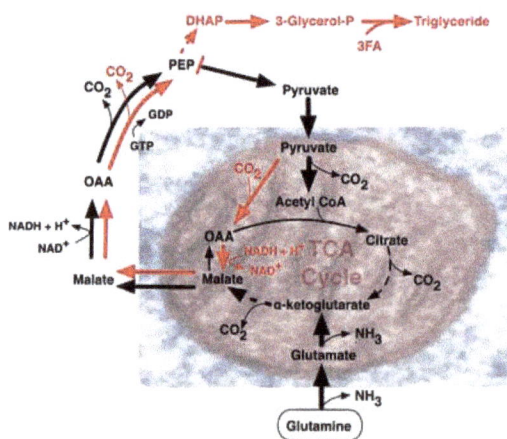

Figure 3. Glutamine metabolism: The influence of anaplerosis (entry of glutamine into the TCA cycle), cataplerosis (removal of glutamine as malate) and glyceroneogenesis (marked red). Mechanisms by which glutamine is metabolized for energy involve entry of glutamine into the TCA cycle (anaplerosis), balanced by its removal (cataplerosis) as malate. Malate can be subsequently converted to oxaloacetate (OAA) and then to phosphoenolpyruvate (PEP) via the cytosolic enzyme phosphenolpyruvate carboxylase (PEPCK). PEP can be converted to pyruvate by pyruvate kinase for entry into the TCA cycle as acetyl-CoA or transamination to alanine. The pathway of glyceroneogenesis in which carbon from sources other than glucose or glycerol contributes to the formation of L-glycerol-3-phosphate (3-Glycerol-P) for conversion to triglycerides, involves a balance of anaplerosis (entry of OAA synthesized from pyruvate via pyruvate carboxylase) and cataplerosis (removal of intermediates to support the synthesis of glyceride-glycerol). Reduction of glycolysis-derived dihydroxyacetone phosphate (DHAP) to 3-Glycerol-P provides cells with the activated glycerol backbone needed to synthesize new triglycerides with fatty acids (FA). Solid arrows, anaplerosis pathways; red arrows, glyceroneogenesis pathway; dotted arrow, mitochondrial cataplerosis pathway.

4. The Warburg Effect in Normal and Cancer Cells; Deriving the Choice, Chain or Chance Perspective

Though conspicuous in cancer, a fundamental question of the original Warburg effect was whether it reflected an aberrant condition or an adaptive feature to be also found in normal cells. Viewed as a "choice", accumulating evidence indicated that the Warburg effect was probably a metabolic pathway advantageously chosen by normal cells during early development. Alternatively, many tumorigenic events are strongly linked to regulation of metabolism, and certain critical mutations may "chain" cancer cells to the Warburg effect pathway. "Chance" events accompanying ageing can include mutations that impair mitochondrial function [23] often encouraging a metabolic drift towards aerobic glycolysis.

Accordingly, influential events can be conceptually considered to reflect A, Environmentally driven metabolic "choice", or B, driving mutational events that "chain" cells to particular metabolic pathways, or C, "chance" deterioration events of ageing (Figure 4). Comparing the Warburg effect in different circumstances, requires appreciation of its influence in contexts that diversely impinge upon an apparently similar phenotype. Metabolic pathways need to be immediately and dynamically responsive to an organism's situation, with the combination of the rate of flux and yield efficiency determining the ultimate outcome.

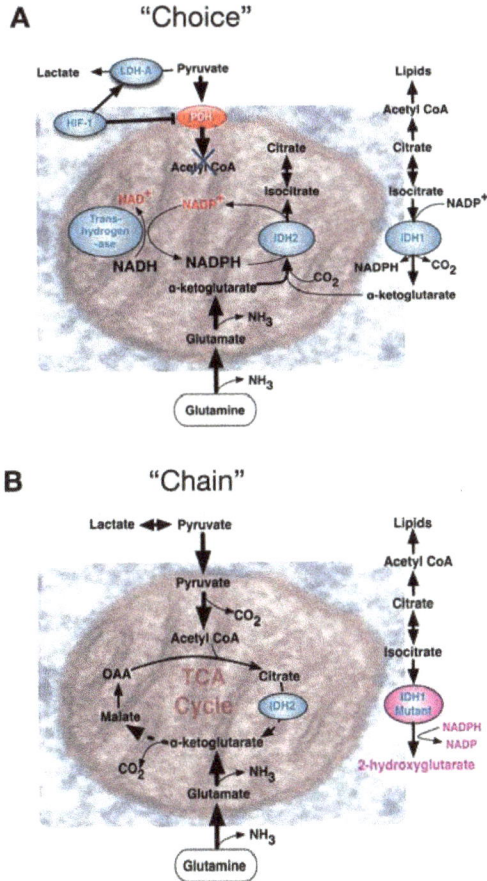

Figure 4. *Cont.*

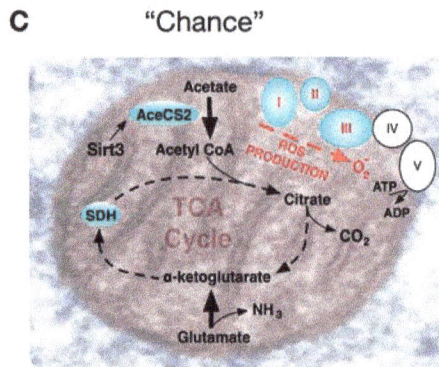

Figure 4. Mechanisms influencing the mitochondrial metabolism and the Warburg effect. (**A**) Environmentally driven metabolic "choice". In cells proliferating under hypoxic pressure with activated HIF-1, the electron transport chain is inhibited because of the lack of oxygen as electron acceptor. HIF-1-induced inactivation of PDH-1 helps ensure that glucose is diverted away from mitochondrial acetyl-CoA-mediated citrate production. The alternative pathway for maintaining citrate synthesis involves reductive carboxylation, thought to rely on a reverse flux of glutamine-derived α-ketogluturate via isocytrate dehydrogenase-2 (IDH2). The reverse flux in mitochondria can be maintained by NADH conversion to NADPH by the mitochondrial transhydrogenase, with the resulting NADPH driving α-ketoglutarate carboxylation. Citrate/isocitrate exported to the cytosol may be metabolized oxidatively by isocytrate dehydrogenase-1 (IDH1), and contributes to the production of cytosolic NADPH. (**B**) Mutational events that "chain" cells to particular metabolic pathways. Oncogenic IDH1 and IDH2 mutations can cause gain of function in cancer cells. Somatic mutation at a crucial arginine residue in cytoplasmic IDH1 or mitochondrial IDH2 are frequent early mutations in glioma and acute myeloid leukaemia which cause an unusual gain of novel enzymatic activity. Instead of isocitrate being converted to α-ketoglutarate, with production of NADPH, α-ketoglutarate is metabolised to 2-hydroxyglutarate with the consumption of NADPH. 2-hydroxyglutarate can compete for α-ketoglutarate-dependent enzymes including histone demethylases and DNA hydroxylases, thereby altering both metabolism and epigenetic phenotypes. (**C**) "Chance" deterioration events in ageing. SIRT3 regulates multiple pathways involved in energy and ROS production. Some mitochondrial metabolic processes activated by SIRT3 include acetate metabolism-mediated direct deacetylation of acetyl-CoA synthetase (AceCS2), the TCA cycle via direct deacetylation of SDH, and ROS production induced by direct acetylation of the respiratory chain complexes I, II and III. Thin solid black arrow, alternative reductive carboxylation pathway; Solid black arrow, TCA cycle pathway; dotted arrow, alternative TCA cycle pathways; red dotted arrow ROS production pathway.

4.1. The Warburg Effect by Choice; an Adaptive Response to Oxygen and Nutrient Restrictions

Supporting the idea that early fetal and placental development is primarily an anaerobic process, the anatomical features of the first trimester gestational sac limit rather than facilitate oxygen transfer to the fetus. Additionally, increased levels of antioxidant molecules are found in the exo-coelomic cavity when oxygen free radicals have the greatest potential for harmful teratogenic effects [24]. An in vivo Xenopus model of embryonic retinal development unequivocally showed that dividing progenitor cells relied more on glycogen to fuel aerobic glycolysis than non-dividing differentiated cells, demonstrating that the Warburg effect was a feature of physiological cell proliferation [25]. Early ex vivo studies in human embryonic stem cells paradoxically showed either increased [26] or decreased oxidative phosphorylation [27] in response to experimentally defined nutrient or oxygen conditions. However, the prevailing view of comprehensive metabolomic studies was that the stem cell redox status changed during differentiation, with the Warburg effect influencing pre-implantation embryo development [28]. Dramatic metabolic differences accompany distinct pluripotent states in early embryogenesis. Embryonic stem cells (ESC) of the inner cell mass of pre-implantation embryos show bivalent energy production, maintaining a relatively high ATP content. In contrast, post-implantation epiblast stem cells (EpiSC) have lower mitochondrial respiratory capacity, despite a more developed mitochondrial content, due to low expression of cytochrome c oxidase. Early mammalian embryo metabolism presents unique features, significantly different from proliferating somatic cells. Pre-implanted cleavage-stage embryos need to mainly replicate DNA and plasma membranes rather than entire cells and during cleavage from a single cell to a blastocyst, cells get progressively smaller. At the blastocyst stage, the somatic cell size is reached, and cells need to replicate fully prior to the next rapid cell division. This requires a unique metabolism that is more closely equivalent to the Warburg effect in cancer. Just as hypoxia has been shown to influence tumorigenesis, HIF-1α sufficed to drive the metabolic switch from an ESC to an EpiSC-like stage [29].

Cancer cells evolve by acquiring a selective advantage for survival and proliferation in a challenging tumor microenvironment. As a tumor outgrows the diffusion limit of local blood supplies, its cells encounter hypoxia. Under such circumstances, a Warburg effect in cancer cells may be considered a chosen reversible adaptation in response to local hypoxia. The stabilized hypoxia inducible transcription factor complex, a heterodimer of oxygen-dependent HIF-1α and constitutively expressed HIF-1β, activates a transcriptional program accommodating cells to hypoxic stress through diverse compensatory mechanisms [30]. Befitting choice for a Warburg effect, both decreased dependence on aerobic respiration and metabolism shifts towards glycolysis become advantageous. These changes are mediated by increased expression of glucose transporters, glycolytic enzymes and inhibitors of mitochondrial metabolism (Figure 4A). At the same time, HIF-induced molecules such as vascular endothelial growth factor (VEGF) stimulate angiogenesis to restore blood flow and oxygen supply. Nonetheless, early blood vessel growth within a growing tumor's poorly organized tissue architecture is often sub-optimal and may not deliver blood effectively, alleviating the oxygen requirements only partially. Adaptive responses of tumor cells to hypoxic conditions in the tumor niche have a persistent influence on tumor growth. HIF-1α has two other isoforms HIF-2α and HIF-3α, as does the HIF-1β subunit, aka the aryl hydrocarbon receptor nuclear translocator (ARNT1, Arnt2 and Arnt3). Under oxygen concentrations below 6% O_2, the HIF α-subunit of short 5-minute half-life, was stabilized upon activation and transport to the nucleus where it dimerized with the β subunit to bind hypoxia responsive elements in target genes [31]. HIF-1α and HIF-2α, with distinct structure within the N-terminal transactivation domain, play key roles in hypoxia-induced cellular responses. Whereas HIF-1α predominantly induces glycolytic pathways, HIF-2α regulates genes important for cell cycle progression and maintenance of stem cell pluripotency, including MYC and the stem cell factor OCT-3/4. Although HIF-1α and HIF-2α both exert similar induction of the angiogenic *VEGF* gene, they may exert opposite effects on other angiogenic factors, e.g., HIF-1α decreases IL-8 expression whereas HIF-2α increases IL-8 mRNA and protein levels [32].

Fluctuating oxygen levels would favor selection of tumors that constitutively upregulate glycolysis, yet the observed heterogeneity of HIF expression within tumor cells would suggest that this signaling pathway is still coupled to oxygen levels in most tumors. MYC and HIF family members both recognize related DNA-binding sequences in the promoters of glycolytic target genes, suggesting a coordinated regulation of expression. But when controlling mitochondrial function, the interactions between MYC and HIF proteins are often antagonistic. HIF-1α can activate expression of pyruvate dehydrogenase kinase 1 (PDK1), which subsequently phosphorylates and inhibits pyruvate dehydrogenase, preventing glucose-derived acetyl-CoA from entering mitochondria and the TCA cycle. Such pathways may restrict the positive role of MYC and thereby restrain cell growth and oxygen consumption in hypoxic conditions. Thus, a complex set of biochemical interactions between MYC and HIF dictates outcomes depending on the promoter context of target genes, the transformed state of the cells and oxygen tension [33].

Emphasizing HIF importance for regulating cell outcomes such as enhanced proliferation, patients with Von Hippel Lindau disease (pVHL) who lack the protein mediating HIF1α degradation, are predisposed to develop a range of highly vascularized tumors [34]. In addition, patients with Chuvash polycythemia, a hereditary disorder of increased sensitization to hypoxia, show hepatic hyper-proliferation and a correspondingly increased organ volume [35]. However, up-regulation of the HIF signaling pathway needn't entirely account for the Warburg effect per se, since many additional events of transformation can directly enforce a Warburg effect metabolism. Nutrient deprived cancer cells can resort to glycolysis for ATP production, evoking pathways involving ROS production and AMPK phosphorylation leading to PDK activation and enhancement of the Warburg effect [36].

4.2. The Warburg Effect Chained to Oncogenic Events

A strong association between cancers and the Warburg effect allows the metabolic label ^{18}F-deoxyglucose to localize tumors via positron emission tomography (FDG-PET). More than simply an indirect metabolic response, the Warburg effect may represent an integral metabolic necessity for cell proliferation towards which multiple oncogenic events conspire [37]. Many well-established irreversible molecular oncogenic changes can be linked to the establishment of a Warburg effect. Therefore, tumor cells may become effectively "chained" to aerobic glycolysis for enhancing proliferation-favorable NADPH production and acetyl-CoA flux to the cytosol for lipid biosynthesis. For example, over-expression of the receptor tyrosine kinase (RTK), human epidermal growth factor 2 (HER-2) by gene amplification, is linked to an aggressive cancer cell subtype by inducing up-regulation of a key lipogenic enzyme, the fatty acid synthase (FASN). This allows rapid response to changes in the flux of lipogenic substrates (e.g., NADPH and acetyl-CoA) and lipogenesis products (e.g., palmitate). In addition, HER-2 can also up-regulate PPARγ. This promotes adipogenesis and lipid storage to avoid endogenous palmitate toxicity by securing palmitate in fat stores rather than allow its negative feedback on FASN function [38]. The K-ras oncogene can modulate the cyclic adenosine monophosphate-dependent protein kinase (cAMP/PKA) signaling pathway, ultimately influencing mitochondria through decreased mitochondrial complex I activity and reduced ATP formation. This makes, *K-ras* mutant cells more sensitive than normal cells to glucose withdrawal [39]. Estrogen receptors directly bind to the promoters of many genes encoding glycolytic enzymes and these can interact synergistically with the proto-oncogene transcription factor MYC to drive the Warburg effect in cancer cells [40].

Few tumorigenic pathways influencing glucose metabolism are more important than the phosphoinositide-3-kinase (PI3K) signaling pathway. Downstream activation of protein kinase AKT induces glucose transporter *GLUT1* gene expression and prevents internalization of GLUT1 protein to maintain cell surface levels for higher glucose uptake. AKT activation also promotes flux through glycolysis, and numerous cancer cell mutations that constitutively activate PI3K, bypassing need for growth signals, ultimately promote aerobic glycolysis [12]. The first reported direct link between oncogene activation and altered glucose metabolism involved MYC induced LDH-A expression [41]. In addition to altering glutamine metabolism, *MYC* oncogene driven cell cycle progression, could induce genomic instability and also promote transcription glycolytic enzymes as well as glucose transporters GLUT-1, GLUT-2 and GLUT-4 [42]. MYC-stimulated LDH-A production could become crucial for proliferation of MYC-dependent tumors [43]. MYC could also influence membrane biosynthesis via enhanced synthesis of fatty acids from glutaminolysis.

An intricate associate between transcription factors and glycolysis, modulated by the carbon source, is coordinated by glucose sensing mediated by MondoA, a basic helix-loop-helix leucine zipper (bHLHZip) transcriptional activator functionally similar to MYC [44]. Resembling MondoA, a further family member Carbohydrate Response Element Binding Protein (ChREBP/MondoB/WBSCR14) can dimerize with yet another bHLHZip protein called Max-like protein x (Mlx). Regulation of their activity is largely controlled by intracellular location with accumulation in the nucleus subject to different glucose-derived metabolites. MondoA:Mlx heterodimers accumulate in the nucleus in response to glucose 6-phosphate (G6P), whereas ChREBP:Mlx accumulates in response to the pentose phosphate intermediate xylulose 5-phosphate [45]. Thus, an adaptive transcriptional response to glucose and

ATP levels is coupled to activation of metabolic genes, including glycolytic enzymes. MondoA and ChREBP also regulate expression of thioredoxin interacting protein (TXNIP) that has a broad range of activities, including glucose homeostasis [46] and inhibition of thioredoxin redox activity, elevating ROS levels that influence key target genes that help control cell growth. Interdependency is such that, MondoA was responsible for most of the glucose-induced transcription in an epithelial cancer cell line [47].

Providing an intriguing link with the highly significant tumor suppressor gene *p53*, MondoA knockdown stimulated cell growth in *p53*-deficient cells, whereas ChREBP knockdown reduced cell growth in *p53* wild-type cells, suggesting that the functional output of MondoA and ChREBP transcriptional responses was subject to p53 status [48]. Indeed, the pleiotropic p53 has been shown to influence and respond to metabolic changes through several cancer cell-antagonizing mechanisms, including promotion of apoptosis, senescence and DNA repair. Moreover, p53 could promote oxidative phosphorylation and dampen glycolysis hence interfering with cell growth and autophagy. As such, *p53* mutations unite oncogenic transformation and altered metabolism [49].

Despite all the above, tumors often retain a degree of metabolic flexibility. In a mass spectrometry-based global metabolomic study, Abu Dawud et al., directly compared human embryonic stem cells (hESCs) and embryonal carcinoma cells (hECCs). As might be expected for a metabolism of choice, undifferentiated hESC utilized aerobic glycolysis rather than OXPHOS, with both normal and cancer cell types expressing equivalent amounts of glycolysis intermediates [50]. Yet the study also highlighted metabolic differences between the two cellular types. Proliferation in hECCs involved higher levels of metabolites. hESCs could be induced to differentiate by depletion of OCT4, that induced gene expression for key mitochondrial respiratory chain proteins, including NDUFC1, UQCRB and COX. Many OXPHOS components were already expressed at the mRNA level in undifferentiated hESC, as though these cells were "poised" to enter the TCA cycle. Moreover, TCA cycle intermediates were enriched in hECC relative to hESC, implying that cancerous hECC may not rely solely on the Warburg effect yet have some dependency on both pathways. Curiously, gene-knockout mitochondrial DNA depleted B16 mouse melanoma cells, restricted to an exclusively glycolytic metabolism, could form primary subcutaneous tumors, yet no longer formed lung tumors when injected intravenously in NOD/SCID recipient mice. This suggested that ROS byproducts of mitochondrial aerobic metabolism may be required for tumor metastasis [51]. This needn't contradict the important contribution of a glycolytic pathway for metastasis [52] but highlights that beyond toxic mutagenic effects, ROS signaling molecules influence adaptive responses to harsh microenvironments, including regulating of cell adhesion via integrins [53]. Certainly, ideas that cancer cells are strictly driven to one particular mode of metabolism is likely to be an over-simplification, and a more complex alternating metabolic scenario seems likely [54]. Surrounding cells that comprise the tumor microenvironment should be also be considered important contributors [55,56].

Though most cancers retain a normal OXPHOS capacity [57], the Warburg Effect can be imposed by links to TCA cycle protein mutations. Fumarate hydratase (FH) catalyzes the hydration of fumarate into malate and FH germline mutations are implicated in hereditary leiomyomatosis and renal cell cancer (HLRCC). FH loss leads to fumarate accumulation that in turn activates hypoxia-inducible factors at normal oxygen tensions. The metabolic consequences of a crippled TCA cycle include a potential NADH shortage, yet *Fh1* deficient cells retain significant mitochondrial bioenergetic activity through compensatory mechanisms involving the biosynthesis and degradation of haem [58]. Other mitochondrial proteins can be mutated in a very specific manner. Frequently, in low-grade gliomas, single amino acid substitution mutations for an arginine residue in the active site of IDH1 and IDH2 evoked production of a new metabolite, 2-hydroxyglutarate (2-HG). Exactly how this exerts oncogenic effects remains to be resolved [59], but it supports the concept of "onco-metabolite" (Figure 4B).

4.3. The Warburg Effect Evolved by Chance during Ageing

Most experimental mitochondrial OXPHOS measurements have employed isolated mitochondrial complexes that needn't entirely correspond with the functioning of mitochondria in vivo. For example, the respiratory chain complex IV enzyme cytochrome oxidase (COX4), with twice the free energy yield of earlier complex I and III respiratory chain steps, represented a rate-limiting step of respiration in intact cells [60], but not in isolated mitochondria [61]. Again, unlike complexes I and III, COX4 converts O_2 into water without forming ROS. Abnormally high levels of ROS can directly induce genomic instability and increase HIF-1α levels, promoting metabolic programming towards the Warburg effect. Notably, Sirtuin gene *Sirt3*, a mammalian NAD-dependent protein deacetylase, homolog of *Saccharomyces cerevisiae Sir2* that regulates yeast life span, deactivates mitochondrial target proteins in critical pathways associated with age-related diseases. Sirtuin deacetylase activity on lysine residues of numerous mitochondrial substrates induces fat oxidation, amino-acid metabolism and electron transport. Transgenic mice lacking *Sirt3* have increased ROS levels, caused by both decreased electron transport and decreased detoxification activity of MnSOD. *Sirt3* reveals a connection between mitochondrial metabolism promoting the Warburg effect, tumorigenesis and induction of oxidative stress associated with randomly accumulated lesions causing degenerative diseases and ageing [62] (Figure 4C).

Notably, several mitochondrial ribosomal and transfer RNAs plus 13 proteins representing subunits of complex I, III, IV and V proton pumps are encoded by mitochondria-specific DNA (mtDNA). mtDNA is transcribed and translated independently from nuclear DNA, with a relatively fragile exclusively maternal inheritance, suffering a ten-fold higher mutation rate than nuclear DNA. The enhanced genetic fragility may explain maternally inherited mitochondrial diseases and ageing may be partly due to somatic mutations of mtDNA leading to impaired COX4 activity, impaired synthesis of ATP and reduced cell energetics with age. Maintaining adequate flux and COX4 activity reduces ROS mediated double strand breaks in mtDNA, with subsequent mtDNA repair vulnerable to introduction of mitochondrial DNA deletions [63]. Accordingly, mitochondrial DNA deletions have been associated with biochemical defects in neurodegenerative diseases like Alzheimer's disease [64]. However, direct evidence for this mechanism as an explanation for ageing and broader spectrum metabolic diseases has been elusive. Recent investigations in cultured human senescent myoblasts [65] and transgenic mouse studies using mtDNA exchange technology [66] lacked evidence that patients with mitochondrial diseases due to pathogenic mtDNA mutations develop premature ageing. Nonetheless, the trans-mitochondrial mice did develop diabetes and lymphomas, so significant mitochondrial respiration defects may lead to over-production of ROS with pathological consequences including ROS impairing glucose incorporation into insulin-targeted organs [67] or insulin secretion from pancreatic β cells [68].

5. The Warburg Effect and Diabetogenesis

A simulated Mars mission over 105 days, as an example of extreme conditions pressuring the human organism, indicated that environmental stress would exert considerable metabolic stress [69], and space flight perturbations in insulin sensitivity led to subclinical type 2 diabetes-like symptoms [70]. However, to what extent chronic pathological effects would be exacerbated during prolonged space flight missions is not known. Nonetheless, reference to what is understood about molecular events and pathways underlying type 2 diabetes is likely to be helpful for assessing the most pertinent measurements for monitoring health in space and preventative measures.

A challenging concern for understanding complex metabolic diseases such as type 2 diabetes mellitus (T2DM) is discrimination between cause and effect, given the close reciprocal relationship between metabolic pathways and disease phenotypes. Nonetheless, foundational metabolic aspects from lessons learned about the Warburg effect in development, cancer and ageing can be applied with a choice, chain or chance perspective to explore relevance in T2DM. Here, we apply this perspective to highlight pivotal events that may serve as useful diagnostic markers or therapeutic targets.

The enigmatic T2DM phenotype reflects multifactorial mechanisms responsible for the phenotypes of impaired insulin secretion from pancreatic β cells and insulin resistance in the major target tissues, such as skeletal muscle. An alarmingly high demographic incidence of T2DM, associated with sedentary lifestyles [71] and inappropriate carbohydrate diets [72] urges need to identify and characterize the repertoire of key regulatory molecules.

5.1. Lessons for T2DM from a Warburg Effect through Normal Cell Metabolic Choice

When the Warburg effect resulted from a contextual choice, gene expression changes following hyperglycemic treatment of early stage bovine blastocysts showed metabolic changes common to diabetes and cancer. This may help explain why diabetic hyperglycemia is often associated with cancer predisposition [73]. In diabetes, hyperglycemia increases intracellular glucose and its conversion to sorbitol by aldose reductase decreases NADPH levels. Hyperglycemic induction of this polyol pathway proves harmful because the lack of NADPH limits glutathione reduction that ordinarily protects cells from oxidative stress [74]. Blastocysts treated with excessive glucose (5 mM) showed increased expression of glutathione peroxidase 8 (GPX8), an antioxidant enzyme that reduces H_2O_2 into water by oxidation of glutathione, implying NADPH depletion may activate a compensatory oxidative stress response. This is similar to the elevation of GPX expression correlated with increased aldose reductase activity found in the pathology of type 1 diabetes [75]. An additional diabetic hallmark of hyperglycemia is activation of the hexosamine pathway. This diverts fructose-6-P from the glycolytic pathway with associated glucosamine-6-P accumulation inducing protein glycosylation of transcription factor Sp1. This induces *SERPINE 1* and *THBS1 gene* expression, both genes being implicated in diabetic pathogenesis. Similar upregulation of *SERPINE1* and *THBS1* in hyperglycemically treated blastocysts likely reflected increased activity of the hexosamine pathway. Underlying diabetic complications is a hyperglycemia-induced uncoupling of OXPHOS, which leads to mitochondrial ROS production inhibiting GAPDH activity, with glycolytic metabolites diverted to stimulate the polyol, AGE, PKC and hexosamine pathways [76]. In agreement, porcine embryos grown in elevated glucose medium showed an early rise in ROS generation and decreased GAPDH activity. Moreover, they also shared the diabetic trait of decreased citrate synthetase activity, responsible for entry of pyruvate-derived acetyl-CoA towards TCA. Decreased TCA activity in high-glucose treated blastocysts can also reflect up-regulation of nuclear receptor PPARγ that regulates lipid metabolism, with subsequent accumulation of oxidized lipids. These are likely to stimulate higher expression of OLR1 (oxidized low-density lipoprotein receptor-1) a scavenger of oxidized lipids, often up-regulated in diabetes, with an important role in pregnancy disorders [77] and tumorigenesis [78].

Matching the growing energy demands of development, blastocysts increase mitochondrial OXPHOS and activate glycolysis and glucose uptake to enhance ATP synthesis. Hyperglycemic conditions increased blastocyst expression of *PDGFC* and *HIF-1α* to promote expression of lactate dehydrogenase A (LDHA) that enhances anaerobic conversion of pyruvate into lactate, with production of NAD^+. Under these conditions, the blastocysts also increased expression of transketolase 1, an enzyme catalyzing the non-oxidative part of the pentose phosphate pathway in order to enhance glucose consumption and lactate production [79]. Thus, under diabetic conditions, the increased anaerobic glycolysis in embryos could limit ROS generation as a beneficial compensation for the impaired energy metabolism. Despite such similarities with the Warburg effect phenotype of cancer cells, in most cases, early development ultimately avoids a tumorigenic fate. In non-tumorigenic cells, the activity of tumor suppressors such as p53 readily activates expression of proteins mediating an apoptotic mitochondrial death pathway in response to a depressed mitochondrial transmembrane potential. Excessive early embryonic lactate production is associated with aborted gestation [80] emphasizing the detrimental impact of early hyperglycemic stress on pre-attachment embryo survival and blastocyst development.

5.2. Lessons for T2DM from an Oncogenically Chained Warburg Effect

A caveat when extending the biochemical events governing the Warburg effect in cancer to other contexts such as diabetes is that cancer cells likely have additional events, such as tumor suppressor mutations, resulting in very different context-dependent metabolic feedback. Thus, key molecular events governing the Warburg effect in tumor cells might not have the same influence in skeletal muscle cells or pancreatic β-cells and vice-versa. Nevertheless, tumor cells can provide an abnormally permissive environment making key control events more significantly apparent. An example is provided by human LDH that catalyzes the reduction of pyruvate into lactate, a watershed event for the switch from aerobic to anaerobic glycolysis. The tetrameric LDH is made up of two subunits, LDHA (aka LDH-M, muscle) and LDHB (aka LDH-H, heart), with isoforms made up of various combinations of these two subunits, e.g., $LDH3=LDHA_2B_2$ whilst $LDH5=LDHA_4$, the form most prevalent in the liver and skeletal muscle. In cancer patients, high plasma hLDH5 levels can serve as a malignant cell biomarker, since it is associated with induction by *HIF-1α* and *MYC*, rather than nonspecific cellular damage. Oncogenic tyrosine kinase fibroblast growth factor receptor (FGFR1) directly phosphorylates LDHA subunits, enhancing the formation of active tetrameric LDH5 that binds to NADH substrate. Thus, oncogenic tyrosine kinases activation, a common event in many cancers, promotes the Warburg effect via LDHA, for enhanced tumor growth [81]. Inhibitors preferentially targeting the LDHA isoform in cellular assays, reduced lactate production in HeLa cells and were particularly effective in blocking cell proliferation in hypoxic conditions [82]. In contrast, pancreatic islet β-cells normally express low levels of LDHA and have high glycerol phosphate dehydrogenase activity. Acute experimental over-expression of LDHA in the murine model MIN6 β-cells perturbed mitochondrial metabolism, interfered with normal glucose-derived pyruvate metabolism in mitochondria and inhibited glucose-stimulated insulin secretion [83]. Thus, an LDHA abnormality highlighted by oncogenically "chained" events remained influential in more normal contexts, being more open to a "choice" of feedback.

Another example from cancer metabolism, is related to pyruvate dehydrogenase kinase 1 (PDHK1) which is also inducible by MYC and HIF-1α. It is again activated by tyrosine phosphorylation that induced the binding of ATP and PDC, promoting the Warburg effect with tumor growth [84]. Complementing tyrosine phosphorylation in glucose homeostasis, protein tyrosine phosphatases (PTPs) are key negative feedback regulators of insulin receptor (IR) signaling that effectively inhibit the action of insulin. Given the predominance of tyrosine kinases in tumorigenesis, it might be anticipated that PTP1B inhibition favors cancer growth, since reduced PTP1B expression was found in hepatocellular carcinomas [85]. Nonetheless, in different cancers its effects remain context dependent, since PTP1B expression driven by androgen receptor signaling, enhanced the progression of prostate cancer [86]. PTP1B expression also plays an important role in T2DM and has been found to be elevated in insulin target tissues of patients. PTP1B can directly interact with the insulin receptor [87] and insulin receptor substrate 1 (IRS-1) [88], dephosphorylating the tyrosine residues in the activation loops of these molecules to down-regulate insulin signaling, thereby contributing to an insulin resistance phenotype in T2DM patients. Transgenic *PTP1B* null mice were healthy, resisted obesity when fed with a high-fat diet and did not evolve diabetes. Restorative effects were observed by treating diabetic mice with PTP1B antisense oligonucleotides [89]. PTP1B is considered an attractive therapeutic target and among the small molecules investigated, antibodies that stabilize the oxidized conformation of PTP1B have been shown to inactivate it [90]. Favoring prospects for preventative astronaut treatment, normal "choice" metabolic responses may present therapeutic targets of more consistent behavior than heterogeneous oncogenically "chained" pathways.

One adaptation to cellular stress observed in tumors is autophagy or "self-eating" of proteins and cell organelles. It is initiated when the autophagosome, a double-membrane vesicle, engulfs cytoplasmic organelles, including mitochondria, before fusion with lysosomes that digest the vesicle's contents. Excessive accumulation of autophagosomes can lead to cell death by apoptosis. This complex process has both tumor-suppressing and tumor-promoting activities depending on whether it is predominantly occurring in epithelial cancer cells or in stromal cells within the tumor niche [91]. Cells undergoing autophagy have fewer mitochondria and shift their metabolism towards glycolysis, producing high-energy mitochondrial fuels e.g., L-lactate, ketone bodies, glutamine and free fatty acids. Neighboring cells may consume these catabolites to fuel mitochondrial OXPHOS. Notably, p53 serves as a key component of the cellular stress response. Among its many activities, nuclear p53 can *trans*-activate numerous autophagy inducers, such as the damage-regulated autophagy modulator (DRAM), a target for silencing by hyper-methylation at its CpG islands in several tumor types [92]. Autophagy is also central to muscle glucose homeostasis in response to exercise. As shown by mice with engineered mutations affecting phosphorylation sites in the mitochondrial anti-apoptotic and anti-autophagy protein BCL2. Exercise-induced autophagy was shown to be dependent on BCL2 activity, and mutant mice defective in autophagy showed impairment of many key metabolic events, including impaired muscle glucose uptake, GLUT4 plasma membrane localization and AMPK activation [93].

Growing awareness of the oncogenic role of epigenetic alterations in genomic DNA encoding metabolically influential proteins led to recognition that fructose-1,6-biphosphatase-1 (FBP1), an antagonist of glycolysis in many cell types, was epigenetically down-regulated by methylation of its promoter sequence, mediated by NF-κB downstream of oncogenic Ras signaling. The incidence of this event sufficed for *FBP1* promoter methylation to be considered a new biomarker for predisposition to gastric cancer [94], implying that an epigenetic mechanism could also account for the Warburg effect [95]. T2DM is associated with low-grade systemic inflammation, and increased NF-κB DNA binding activity was found to be one of the characteristics of muscle tissue from T2DM patients [96], in agreement with suggestions that the NF-κB pathway may represent a therapeutic target for insulin resistance [97]. Comprehensive DNA methylation profiling in pancreatic islets of T2DM patients versus non-diabetic donors revealed aberrantly methylated disease-specific genes that are impacting pathways implicated in pancreatic β-cell survival and function. Further studies would be required to confirm to what extent this reflects causal or consequential changes [98]. Epigenetic DNA methylation is very relevant for space travel, since an irradiation response in brain tissue and inhibition of DNA methylation can be an effective neuroprotective strategy to avoid radiation-induced cognitive deficits [99].

A further level of molecular regulation thought to influence 74–92% of all protein-encoding mRNAs involves microRNA (miRNA) molecules that bind to the $3'$ untranslated region of their target mRNA through imperfect pair bonding and effect their down-regulation. Actively involved in tumorigenesis, they can serve as oncogenes or tumor suppressor genes, depending on the context [100] and are subject to regulation by oncogenes also. Pertinently, miRNAs are implicated directly in enhancing glutamine metabolism and the Warburg effect in cancer cells [101] with several miRNAs implicated in T2DM [102]. As a specific example, miR-143 down-regulation, found in a number of tumors, can up-regulate *hexokinase 2 (HK2)* expression to promote a shift towards aerobic glycolysis in cancer cells [103]. A Warburg effect and hyperplastic metabolism via miR-143 down-regulation and *HK2* overexpression characterized esophageal squamous cell carcinomas induced by a dietary zinc deficiency [104]. It would be premature to suggest this fully defines the outcome of miR-143 expression levels, since it was up-regulated in the liver of T2DM obese mouse models, where it also impaired glucose metabolism, but through induction of insulin resistance via miR-143-ORP8-dependent inhibition of insulin-stimulated AKT activation [105]. Linking molecular metabolic mediators to dietary zinc is consistent with observations that metal elements can facilitate tissue repair processes both pre-emptively and after recovery from otherwise lethal radiation injury [106] emphasizing the importance of nutrition for protection against radiation exposure, including radiotherapy and space flight [107].

5.3. Lessons for T2DM from a Warburg Effect by Chance Whilst Ageing

The diabetic myotube phenotypes of increased basal glucose oxidation and incomplete lipid oxidation could be reproduced by using malonate to inhibit the TCA cycle enzyme succinate dehydrogenase. This impaired TCA cycle flux in skeletal muscle may causally contribute to a diabetic phenotype. It underscored the need for a detailed understanding of structure, regulation, modification and expression of the multifunctional TCA cycle enzymes [108]. However, with regard to an age-induced Warburg effect, the functional capacity of the mitochondria was retained in cultured senescent myotubes. Cells at late passage numbers showed a reduced mitochondrial mass and decreased whole cell ATP level, yet retained full mitochondrial ATP production capacity and increased ROS production [65]. This does not contradict studies showing that an alteration in mitochondrial oxidative and ATP generating activity was important for declined muscle function in ageing, but suggests that an indirectly evolved phenotype with ageing is per se not necessarily a direct cause of dysfunction in the TCA cycle enzymes. An age-associated lower mitochondrial mass may require the remaining mitochondria to compensate with elevated energy generation, yet at the cost of increased harmful ROS production. Notably, consistent with an age-related loss of mitochondrial mass, muscle-specific mutations that accumulated with aging were found in critical human mtDNA regions near the site of mtDNA attachment to the inner mitochondrial membrane, most likely aiming to control mtDNA replication and/or copy number [109].

Diabetes usually occurs later in life, making ageing an important risk factor. Curiously, several genes that influence the Warburg effect probably have a role in longevity. For example, sirtuins (SIRT1-7), a member of the highly conserved protein family with seven transmembrane domains, can extend the lifespan of model organisms. Targets for SIRT1 mediated deacetylation include a broad range of transcription factors and co-activators governing many central metabolic pathways, and several different models of *SIRT1* over-expression in transgenic mice were shown to prevent diabetes [110]. Notably, human fibroblasts with inactivated SIRT2 showed Warburg effect-like decreased OXPHOS and increased glycolysis [111]. SIRT3, like SIRT4 and SIRT5, resides within mitochondria and deacetylates and regulates many mitochondrial proteins. Compounding its metabolic significance, SIRT1 can deacetylate SIRT3 which becomes hyper-acetylated in aged and obese mice [112].

Genetic polymorphisms in the *SIRT3* gene promoter can influence levels of gene expression. Polymorphisms in the *PTP1B* gene have also been associated with improved health in old age. Direct comparison of relatively young (3 month) and aged (16 month) mice with a wild-type or $PTP1B^{-/-}$ genetic background revealed a significant difference. Mice with *PTP1B* deficiency were protected from the development of peripheral insulin resistance, adiposity, hyperinsulinemia and islet hyperplasia found in the wild type mice at 16 months of age, indicating that this enzyme had a critical role in age-dependent onset of the murine T2DM phenotype [113]. Notably, the increase in mRNA, protein levels and PTP1B activity with obesity was confined to the liver and muscle of 16-month old wild-type mice indicating a tissue-specific response. Though upstream events elevating *PTP1B* activity were not fully defined, the transcriptional activators NF-κB, and p53 were co-elevated in the liver and muscle tissue of the 16-month old wild-type mice. p53 is known to interact with several longevity pathways, including activation of AMPK and repression of the insulin IGF1 and mTOR pathways. How tissue specific discriminatory effects may arise is far from clear, but low levels of antioxidant enzymes such as glutathione peroxidase in pancreatic β-cells compared to other tissues, may explain the increased susceptibility to oxidative stress-mediated tissue damage [114]. Appreciating an important interrelationship between p53 and mitochondria and telomeres is certainly likely to be helpful to resolve key mechanisms influencing oxidative stress metabolism that are altered by ageing [115].

Ageing is associated with a reduced muscle mass and strength, that can be influenced by exercise. Resistance training was shown to increase muscle strength and function even in older adults, reducing biomarkers of oxidative stress with improved mitochondrial function. Notably, sarcopenia patients, exhibited mitochondrial DNA mutations and deletions in mature myocytes, but not in their muscle satellite cells. Post-injury regeneration of skeletal muscle following strenuous exercise involved activation, proliferation and differentiation of the resident satellite stem cells and endothelial precursor cells. Tunneling nanotubes that connect myogenic skeletal muscle satellite cells to muscle fibers may serve as conduits for intracellular material, including mitochondria [116]. The concept of mtDNA shifting suggests that resistance-exercise training is leading to a replacement of defective mitochondria bearing DNA mutation and deletions with healthier mitochondria from satellite cells [117]. Though not entirely clear whether this reflected mitochondrial DNA transfer from satellite cells, or regeneration of new fibers from satellite cells with non-deleted mtDNA, recent evidence confirms that RNA-mediated improvement in mitochondrial activity could mitigate oxidative stress to drive enhanced regeneration of injured muscle [118]. In addition, bone marrow derived stromal cells can also repair tissue injury through the transfer of mitochondria via nanotubes and microvesicles [119]. Therefore, intercellular interaction is likely involved in controlling the Warburg effect, and mitochondrial transfer may be an important process in many diseases. Given that mitochondria regulate cytoplasmic radiation induced genotoxic damage [120] intercellular tunneling nanotube connections may have a role in the design of countermeasures against therapeutic and environmental radiation exposure.

6. Not Forgetting the Reverse and Inverse Warburg Hypothesis

In multicellular tumor microenvironment models, cancer cell acquired mutations that can directly or indirectly favor ROS generation [121,122]. The "reverse Warburg" hypothesis proposes that in response to increased local hydrogen peroxide levels, it is the adjacent cancer-associated fibroblasts rather than cancer cells themselves [123] that undergo most of the aerobic glycolysis [124]. In turn, ATP produced by stromal cells may help cross-feed cancer cells to compensate mutations and maintain mitochondrial activity. This would allow cancer cells to proceed with an active TCA cycle and with oxidative phosphorylation pathways to meet the enhanced ATP and anabolic demand of proliferating tumor cells.

Helping to support this scenario, cancer cells express cytoprotective genes that sustain anabolic metabolism despite the oxidative pressure in their microenvironment. Pivotal for both enhancing metabolism and protection, the transcription factor Nuclear erythroid factor 2-like (Nrf2), constitutively active in many cancer cell types, can redirect glucose and glutamine into anabolic pathways, including purine nucleotide generation in the presence of PI3K-Akt signaling [125]. The broad repertoire of transcriptional targets directed by Nrf2 include at least six metabolic regulator genes involved in the PPP and NADPH production pathways, plus numerous anti-oxidant enzymes and cytoprotective factors [126]. Somatic mutations inactivating Kelch-like ECH-associated protein 1 (Keap1), the canonical inhibitor of Nrf2, are relatively frequent, preventing Nrf2-Keap1 interaction in the cytoplasm and promoting proteasomal degradation of Nrf2. Mutations in *Nrf2* itself are frequently disrupting discs-large (DLG) or glutamate-threonine-glycine-glutamate (ETGE) motifs associated with low or high affinity Keap1 interactions, thus favoring constitutive activation of Nrf2 [127]. One may regard Nrf2 as a nefarious complement to oxidative and hypoxic environments that facilitates tumor growth [128]. However, in normal cell metabolism contexts, the repressive action of Nrf2 on ATP citrate lyase, Acc1 and Fasn, can suppress *de novo* lipogenesis which is required for tumor growth. Accordingly, Nrf2 has a highly context-dependent role, permissive for already initiated cancer growth yet also beneficial for cancer prevention, since absence of Nrf2 also reduces DNA repair in normal cells [129], increasing cell susceptibility to carcinogenic agents [130]. Astronauts resident in the international space station for six months showed ROS alterations thought to reflect reduced mitochondrial synthesis and limited expression of genes including *Nrf2*, highlighting a potentially increased cancer risk in astronauts [131].

Besides cancer and diabetes, the Warburg effect influences numerous morbidities. Although it improves host responses to pathogens through T cell activation [132], the Warburg effect is generally associated with progression of chronic disease. It is implicated in multiple sclerosis, pulmonary hypertension and idiopathic pulmonary fibrosis, cardiac hypertrophy, atherosclerosis, polycystic kidney disease and Alzheimer's disease (AD) [133]. Notably, for the latter, a hypothesis termed "inverse Warburg" effect has been proposed. Distinct from the "reverse Warburg" hypothesis of tumorigenesis, it nevertheless involves a reciprocal metabolic coupling between adjacent cell types [134], namely neurons and microglia. Proposed by Demetrius et al., the bioenergetics-based inverse Warburg model for sporadic forms of AD implicates energy generation and age as critical elements in the origin of this neurodegenerative disease. As a first step, age-induced chance inefficiency in the mitochondrial activity [135], such as that mediated by DNA protein kinase within certain neurons [136] causes a compensatory ramped up-regulation of oxidative phosphorylation to maintain cell viability. In effect, this is a reversal of Warburg effect pathways that replace oxidative phosphorylation with aerobic glycolysis. Subsequently, a cascade of events evokes a selective advantage for neurons with mild mitochondrial impairment and increased OXPHOS activity over healthy neurons. Neurons adopt this option due to low expression levels of *PFKFB3* encoding the 6-phosphofructo-2-kinase/fructose-2,6-biphosphatase 3 enzyme that is critical for the regulation of glycolysis [137]. Thus, neurons have a limited capacity to up-regulate glycolysis for compensating the increased energy demands. In contrast, microglial astrocytes express high levels of PFKFB3 and also show deficiencies in pyruvate dehydrogenase activity and elements for shuttling cytosolic NADH within mitochondria for pyruvate oxidation instead of lactate formation. Accordingly, neurons are essentially oxidative cells, whereas astrocytes have high glycolytic capacity [138] and use this alternative metabolic route to generate lactate that can fuel adjacent neurons. Defective neurons with excessive OXPHOS activity produce ROS that promote pathological ageing, with abnormal cell cycle entry and eventual neuronal loss. At the same time, production of lactate or ketone bodies can be considered a compensatory response in adjacent microglia to meet the high energy demands of defective neurons. The proposal that intercellular Warburg effect-mediated interactions may modulate disease progression [139] can help account for idiosyncrasies surrounding a metabolic amyloid cascade hypothesis, whereby aggregation of neurite plaques, with consequential disruption of synaptic connections, underlies the death of neurons and the overwhelming forgetfulness of dementia [140].

7. New Light on Redox Metabolism Pathways in Extreme Conditions

Historically, numerous medical advances trace origin to causal relationships from extreme examples providing unique peculiarities that highlighted key functional mediators, e.g., the occupational hazard of early 20th century watch dial painters, who pointed brushes in their mouths and contracted malignancies from radioluminescent radium poisoning [141]. A current extreme occupational endeavor inevitably associated with numerous metabolic risks is human spaceflight, with astronauts being subjected to unique microgravity and ionizing radiation conditions [142].

Weightlessness from microgravity drives dramatic physiological changes, including an altered cardiovascular system and circulation, immunosuppression and changes to calcium, sodium and bone metabolism [143]. Nonetheless, it is notable that an additional higher cardiovascular disease mortality risk can be found in Apollo lunar astronauts, the only humans to have travelled beyond Earth's radiation-shielding magnetosphere, versus astronauts that have only flown in low Earth orbit. Advances in systems biology provide an improved interpretation of the molecular signatures and networks altered in human cells by microgravity and ionizing radiation-induced oxidative stress [144], highlighting a close association among oxidative phosphorylation/respiratory electron transport, ubiquitin-proteasome system and neurodegenerative conditions (e.g., Alzheimer's disease) across multiple datasets. Mice genetically modelling AD exposed to 56Fe ion particle radiation showed an accelerated age-associated accumulation of amyloid beta (Aβ) plaques and increased cognitive impairment [145]. This concurred with later studies showing neuronal network fragility to the combination of simulated microgravity and chronic exposure to radiation [146]. Extending repercussions beyond generation of ROS, immune dysfunction induced by spaceflight has been closely associated with metabolic change in fatty acid oxidation and decreased glycolysis-related pathways [147].

Fortunately, experimental methods applicable to the analysis of the Warburg effect and mitochondria are advancing. Refined isolation of mitochondria and measurement of mitochondrial membrane potential and metabolites has accelerated screening of potential new drugs that shift energy metabolism from mitochondrial OXPHOS to aerobic glycolysis [148]. Powerful proteomic approaches allow broad scale analysis of metabolite profiles, highlighting associations between amino acid metabolites and the risk of developing diabetes [149]. A comprehensive metabolic signature has been obtained for defining the Warburg effect in situations such as pancreatic cancer [150]. Spectrophotometric protocols allow simple and reliable assessment of respiratory chain function to be applied to minute quantities of muscle tissue, cultured cells and isolated mitochondria [151].

To date, the current repertoire of nutritional supplements has done little to provide true clinical benefit [152]. However, improved, more sophisticated transgenic mouse models will allow better discrimination between cause and effect when exploring metabolic disorders. Despite only partial mimicry, simulated microgravity ground experiments can suffice to potentiate the effect of ROS [153] shown to exacerbate the effects of heavy ion radiation in human B lymphoblasts [154]. Dedicated heavy-ion radiobiology research centers, e.g., the NASA Space radiation laboratory at Brookhaven, are pivotal for improving space-relevant radiobiology knowledge and exploration of how heavy-ion species might be exploited for focused anti-cancer therapy [155]. Combining heavy-ion irradiation from a carbon ion combined with a clinostat to simulate microgravity allowed exploration of synchronous effects pertinent to space radiation research [156]. Additional technological advances include laser-plasma-accelerators representing advantageous tools for accurately reproducing broadband radiation particle flux similar to conditions in space [157]. Prospective research adopting Extreme Light Infrastructure-Nuclear Physics (ELI-NP) at the Center for Advanced Laster Technologies (CETAL) facility, will provide two very high intensity 10 PW lasers and a very intense (10^{13} γ/s) brilliant γ beam with which to explore appropriate shielding methods [158] for deep space flights and for developing relevant radiobiology investigations.

8. Conclusions

The Warburg effect hypothesis, originated from findings in cancer cells, has proved helpful to better understand key mechanisms underlying a broad range of metabolic diseases. Certainly, the situation is extremely complex, and many examples of anaerobic glycolysis co-existing with normal TCA function and OXPHOS pathways suggest a dynamically regulated shared metabolic balance rather than strict adoption of one mode of respiration at the expense of the other. Biomarkers for oxidative stress assessed in broader population studies of environmental metabolic stressors may have a direct causal role and represent indirect indicators of questionable therapeutic value or serve as good intervention targets [159]. Assumptions that measurements made ex vivo apply to in vivo microenvironments need confirmation [160] because, in a heterogeneous tissue, clusters of abnormality become risk factors in their own right. The impact of an event such as mtDNA mutation with regard to disease conditions may vary between different cell types, different tissues and even different human populations [161]. Molecules exerting significant phenotypic effects in more than one disease state are being discovered, and a broader network view can improve our interpretation of how metabolic mechanisms are coordinated. Applying a choice, chain, chance perspective to integrate molecular metabolic pathways influenced by the Warburg effect in developmental biology, cancer and type 2 diabetes mellitus can provide unifying insights.

The scope of relevance for the Warburg effect has extended well beyond its origins as an anomaly of cancer. It is clear that, for any long-term space mission, astronauts will need appropriate physical shielding and additional medicinal countermeasures to limit both excessive oxidative stress triggered by ionizing radiation and the associated inflammatory processes mediated by the NF-κB pathway [162]. Use of cytoprotective agents, such as Amifostine, can provide radioprotection to normal tissues by triggering a metabolic shift with induction of glycolysis and blockage of mitochondrial pyruvate usage through a Warburg effect pathway that reduces ROS production [163]. A possible therapeutic solution might be the use of Nrf2 activators, such as dimethyl fumarate and sulforaphane, for enhancing the endogenous antioxidant protection against the oxidative pressure of galactic cosmic radiation [164]. Quantitative analysis of low-molecular mass endogenous metabolites is becoming ever-more achievable with modern analytical techniques that are well-suited to complex pathologies [165]. It is anticipated that space exploration will continue to be a major stimulus for advancing our understanding of human metabolism [166] and reciprocate with unique experiments that extend our capacity to overcome even the most challenging complex human diseases and environmental pressures.

Acknowledgments: Work was partly supported by the M-Era.Net program through the NANOTHER project, grant 52/2016, and by the ELI-RO program through the grant 13 ELI/2016.

Conflicts of Interest: The authors declare no conflict of interest.

References

1. Ehlmann, B.L.; Chowdhury, J.; Marzullo, T.C.; Collins, R.E.; Litzenberger, J.; Ibsen, S.; Krauser, W.R.; DeKock, B.; Hannon, M.; Kinnevan, J.; et al. Humans to Mars: A feasibility and cost–benefit analysis. *Acta Astronaut.* **2005**, *56*, 851–858. [CrossRef] [PubMed]
2. Wells, H.G. *The War of the Worlds*; William Heinemann: Portsmouth, NH, USA, 2003.
3. Roger, A.J.; Muñoz-Gómez, S.A.; Kamikawa, R. The Origin and Diversification of Mitochondria. *Curr. Biol.* **2017**, *27*, R1177–R1192. [CrossRef] [PubMed]
4. Kadenbach, B. Introduction to mitochondrial oxidative phosphorylation. *Adv. Exp. Med. Biol.* **2012**, *748*, 1–11. [PubMed]
5. Warburg, O.; Posener, K.; Negelein, E. Über den stoffwechsel der Carcinomzelle. *Biochem. Z.* **1924**, *152*, 319–344. [CrossRef]
6. Wong, J.L. From fertilization to cancer: A lifelong pursuit into how cells use oxygen. Otto Heinrich Warburg (October 8, 1883–August 1, 1970). *Mol. Reprod. Dev.* **2011**, *78*. [CrossRef] [PubMed]
7. Levine, A.J.; Puzio-Kuter, A.M. The control of the metabolic switch in cancers by oncogenes and tumor suppressor genes. *Science* **2010**, *330*, 1340–1344. [CrossRef] [PubMed]
8. Zhan, C.; Yan, L.; Wang, L.; Ma, J.; Jiang, W.; Zhang, Y.; Shi, Y.; Wang, Q. Isoform switch of pyruvate kinase M1 indeed occurs but not to pyruvate kinase M2 in human tumorigenesis. *PLoS ONE* **2015**, *10*, e0118663. [CrossRef] [PubMed]
9. Vander Heiden, M.G.; Locasale, J.W.; Swanson, K.D.; Sharfi, H.; Heffron, G.J.; Amador-Noguez, D.; Christofk, H.R.; Wagner, G.; Rabinowitz, J.D.; Asara, J.M.; et al. Evidence for an alternative glycolytic pathway in rapidly proliferating cells. *Science* **2010**, *329*, 1492–1499. [CrossRef] [PubMed]
10. Tran, Q.; Lee, H.; Park, J.; Kim, S.H.; Park, J. Targeting cancer metabolism—Revisiting the Warburg effects. *Toxicol. Res.* **2016**, *32*, 177–193. [CrossRef] [PubMed]
11. Redel, B.K.; Brown, A.N.; Spate, L.D.; Whitworth, K.M.; Green, J.A.; Prather, R.S. Glycolysis in preimplantation development is partially controlled by the Warburg Effect. *Mol. Reprod. Dev.* **2012**, *79*, 262–271. [CrossRef] [PubMed]
12. DeBerardinis, R.J.; Mancuso, A.; Daikhin, E.; Nissim, I.; Yudkoff, M.; Wehrli, S.; Thompson, C.B. Beyond aerobic glycolysis: Transformed cells can engage in glutamine metabolism that exceeds the requirement for protein and nucleotide synthesis. *Proc. Natl. Acad. Sci. USA* **2007**, *104*, 19345–19350. [CrossRef] [PubMed]
13. Schuster, S.; de Figueiredo, L.F.; Schroeter, A.; Kaleta, C. Combining metabolic pathway analysis with Evolutionary Game Theory: Explaining the occurrence of low-yield pathways by an analytic optimization approach. *Biosystems* **2011**, *105*, 147–153. [CrossRef] [PubMed]
14. Golpour, M.; Akhavan Niaki, H.; Khorasani, H.R.; Hajian, A.; Mehrasa, R.; Mostafazadeh, A. Human fibroblast switches to anaerobic metabolic pathway in response to serum starvation: A mimic of warburg effect. *Int. J. Mol. Cell. Med.* **2014**, *3*, 74–80. [PubMed]
15. Jeon, S.M.; Chandel, N.S.; Hay, N. AMPK regulates NADPH homeostasis to promote tumour cell survival during energy stress. *Nature* **2012**, *485*, 661–665. [CrossRef] [PubMed]

16. Choi, H.S.; Hwang, C.K.; Song, K.Y.; Law, P.Y.; Wei, L.N.; Loh, H.H. Poly(C)-binding proteins as transcriptional regulators of gene expression. *Biochem. Biophys. Res. Commun.* **2009**, *380*, 431–436. [CrossRef] [PubMed]

17. Rutter, J.; Reick, M.; Wu, L.C.; McKnight, S.L. Regulation of clock and NPAS2 DNA binding by the redox state of NAD cofactors. *Science* **2001**, *293*, 510–514. [CrossRef] [PubMed]

18. Ward, P.S.; Thompson, C.B. Metabolic reprogramming: A cancer hallmark even warburg did not anticipate. *Cancer Cell* **2012**, *21*, 297–308. [CrossRef] [PubMed]

19. Dang, C.V.; Le, A.; Gao, P. MYC-induced cancer cell energy metabolism and therapeutic opportunities. *Clin. Cancer Res.* **2009**, *15*, 6479–6483. [CrossRef] [PubMed]

20. Metallo, C.M.; Gameiro, P.A.; Bell, E.L.; Mattaini, K.R.; Yang, J.; Hiller, K.; Jewell, C.M.; Johnson, Z.R.; Irvine, D.J.; Guarente, L.; et al. Reductive glutamine metabolism by IDH1 mediates lipogenesis under hypoxia. *Nature* **2011**, *481*, 380–384. [CrossRef] [PubMed]

21. MacDonald, V.E.; Howe, L.J. Histone acetylation: Where to go and how to get there. *Epigenetics* **2009**, *4*, 139–143. [CrossRef] [PubMed]

22. Zielke, H.R.; Ozand, P.T.; Tildon, J.T.; Sevdalian, D.A.; Cornblath, M. Reciprocal regulation of glucose and glutamine utilization by cultured human diploid fibroblasts. *J. Cell. Physiol.* **1978**, *95*, 41–48. [CrossRef] [PubMed]

23. Vafai, S.B.; Mootha, V.K. Mitochondrial disorders as windows into an ancient organelle. *Nature* **2012**, *491*, 374–383. [CrossRef] [PubMed]

24. Jauniaux, E.; Gulbis, B.; Burton, G.J. The human first trimester gestational sac limits rather than facilitates oxygen transfer to the foetus—A review. *Placenta* **2003**, *24*, S86–S93. [CrossRef] [PubMed]

25. Burton, G.J.; Jauniaux, E.; Murray, A.J. Oxygen and placental development; parallels and differences with tumour biology. *Placenta* **2017**, *56*, 14–18. [CrossRef] [PubMed]

26. Birket, M.J.; Orr, A.L.; Gerencser, A.A.; Madden, D.T.; Vitelli, C.; Swistowski, A.; Brand, M.D.; Zeng, X. A reduction in ATP demand and mitochondrial activity with neural differentiation of human embryonic stem cells. *J. Cell Sci.* **2011**, *124*, 348–358. [CrossRef] [PubMed]

27. Facucho-Oliveira, J.M.; St John, J.C. The relationship between pluripotency and mitochondrial DNA proliferation during early embryo development and embryonic stem cell differentiation. *Stem Cell Rev.* **2009**, *5*, 140–158. [CrossRef] [PubMed]

28. Krisher, R.L.; Prather, R.S. A role for the Warburg effect in preimplantation embryo development: Metabolic modification to support rapid cell proliferation. *Mol. Reprod. Dev.* **2012**, *79*, 311–320. [CrossRef] [PubMed]

29. Zhou, W.; Choi, M.; Margineantu, D.; Margaretha, L.; Hesson, J.; Cavanaugh, C.; Blau, C.A.; Horwitz, M.S.; Hockenbery, D.; Ware, C.; et al. HIF1α induced switch from bivalent to exclusively glycolytic metabolism during ESC-to-EpiSC/hESC transition. *EMBO J.* **2012**, *31*, 2103–2116. [CrossRef] [PubMed]

30. Kaelin, W.G.; Ratcliffe, P.J. Oxygen sensing by metazoans: The central role of the HIF hydroxylase pathway. *Mol. Cell* **2008**, *30*, 393–402. [CrossRef] [PubMed]

31. Chan, M.C.; Ilott, N.E.; Schödel, J.; Sims, D.; Tumber, A.; Lippl, K.; Mole, D.R.; Pugh, C.W.; Ratcliffe, P.J.; Ponting, C.P.; et al. Tuning the Transcriptional Response to Hypoxia by Inhibiting Hypoxia-inducible Factor (HIF) Prolyl and Asparaginyl Hydroxylases. *J. Biol. Chem.* **2016**, *291*, 20661–20673. [CrossRef] [PubMed]

32. Loboda, A.; Jozkowicz, A.; Dulak, J. HIF-1 versus HIF-2—Is one more important than the other. *Vasc. Pharmacol.* **2012**, *56*, 245–251. [CrossRef] [PubMed]

33. Doe, M.R.; Ascano, J.M.; Kaur, M.; Cole, M.D. Myc posttranscriptionally induces HIF1 protein and target gene expression in normal and cancer cells. *Cancer Res.* **2012**, *72*, 949–957. [CrossRef] [PubMed]

34. Richard, S.; Gardie, B.; Couvé, S.; Gad, S. Von Hippel-Lindau: How a rare disease illuminates cancer biology. *Semin. Cancer Biol.* **2013**, *23*, 26–37. [CrossRef] [PubMed]

35. Yoon, D.; Okhotin, D.V.; Kim, B.; Okhotina, Y.; Okhotin, D.J.; Miasnikova, G.Y.; Sergueeva, A.I.; Polyakova, L.A.; Maslow, A.; Lee, Y.; et al. Increased size of solid organs in patients with Chuvash polycythemia and in mice with altered expression of HIF-1α and HIF-2α. *J. Mol. Med.* **2010**, *88*, 523–530. [CrossRef] [PubMed]

36. Wu, C.A.; Chao, Y.; Shiah, S.G.; Lin, W.W. Nutrient deprivation induces the Warburg effect through ROS/AMPK-dependent activation of pyruvate dehydrogenase kinase. *Biochim. Biophys. Acta* **2013**, *1833*, 1147–1156. [CrossRef] [PubMed]

37. Ward, P.S.; Thompson, C.B. Signaling in control of cell growth and metabolism. *Cold Spring Harb. Perspect. Biol.* **2012**, *4*, a006783. [CrossRef] [PubMed]

38. Menendez, J.A. Fine-tuning the lipogenic/lipolytic balance to optimize the metabolic requirements of cancer cell growth: Molecular mechanisms and therapeutic perspectives. *Biochim. Biophys. Acta* **2010**, *1801*, 381–391. [CrossRef] [PubMed]

39. Palorini, R.; De Rasmo, D.; Gaviraghi, M.; Sala Danna, L.; Signorile, A.; Cirulli, C.; Chiaradonna, F.; Alberghina, L.; Papa, S. Oncogenic K-ras expression is associated with derangement of the cAMP/PKA pathway and forskolin-reversible alterations of mitochondrial dynamics and respiration. *Oncogene* **2013**, *32*, 352–362. [CrossRef] [PubMed]

40. Cai, Q.; Lin, T.; Kamarajugadda, S.; Lu, J. Regulation of glycolysis and the Warburg effect by estrogen-related receptors. *Oncogene* **2013**, *32*, 2079–2086. [CrossRef] [PubMed]

41. Shim, H.; Dolde, C.; Lewis, B.C.; Wu, C.S.; Dang, G.; Jungmann, R.A.; Dalla-Favera, R.; Dang, C.V. c-Myc transactivation of LDH-A: Implications for tumor metabolism and growth. *Proc. Natl. Acad. Sci. USA* **1997**, *94*, 6658–6663. [CrossRef] [PubMed]

42. Ahuja, P.; Zhao, P.; Angelis, E.; Ruan, H.; Korge, P.; Olson, A.; Wang, Y.; Jin, E.S.; Jeffrey, F.M.; Portman, M.; et al. Myc controls transcriptional regulation of cardiac metabolism and mitochondrial biogenesis in response to pathological stress in mice. *J. Clin. Investig.* **2010**, *120*, 1494–1505. [CrossRef] [PubMed]

43. Le, A.; Cooper, C.R.; Gouw, A.M.; Dinavahi, R.; Maitra, A.; Deck, L.M.; Royer, R.E.; Vander Jagt, D.L.; Semenza, G.L.; Dang, C.V. Inhibition of lactate dehydrogenase A induces oxidative stress and inhibits tumor progression. *Proc. Natl. Acad. Sci. USA* **2010**, *107*, 2037–2042. [CrossRef] [PubMed]

44. Sloan, E.J.; Ayer, D.E. Myc, mondo, and metabolism. *Genes Cancer* **2010**, *1*, 587–596. [CrossRef] [PubMed]

45. Sans, C.L.; Satterwhite, D.J.; Stoltzman, C.A.; Breen, K.T.; Ayer, D.E. MondoA-Mlx heterodimers are candidate sensors of cellular energy status: Mitochondrial localization and direct regulation of glycolysis. *Mol. Cell. Biol.* **2006**, *26*, 4863–4871. [CrossRef] [PubMed]

46. Wilde, B.R.; Ayer, D.E. Interactions between Myc and MondoA transcription factors in metabolism and tumourigenesis. *Br. J. Cancer* **2015**, *113*, 1529–1533. [CrossRef] [PubMed]

47. Stoltzman, C.A.; Peterson, C.W.; Breen, K.T.; Muoio, D.M.; Billin, A.N.; Ayer, D.E. Glucose sensing by MondoA:Mlx complexes: A role for hexokinases and direct regulation of thioredoxin-interacting protein expression. *Proc. Natl. Acad. Sci. USA* **2008**, *105*, 6912–6917. [CrossRef] [PubMed]

48. Tong, X.; Zhao, F.; Mancuso, A.; Gruber, J.J.; Thompson, C.B. The glucose-responsive transcription factor ChREBP contributes to glucose-dependent anabolic synthesis and cell proliferation. *Proc. Natl. Acad. Sci. USA* **2009**, *106*, 21660–21665. [CrossRef] [PubMed]

49. Vousden, K.H.; Ryan, K.M. p53 and metabolism. *Nat. Rev. Cancer* **2009**, *9*, 691–700. [CrossRef] [PubMed]

50. Abu Dawud, R.; Schreiber, K.; Schomburg, D.; Adjaye, J. Human embryonic stem cells and embryonal carcinoma cells have overlapping and distinct metabolic signatures. *PLoS ONE* **2012**, *7*, e39896. [CrossRef] [PubMed]

51. Berridge, M.V.; Tan, A.S. Effects of mitochondrial gene deletion on tumorigenicity of metastatic melanoma: Reassessing the Warburg effect. *Rejuvenation Res.* **2010**, *13*, 139–141. [CrossRef] [PubMed]

52. Sottnik, J.L.; Lori, J.C.; Rose, B.J.; Thamm, D.H. Glycolysis inhibition by 2-deoxy-D-glucose reverts the metastatic phenotype in vitro and in vivo. *Clin. Exp. Metastasis* **2011**, *28*, 865–875. [CrossRef] [PubMed]

53. Goitre, L.; Pergolizzi, B.; Ferro, E.; Trabalzini, L.; Retta, S.F. Molecular Crosstalk between Integrins and Cadherins: Do Reactive Oxygen Species Set the Talk? *J. Signal Transduct.* **2012**, *2012*, 807682. [CrossRef] [PubMed]

54. Smolková, K.; Plecitá-Hlavatá, L.; Bellance, N.; Benard, G.; Rossignol, R.; Ježek, P. Waves of gene regulation suppress and then restore oxidative phosphorylation in cancer cells. *Int. J. Biochem. Cell Biol.* **2011**, *43*, 950–968. [CrossRef] [PubMed]

55. Ohkouchi, S.; Block, G.J.; Katsha, A.M.; Kanehira, M.; Ebina, M.; Kikuchi, T.; Saijo, Y.; Nukiwa, T.; Prockop, D.J. Mesenchymal stromal cells protect cancer cells from ROS-induced apoptosis and enhance the Warburg effect by secreting STC1. *Mol. Ther.* **2012**, *20*, 417–423. [CrossRef] [PubMed]

56. Gupta, S.; Roy, A.; Dwarakanath, B.S. Metabolic Cooperation and Competition in the Tumor Microenvironment: Implications for Therapy. *Front. Oncol.* **2017**, *7*, 68. [CrossRef] [PubMed]

57. Lim, H.Y.; Ho, Q.S.; Low, J.; Choolani, M.; Wong, K.P. Respiratory competent mitochondria in human ovarian and peritoneal cancer. *Mitochondrion* **2011**, *11*, 437–443. [CrossRef] [PubMed]

58. Frezza, C.; Zheng, L.; Folger, O.; Rajagopalan, K.N.; MacKenzie, E.D.; Jerby, L.; Micaroni, M.; Chaneton, B.; Adam, J.; Hedley, A.; et al. Haem oxygenase is synthetically lethal with the tumour suppressor fumarate hydratase. *Nature* **2011**, *477*, 225–228. [CrossRef] [PubMed]

59. Frezza, C.; Tennant, D.A.; Gottlieb, E. IDH1 mutations in gliomas: When an enzyme loses its grip. *Cancer Cell* **2010**, *17*, 7–9. [CrossRef] [PubMed]

60. Pacelli, C.; Latorre, D.; Cocco, T.; Capuano, F.; Kukat, C.; Seibel, P.; Villani, G. Tight control of mitochondrial membrane potential by cytochrome c oxidase. *Mitochondrion* **2011**, *11*, 334–341. [CrossRef] [PubMed]

61. Groen, A.K.; Wanders, R.J.; Westerhoff, H.V.; van der Meer, R.; Tager, J.M. Quantification of the contribution of various steps to the control of mitochondrial respiration. *J. Biol. Chem.* **1982**, *257*, 2754–2757. [PubMed]

62. Haigis, M.C.; Deng, C.X.; Finley, L.W.; Kim, H.S.; Gius, D. SIRT3 is a mitochondrial tumor suppressor: A scientific tale that connects aberrant cellular ROS, the Warburg effect, and carcinogenesis. *Cancer Res.* **2012**, *72*, 2468–2472. [CrossRef] [PubMed]

63. Krishnan, K.J.; Reeve, A.K.; Samuels, D.C.; Chinnery, P.F.; Blackwood, J.K.; Taylor, R.W.; Wanrooij, S.; Spelbrink, J.N.; Lightowlers, R.N.; Turnbull, D.M. What causes mitochondrial DNA deletions in human cells. *Nat. Genet.* **2008**, *40*, 275–279. [CrossRef] [PubMed]

64. Krishnan, K.J.; Ratnaike, T.E.; De Gruyter, H.L.; Jaros, E.; Turnbull, D.M. Mitochondrial DNA deletions cause the biochemical defect observed in Alzheimer's disease. *Neurobiol. Aging* **2012**, *33*, 2210–2214. [CrossRef] [PubMed]

65. Minet, A.D.; Gaster, M. Cultured senescent myoblasts derived from human vastus lateralis exhibit normal mitochondrial ATP synthesis capacities with correlating concomitant ROS production while whole cell ATP production is decreased. *Biogerontology* **2012**, *13*, 277–285. [CrossRef] [PubMed]

66. Hashizume, O.; Shimizu, A.; Yokota, M.; Sugiyama, A.; Nakada, K.; Miyoshi, H.; Itami, M.; Ohira, M.; Nagase, H.; Takenaga, K.; et al. Specific mitochondrial DNA mutation in mice regulates diabetes and lymphoma development. *Proc. Natl. Acad. Sci. USA* **2012**, *109*, 10528–10533. [CrossRef] [PubMed]

67. Hirabara, S.M.; Curi, R.; Maechler, P. Saturated fatty acid-induced insulin resistance is associated with mitochondrial dysfunction in skeletal muscle cells. *J. Cell. Physiol.* **2010**, *222*, 187–194. [CrossRef] [PubMed]

68. Maechler, P.; Wollheim, C.B. Mitochondrial function in normal and diabetic β-cells. *Nature* **2001**, *414*, 807–812. [CrossRef] [PubMed]

69. Strollo, F.; Vassilieva, G.; Ruscica, M.; Masini, M.; Santucci, D.; Borgia, L.; Magni, P.; Celotti, F.; Nikiporuc, I. Changes in stress hormones and metabolism during a 105-day simulated Mars mission. *Aviat. Space Environ. Med.* **2014**, *85*, 793–797. [CrossRef] [PubMed]

70. Tobin, B.W.; Uchakin, P.N.; Leeper-Woodford, S.K. Insulin secretion and sensitivity in space flight: Diabetogenic effects. *Nutrition* **2002**, *18*, 842–848. [CrossRef]

71. Rejeski, W.J.; Ip, E.H.; Bertoni, A.G.; Bray, G.A.; Evans, G.; Gregg, E.W.; Zhang, Q.; Look AHEAD Research Group. Lifestyle change and mobility in obese adults with type 2 diabetes. *N. Engl. J. Med.* **2012**, *366*, 1209–1217. [CrossRef] [PubMed]

72. Lustig, R.H.; Schmidt, L.A.; Brindis, C.D. Public health: The toxic truth about sugar. *Nature* **2012**, *482*, 27–29. [CrossRef] [PubMed]

73. Kellenberger, L.D.; Bruin, J.E.; Greenaway, J.; Campbell, N.E.; Moorehead, R.A.; Holloway, A.C.; Petrik, J. The role of dysregulated glucose metabolism in epithelial ovarian cancer. *J. Oncol.* **2010**, *2010*, 514310. [CrossRef] [PubMed]

74. Lee, B.R.; Um, H.D. Hydrogen peroxide suppresses U937 cell death by two different mechanisms depending on its concentration. *Exp. Cell Res.* **1999**, *248*, 430–438. [CrossRef] [PubMed]

75. Hodgkinson, A.D.; Bartlett, T.; Oates, P.J.; Millward, B.A.; Demaine, A.G. The response of antioxidant genes to hyperglycemia is abnormal in patients with type 1 diabetes and diabetic nephropathy. *Diabetes* **2003**, *52*, 846–851. [CrossRef] [PubMed]

76. Nishikawa, T.; Edelstein, D.; Brownlee, M. The missing link: A single unifying mechanism for diabetic complications. *Kidney Int. Suppl.* **2000**, *77*, S26–S30. [CrossRef] [PubMed]

77. Ethier-Chiasson, M.; Forest, J.C.; Giguère, Y.; Masse, A.; Marseille-Tremblay, C.; Lévy, E.; Lafond, J. Modulation of placental protein expression of OLR1: Implication in pregnancy-related disorders or pathologies. *Reproduction* **2008**, *136*, 491–502. [CrossRef] [PubMed]

78. Khaidakov, M.; Mitra, S.; Kang, B.Y.; Wang, X.; Kadlubar, S.; Novelli, G.; Raj, V.; Winters, M.; Carter, W.C.; Mehta, J.L. Oxidized LDL receptor 1 (OLR1) as a possible link between obesity, dyslipidemia and cancer. *PLoS ONE* **2011**, *6*, e20277. [CrossRef] [PubMed]

79. Cagnone, G.L.; Dufort, I.; Vigneault, C.; Sirard, M.A. Differential gene expression profile in bovine blastocysts resulting from hyperglycemia exposure during early cleavage stages. *Biol. Reprod.* **2012**, *86*, 50. [CrossRef] [PubMed]

80. Gardner, D.K.; Wale, P.L.; Collins, R.; Lane, M. Glucose consumption of single post-compaction human embryos is predictive of embryo sex and live birth outcome. *Hum. Reprod.* **2011**, *26*, 1981–1986. [CrossRef] [PubMed]

81. Fan, J.; Hitosugi, T.; Chung, T.W.; Xie, J.; Ge, Q.; Gu, T.L.; Polakiewicz, R.D.; Chen, G.Z.; Boggon, T.J.; Lonial, S.; et al. Tyrosine phosphorylation of lactate dehydrogenase A is important for NADH/NAD+ redox homeostasis in cancer cells. *Mol. Cell. Biol.* **2011**, *31*, 4938–4950. [CrossRef] [PubMed]

82. Granchi, C.; Roy, S.; Giacomelli, C.; Macchia, M.; Tuccinardi, T.; Martinelli, A.; Lanza, M.; Betti, L.; Giannaccini, G.; Lucacchini, A.; et al. Discovery of N-hydroxyindole-based inhibitors of human lactate dehydrogenase isoform A (LDH-A) as starvation agents against cancer cells. *J. Med. Chem.* **2011**, *54*, 1599–1612. [CrossRef] [PubMed]

83. Malmgren, S.; Nicholls, D.G.; Taneera, J.; Bacos, K.; Koeck, T.; Tamaddon, A.; Wibom, R.; Groop, L.; Ling, C.; Mulder, H.; et al. Tight coupling between glucose and mitochondrial metabolism in clonal β-cells is required for robust insulin secretion. *J. Biol. Chem.* **2009**, *284*, 32395–32404. [CrossRef] [PubMed]

84. Hitosugi, T.; Fan, J.; Chung, T.W.; Lythgoe, K.; Wang, X.; Xie, J.; Ge, Q.; Gu, T.L.; Polakiewicz, R.D.; Roesel, J.L.; et al. Tyrosine phosphorylation of mitochondrial pyruvate dehydrogenase kinase 1 is important for cancer metabolism. *Mol. Cell* **2011**, *44*, 864–877. [CrossRef] [PubMed]

85. Zheng, L.Y.; Zhou, D.X.; Lu, J.; Zhang, W.J.; Zou, D.J. Down-regulated expression of the protein-tyrosine phosphatase 1B (PTP1B) is associated with aggressive clinicopathologic features and poor prognosis in hepatocellular carcinoma. *Biochem. Biophys. Res. Commun.* **2012**, *420*, 680–684. [CrossRef] [PubMed]

86. Lessard, L.; Labbé, D.P.; Deblois, G.; Bégin, L.R.; Hardy, S.; Mes-Masson, A.M.; Saad, F.; Trotman, L.C.; Giguère, V.; Tremblay, M.L. PTP1B is an androgen receptor-regulated phosphatase that promotes the progression of prostate cancer. *Cancer Res.* **2012**, *72*, 1529–1537. [CrossRef] [PubMed]

87. Li, S.; Depetris, R.S.; Barford, D.; Chernoff, J.; Hubbard, S.R. Crystal structure of a complex between protein tyrosine phosphatase 1B and the insulin receptor tyrosine kinase. *Structure* **2005**, *13*, 1643–1651. [CrossRef] [PubMed]

88. Goldstein, B.J.; Bittner-Kowalczyk, A.; White, M.F.; Harbeck, M. Tyrosine dephosphorylation and deactivation of insulin receptor substrate-1 by protein-tyrosine phosphatase 1B. Possible facilitation by the formation of a ternary complex with the Grb2 adaptor protein. *J. Biol. Chem.* **2000**, *275*, 4283–4289. [CrossRef] [PubMed]

89. Zinker, B.A.; Rondinone, C.M.; Trevillyan, J.M.; Gum, R.J.; Clampit, J.E.; Waring, J.F.; Xie, N.; Wilcox, D.; Jacobson, P.; Frost, L.; et al. PTP1B antisense oligonucleotide lowers PTP1B protein, normalizes blood glucose, and improves insulin sensitivity in diabetic mice. *Proc. Natl. Acad. Sci. USA* **2002**, *99*, 11357–11362. [CrossRef] [PubMed]

90. Haque, A.; Andersen, J.N.; Salmeen, A.; Barford, D.; Tonks, N.K. Conformation-sensing antibodies stabilize the oxidized form of PTP1B and inhibit its phosphatase activity. *Cell* **2011**, *147*, 185–198. [CrossRef] [PubMed]

91. Salem, A.F.; Whitaker-Menezes, D.; Lin, Z.; Martinez-Outschoorn, U.E.; Tanowitz, H.B.; Al-Zoubi, M.S.; Howell, A.; Pestell, R.G.; Sotgia, F.; Lisanti, M.P. Two-compartment tumor metabolism: Autophagy in the tumor microenvironment and oxidative mitochondrial metabolism (OXPHOS) in cancer cells. *Cell Cycle* **2012**, *11*, 2545–2556. [CrossRef] [PubMed]

92. Crighton, D.; Wilkinson, S.; O'Prey, J.; Syed, N.; Smith, P.; Harrison, P.R.; Gasco, M.; Garrone, O.; Crook, T.; Ryan, K.M. DRAM, a p53-induced modulator of autophagy, is critical for apoptosis. *Cell* **2006**, *126*, 121–134. [CrossRef] [PubMed]

93. He, C.; Bassik, M.C.; Moresi, V.; Sun, K.; Wei, Y.; Zou, Z.; An, Z.; Loh, J.; Fisher, J.; Sun, Q.; et al. Exercise-induced BCL2-regulated autophagy is required for muscle glucose homeostasis. *Nature* **2012**, *481*, 511–515. [CrossRef] [PubMed]

94. Liu, X.; Wang, X.; Zhang, J.; Lam, E.K.; Shin, V.Y.; Cheng, A.S.; Yu, J.; Chan, F.K.; Sung, J.J.; Jin, H.C. Warburg effect revisited: An epigenetic link between glycolysis and gastric carcinogenesis. *Oncogene* **2010**, *29*, 442–450. [CrossRef] [PubMed]

95. Wang, X.; Jin, H. The epigenetic basis of the Warburg effect. *Epigenetics* **2010**, *5*, 566–568. [CrossRef] [PubMed]

96. Andreasen, A.S.; Kelly, M.; Berg, R.M.; Møller, K.; Pedersen, B.K. Type 2 diabetes is associated with altered NF-κB DNA binding activity, JNK phosphorylation, and AMPK phosphorylation in skeletal muscle after LPS. *PLoS ONE* **2011**, *6*, e23999. [CrossRef] [PubMed]

97. Ruan, H.; Pownall, H.J. The adipocyte IKK/NF-κB pathway: A therapeutic target for insulin resistance. *Curr. Opin. Investig. Drugs* **2009**, *10*, 346–352. [PubMed]

98. Volkmar, M.; Dedeurwaerder, S.; Cunha, D.A.; Ndlovu, M.N.; Defrance, M.; Deplus, R.; Calonne, E.; Volkmar, U.; Igoillo-Esteve, M.; Naamane, N.; et al. DNA methylation profiling identifies epigenetic dysregulation in pancreatic islets from type 2 diabetic patients. *EMBO J.* **2012**, *31*, 1405–1426. [CrossRef] [PubMed]

99. Acharya, M.M.; Baddour, A.A.; Kawashita, T.; Allen, B.D.; Syage, A.R.; Nguyen, T.H.; Yoon, N.; Giedzinski, E.; Yu, L.; Parihar, V.K.; et al. Epigenetic determinants of space radiation-induced cognitive dysfunction. *Sci. Rep.* **2017**, *7*, 42885. [CrossRef] [PubMed]

100. Lujambio, A.; Lowe, S.W. The microcosmos of cancer. *Nature* **2012**, *482*, 347–355. [CrossRef] [PubMed]

101. Gao, P.; Tchernyshyov, I.; Chang, T.C.; Lee, Y.S.; Kita, K.; Ochi, T.; Zeller, K.I.; De Marzo, A.M.; Van Eyk, J.E.; Mendell, J.T.; et al. c-Myc suppression of miR-23a/b enhances mitochondrial glutaminase expression and glutamine metabolism. *Nature* **2009**, *458*, 762–765. [CrossRef] [PubMed]

102. Ferland-McCollough, D.; Ozanne, S.E.; Siddle, K.; Willis, A.E.; Bushell, M. The involvement of microRNAs in Type 2 diabetes. *Biochem. Soc. Trans.* **2010**, *38*, 1565–1570. [CrossRef] [PubMed]

103. Zhao, S.; Liu, H.; Liu, Y.; Wu, J.; Wang, C.; Hou, X.; Chen, X.; Yang, G.; Zhao, L.; Che, H.; et al. miR-143 inhibits glycolysis and depletes stemness of glioblastoma stem-like cells. *Cancer Lett.* **2013**, *333*, 253–260. [CrossRef] [PubMed]

104. Fong, L.Y.; Jing, R.; Smalley, K.J.; Taccioli, C.; Fahrmann, J.; Barupal, D.K.; Alder, H.; Farber, J.L.; Fiehn, O.; Croce, C.M. Integration of metabolomics, transcriptomics, and microRNA expression profiling reveals a miR-143-HK2-glucose network underlying zinc-deficiency-associated esophageal neoplasia. *Oncotarget* **2017**, *8*, 81910–81925. [CrossRef] [PubMed]

105. Jordan, S.D.; Krüger, M.; Willmes, D.M.; Redemann, N.; Wunderlich, F.T.; Brönneke, H.S.; Merkwirth, C.; Kashkar, H.; Olkkonen, V.M.; Böttger, T.; et al. Obesity-induced overexpression of miRNA-143 inhibits insulin-stimulated AKT activation and impairs glucose metabolism. *Nat. Cell Biol.* **2011**, *13*, 434–446. [CrossRef] [PubMed]

106. Sorenson, J.R. Cu, Fe, Mn, and Zn chelates offer a medicinal chemistry approach to overcoming radiation injury. *Curr. Med. Chem.* **2002**, *9*, 639–662. [CrossRef] [PubMed]

107. Bergouignan, A.; Stein, T.P.; Habold, C.; Coxam, V.; O' Gorman, D.; Blanc, S. Towards human exploration of space: The THESEUS review series on nutrition and metabolism research priorities. *NPJ Microgravity* **2016**, *2*, 16029. [CrossRef] [PubMed]

108. Gaster, M.; Nehlin, J.O.; Minet, A.D. Impaired TCA cycle flux in mitochondria in skeletal muscle from type 2 diabetic subjects: Marker or maker of the diabetic phenotype. *Arch. Physiol. Biochem.* **2012**, *118*, 156–189. [CrossRef] [PubMed]

109. Wang, Y.; Michikawa, Y.; Mallidis, C.; Bai, Y.; Woodhouse, L.; Yarasheski, K.E.; Miller, C.A.; Askanas, V.; Engel, W.K.; Bhasin, S.; et al. Muscle-specific mutations accumulate with aging in critical human mtDNA control sites for replication. *Proc. Natl. Acad. Sci. USA* **2001**, *98*, 4022–4027. [CrossRef] [PubMed]

110. Guarente, L. Sirtuins, Aging, and Medicine. *N. Engl. J. Med.* **2011**, *364*, 2235–2244. [CrossRef] [PubMed]

111. Cha, Y.; Han, M.J.; Cha, H.J.; Zoldan, J.; Burkart, A.; Jung, J.H.; Jang, Y.; Kim, C.H.; Jeong, H.C.; Kim, B.G.; et al. Metabolic control of primed human pluripotent stem cell fate and function by the miR-200c-SIRT2 axis. *Nat. Cell Biol.* **2017**, *19*, 445–456. [CrossRef] [PubMed]

112. Kwon, S.; Seok, S.; Yau, P.; Li, X.; Kemper, B.; Kemper, J.K. Obesity and aging diminish sirtuin 1 (SIRT1)-mediated deacetylation of SIRT3, leading to hyperacetylation and decreased activity and stability of SIRT3. *J. Biol. Chem.* **2017**, *292*, 17312–17323. [CrossRef] [PubMed]

113. González-Rodríguez, Á.; Más-Gutierrez, J.A.; Mirasierra, M.; Fernandez-Pérez, A.; Lee, Y.J.; Ko, H.J.; Kim, J.K.; Romanos, E.; Carrascosa, J.M.; Ros, M.; et al. Essential role of protein tyrosine phosphatase 1B in obesity-induced inflammation and peripheral insulin resistance during aging: Role of PTP1B in aging-related insulin resistance. *Aging Cell* **2012**, *11*, 284–296. [CrossRef] [PubMed]

114. Tanaka, Y.; Tran, P.O.T.; Harmon, J.; Robertson, R.P. A role for glutathione peroxidase in protecting pancreatic cells against oxidative stress in a model of glucose toxicity. *Proc. Natl. Acad. Sci. USA* **2002**, *99*, 12363–12368. [CrossRef] [PubMed]

115. Sahin, E.; DePinho, R.A. Axis of ageing: Telomeres, p53 and mitochondria. *Nat. Rev. Mol. Cell Biol.* **2012**, *13*, 397–404. [CrossRef] [PubMed]

116. Tavi, P.; Korhonen, T.; Hänninen, S.L.; Bruton, J.D.; Lööf, S.; Simon, A.; Westerblad, H. Myogenic skeletal muscle satellite cells communicate by tunnelling nanotubes. *J. Cell. Physiol.* **2010**, *223*, 376–383. [CrossRef] [PubMed]

117. Tarnopolsky, M.A. Mitochondrial DNA shifting in older adults following resistance exercise training. *Appl. Physiol. Nutr. Metab.* **2009**, *34*, 348–354. [CrossRef] [PubMed]

118. Jash, S.; Adhya, S. Induction of muscle regeneration by RNA-mediated mitochondrial restoration. *FASEB J.* **2012**, *26*, 4187–4197. [CrossRef] [PubMed]

119. Islam, M.N.; Das, S.R.; Emin, M.T.; Wei, M.; Sun, L.; Westphalen, K.; Rowlands, D.J.; Quadri, S.K.; Bhattacharya, S.; Bhattacharya, J. Mitochondrial transfer from bone-marrow-derived stromal cells to pulmonary alveoli protects against acute lung injury. *Nat. Med.* **2012**, *18*, 759–765. [CrossRef] [PubMed]

120. Zhang, B.; Davidson, M.M.; Hei, T.K. Mitochondria regulate DNA damage and genomic instability induced by high LET radiation. *Life Sci. Space Res.* **2014**, *1*, 80–88. [CrossRef] [PubMed]

121. López-Lázaro, M. Dual role of hydrogen peroxide in cancer: Possible relevance to cancer chemoprevention and therapy. *Cancer Lett.* **2007**, *252*, 1–8. [CrossRef] [PubMed]

122. Martinez-Outschoorn, U.E.; Curry, J.M.; Ko, Y.H.; Lin, Z.; Tuluc, M.; Cognetti, D.; Birbe, R.C.; Pribitkin, E.; Bombonati, A.; Pestell, R.G.; et al. Oncogenes and inflammation rewire host energy metabolism in the tumor microenvironment: RAS and NFκB target stromal MCT4. *Cell Cycle* **2013**, *12*, 2580–2597. [CrossRef] [PubMed]

123. Gordon, N.; Skinner, A.M.; Pommier, R.F.; Schillace, R.V.; O'Neill, S.; Peckham, J.L.; Muller, P.; Condron, M.E.; Donovan, C.; Naik, A.; et al. Gene expression signatures of breast cancer stem and progenitor cells do not exhibit features of Warburg metabolism. *Stem Cell Res. Ther.* **2015**, *6*, 157. [CrossRef] [PubMed]

124. Xu, X.D.; Shao, S.X.; Jiang, H.P.; Cao, Y.W.; Wang, Y.H.; Yang, X.C.; Wang, Y.L.; Wang, X.S.; Niu, H.T. Warburg effect or reverse Warburg effect? A review of cancer metabolism. *Oncol. Res. Treat.* **2015**, *38*, 117–122. [CrossRef] [PubMed]

125. Mitsuishi, Y.; Taguchi, K.; Kawatani, Y.; Shibata, T.; Nukiwa, T.; Aburatani, H.; Yamamoto, M.; Motohashi, H. Nrf2 redirects glucose and glutamine into anabolic pathways in metabolic reprogramming. *Cancer Cell* **2012**, *22*, 66–79. [CrossRef] [PubMed]

126. Manda, G.; Isvoranu, G.; Comanescu, M.V.; Manea, A.; Debelec Butuner, B.; Korkmaz, K.S. The redox biology network in cancer pathophysiology and therapeutics. *Redox Biol.* **2015**, *5*, 347–357. [CrossRef] [PubMed]

127. Knatko, E.V.; Higgins, M.; Fahey, J.W.; Dinkova-Kostova, A.T. Loss of Nrf2 abrogates the protective effect of Keap1 downregulation in a preclinical model of cutaneous squamous cell carcinoma. *Sci. Rep.* **2016**, *6*, 25804. [CrossRef] [PubMed]

128. Toth, R.K.; Warfel, N.A. Strange Bedfellows: Nuclear Factor, Erythroid 2-Like 2 (Nrf2) and Hypoxia-Inducible Factor 1 (HIF-1) in Tumor Hypoxia. *Antioxidants* **2017**, *6*, 27. [CrossRef] [PubMed]

129. Jayakumar, S.; Pal, D.; Sandur, S.K. Nrf2 facilitates repair of radiation induced DNA damage through homologous recombination repair pathway in a ROS independent manner in cancer cells. *Mutat. Res.* **2015**, *779*, 33–45. [CrossRef] [PubMed]

130. Chartoumpekis, D.V.; Wakabayashi, N.; Kensler, T.W. Keap1/Nrf2 pathway in the frontiers of cancer and non-cancer cell metabolism. *Biochem. Soc. Trans.* **2015**, *43*, 639–644. [CrossRef] [PubMed]

131. Indo, H.P.; Majima, H.J.; Terada, M.; Suenaga, S.; Tomita, K.; Yamada, S.; Higashibata, A.; Ishioka, N.; Kanekura, T.; Nonaka, I.; et al. Changes in mitochondrial homeostasis and redox status in astronauts following long stays in space. *Sci. Rep.* **2016**, *6*, 39015. [CrossRef] [PubMed]

132. Peng, M.; Yin, N.; Chhangawala, S.; Xu, K.; Leslie, C.S.; Li, M.O. Aerobic glycolysis promotes T helper 1 cell differentiation through an epigenetic mechanism. *Science* **2016**, *354*, 481–484. [CrossRef] [PubMed]

133. Chen, Z.; Liu, M.; Li, L.; Chen, L. Involvement of the Warburg effect in non-tumor diseases processes. *J. Cell. Physiol.* **2017**. [CrossRef] [PubMed]
134. Pavlides, S.; Tsirigos, A.; Vera, I.; Flomenberg, N.; Frank, P.G.; Casimiro, M.C.; Wang, C.; Pestell, R.G.; Martinez-Outschoorn, U.E.; Howell, A.; et al. Transcriptional evidence for the "Reverse Warburg Effect" in human breast cancer tumor stroma and metastasis: Similarities with oxidative stress, inflammation, Alzheimer's disease, and "Neuron-Glia Metabolic Coupling". *Aging* **2010**, *2*, 185–199. [CrossRef] [PubMed]
135. Grimm, A.; Friedland, K.; Eckert, A. Mitochondrial dysfunction: The missing link between aging and sporadic Alzheimer's disease. *Biogerontology* **2016**, *17*, 281–296. [CrossRef] [PubMed]
136. Park, S.J.; Gavrilova, O.; Brown, A.L.; Soto, J.E.; Bremner, S.; Kim, J.; Xu, X.; Yang, S.; Um, J.H.; Koch, L.G.; et al. DNA-PK promotes the mitochondrial, metabolic, and physical decline that occurs during aging. *Cell Metab.* **2017**, *25*, 1135–1146. [CrossRef] [PubMed]
137. Herrero-Mendez, A.; Almeida, A.; Fernández, E.; Maestre, C.; Moncada, S.; Bolaños, J.P. The bioenergetic and antioxidant status of neurons is controlled by continuous degradation of a key glycolytic enzyme by APC/C-Cdh1. *Nat. Cell Biol.* **2009**, *11*, 747–752. [CrossRef] [PubMed]
138. Neves, A.; Costalat, R.; Pellerin, L. Determinants of brain cell metabolic phenotypes and energy substrate utilization unraveled with a modeling approach. *PLoS Comput. Biol.* **2012**, *8*, e1002686. [CrossRef] [PubMed]
139. Atlante, A.; de Bari, L.; Bobba, A.; Amadoro, G. A disease with a sweet tooth: Exploring the Warburg effect in Alzheimer's disease. *Biogerontology* **2017**, *18*, 301–319. [CrossRef] [PubMed]
140. Karran, E.; de Strooper, B. The amyloid cascade hypothesis: Are we poised for success or failure. *J. Neurochem.* **2016**, *139*, 237–252. [CrossRef] [PubMed]
141. Gunderman, R.B.; Gonda, A.S. Radium girls. *Radiology* **2015**, *274*, 314–318. [CrossRef] [PubMed]
142. Setlow, R.B. The hazards of space travel. *EMBO Rep.* **2003**, *4*, 1013–1016. [CrossRef] [PubMed]
143. Grimm, D.; Grosse, J.; Wehland, M.; Mann, V.; Reseland, J.E.; Sundaresan, A.; Corydon, T.J. The impact of microgravity on bone in humans. *Bone* **2016**, *87*, 44–56. [CrossRef] [PubMed]
144. Mukhopadhyay, S.; Saha, R.; Palanisamy, A.; Ghosh, M.; Biswas, A.; Roy, S.; Pal, A.; Sarkar, K.; Bagh, S. A systems biology pipeline identifies new immune and disease related molecular signatures and networks in human cells during microgravity exposure. *Sci. Rep.* **2016**, *6*, 25975. [CrossRef] [PubMed]
145. Cherry, J.D.; Liu, B.; Frost, J.L.; Lemere, C.A.; Williams, J.P.; Olschowka, J.A.; O'Banion, M.K. Galactic cosmic radiation leads to cognitive impairment and increased aβ plaque accumulation in a mouse model of Alzheimer's disease. *PLoS ONE* **2012**, *7*, e53275. [CrossRef] [PubMed]
146. Pani, G.; Verslegers, M.; Quintens, R.; Samari, N.; de Saint-Georges, L.; van Oostveldt, P.; Baatout, S.; Benotmane, M.A. Combined exposure to simulated microgravity and acute or chronic radiation reduces neuronal network integrity and survival. *PLoS ONE* **2016**, *11*, e0155260. [CrossRef] [PubMed]
147. Pecaut, M.J.; Mao, X.W.; Bellinger, D.L.; Jonscher, K.R.; Stodieck, L.S.; Ferguson, V.L.; Bateman, T.A.; Mohney, R.P.; Gridley, D.S. Is spaceflight-induced immune dysfunction linked to systemic changes in metabolism. *PLoS ONE* **2017**, *12*, e0174174. [CrossRef] [PubMed]
148. Gohil, V.M.; Sheth, S.A.; Nilsson, R.; Wojtovich, A.P.; Lee, J.H.; Perocchi, F.; Chen, W.; Clish, C.B.; Ayata, C.; Brookes, P.S.; et al. Nutrient-sensitized screening for drugs that shift energy metabolism from mitochondrial respiration to glycolysis. *Nat. Biotechnol.* **2010**, *28*, 249–255. [CrossRef] [PubMed]
149. Wang, T.J.; Larson, M.G.; Vasan, R.S.; Cheng, S.; Rhee, E.P.; McCabe, E.; Lewis, G.D.; Fox, C.S.; Jacques, P.F.; Fernandez, C.; et al. Metabolite profiles and the risk of developing diabetes. *Nat. Med.* **2011**, *17*, 448–453. [CrossRef] [PubMed]
150. Zhou, W.; Capello, M.; Fredolini, C.; Racanicchi, L.; Piemonti, L.; Liotta, L.A.; Novelli, F.; Petricoin, E.F. Proteomic analysis reveals Warburg effect and anomalous metabolism of glutamine in pancreatic cancer cells. *J. Proteome Res.* **2012**, *11*, 554–563. [CrossRef] [PubMed]
151. Spinazzi, M.; Casarin, A.; Pertegato, V.; Salviati, L.; Angelini, C. Assessment of mitochondrial respiratory chain enzymatic activities on tissues and cultured cells. *Nat. Protoc.* **2012**, *7*, 1235–1246. [CrossRef] [PubMed]
152. Pfeffer, G.; Majamaa, K.; Turnbull, D.M.; Thorburn, D.; Chinnery, P.F. Treatment for mitochondrial disorders. *Cochrane Database Syst. Rev.* **2012**, CD004426. [CrossRef]
153. Ran, F.; An, L.; Fan, Y.; Hang, H.; Wang, S. Simulated microgravity potentiates generation of reactive oxygen species in cells. *Biophys. Rep.* **2016**, *2*, 100–105. [CrossRef] [PubMed]

154. Dang, B.; Yang, Y.; Zhang, E.; Li, W.; Mi, X.; Meng, Y.; Yan, S.; Wang, Z.; Wei, W.; Shao, C.; et al. Simulated microgravity increases heavy ion radiation-induced apoptosis in human B lymphoblasts. *Life Sci.* **2014**, *97*, 123–128. [CrossRef] [PubMed]

155. Held, K.D.; Blakely, E.A.; Story, M.D.; Lowenstein, D.I. Use of the NASA space radiation laboratory at Brookhaven national laboratory to conduct charged particle radiobiology studies relevant to ion therapy. *Radiat. Res.* **2016**, *185*, 563–567. [CrossRef] [PubMed]

156. Ikeda, H.; Souda, H.; Puspitasari, A.; Held, K.D.; Hidema, J.; Nikawa, T.; Yoshida, Y.; Kanai, T.; Takahashi, A. Development and performance evaluation of a three-dimensional clinostat synchronized heavy-ion irradiation system. *Life Sci. Space Res.* **2017**, *12*, 51–60. [CrossRef] [PubMed]

157. Hidding, B.; Karger, O.; Königstein, T.; Pretzler, G.; Manahan, G.G.; McKenna, P.; Gray, R.; Wilson, R.; Wiggins, S.M.; Welsh, G.H.; et al. Laser-plasma-based space radiation reproduction in the laboratory. *Sci. Rep.* **2017**, *7*, 42354. [CrossRef] [PubMed]

158. Popovici, M.A.; Mitu, I.O.; Căta-Danil, G.; Negoiṭ Ă, F.; Ivan, C. Shielding assessment of high field (QED) experiments at the ELI-NP 10 PW laser system. *J. Radiol. Prot.* **2017**, *37*, 176–188. [CrossRef] [PubMed]

159. Ghezzi, P.; Floridi, L.; Boraschi, D.; Cuadrado, A.; Manda, G.; Levic, S.; D'Acquisto, F.; Hamilton, A.; Athersuch, T.J.; Selley, L. Oxidative stress and inflammation induced by environmental and psychological stressors: A biomarker perspective. *Antioxid. Redox Signal.* **2017**. [CrossRef] [PubMed]

160. Neveu, M.A.; De Preter, G.; Marchand, V.; Bol, A.; Brender, J.R.; Saito, K.; Kishimoto, S.; Porporato, P.E.; Sonveaux, P.; Grégoire, V.; et al. Multimodality imaging identifies distinct metabolic profiles in vitro and in vivo. *Neoplasia* **2016**, *18*, 742–752. [CrossRef] [PubMed]

161. Lam, C.W. Mutation not universally linked with diabetes. *Nature* **2002**, *416*, 677. [CrossRef] [PubMed]

162. Zhang, Y.; Moreno-Villanueva, M.; Krieger, S.; Ramesh, G.T.; Neelam, S.; Wu, H. Transcriptomics, NF-κB pathway, and their potential spaceflight-related health consequences. *Int. J. Mol. Sci.* **2017**, *18*, 1166. [CrossRef] [PubMed]

163. Koukourakis, M.I.; Giatromanolaki, A.; Zois, C.E.; Kalamida, D.; Pouliliou, S.; Karagounis, I.V.; Yeh, T.; Abboud, M.I.; Claridge, T.D.W.; Schofield, C.J.; et al. Normal tissue radioprotection by amifostine via Warburg-type effects. *Sci. Rep.* **2016**, *6*, 30986. [CrossRef] [PubMed]

164. Hellweg, C.E.; Spitta, L.F.; Henschenmacher, B.; Diegeler, S.; Baumstark-Khan, C. Transcription Factors in the Cellular Response to Charged Particle Exposure. *Front. Oncol.* **2016**, *6*, 61. [CrossRef] [PubMed]

165. Yi, L.; Liu, W.; Wang, Z.; Ren, D.; Peng, W. Characterizing Alzheimer's disease through metabolomics and investigating anti-Alzheimer's disease effects of natural products. *Ann. N. Y. Acad. Sci.* **2017**, *1398*, 130–141. [CrossRef] [PubMed]

166. Schmidt, M.A.; Goodwin, T.J.; Pelligra, R. Incorporation of omics analyses into artificial gravity research for space exploration countermeasure development. *Metabolomics* **2016**, *12*, 36. [CrossRef] [PubMed]

International Journal of
Molecular Sciences

MDPI

Article

Neuroprotection by Caffeine in Hyperoxia-Induced Neonatal Brain Injury

Stefanie Endesfelder [1,*], Ulrike Weichelt [2], Evelyn Strauß [1], Anja Schlör [3], Marco Sifringer [4], Till Scheuer [1], Christoph Bührer [1] and Thomas Schmitz [1]

[1] Department of Neonatology, Charité, Universitätsmedizin Berlin, 13353 Berlin, Germany; evelyn.strauss@charite.de (E.S.); till.scheuer@charite.de (T.S.); christoph.buehrer@charite.de (C.B.); thomas.schmitz@charite.de (T.S.)
[2] Department of Physiology, Charité, Universitätsmedizin Berlin, 10117 Berlin, Germany; ulrike.weichelt@charite.de
[3] Department of Biochemistry and Biology, University of Potsdam, 14476 Potsdam, Germany; anja.schloer@uni-potsdam.de
[4] Department of Anesthesiology and Intensive Care Medicine, Charité, Universitätsmedizin Berlin, 13353 Berlin, Germany; marco.sifringer@charite.de
* Correspondence: stefanie.endesfelder@charite.de; Tel.: +49-30-450-559548; Fax: +49-30-450-559979

Academic Editors: Melpo Christofidou-Solomidou and Thomas J. Goodwin
Received: 14 November 2016; Accepted: 12 January 2017; Published: 18 January 2017

Abstract: Sequelae of prematurity triggered by oxidative stress and free radical-mediated tissue damage have coined the term "oxygen radical disease of prematurity". Caffeine, a potent free radical scavenger and adenosine receptor antagonist, reduces rates of brain damage in preterm infants. In the present study, we investigated the effects of caffeine on oxidative stress markers, anti-oxidative response, inflammation, redox-sensitive transcription factors, apoptosis, and extracellular matrix following the induction of hyperoxia in neonatal rats. The brain of a rat pups at postnatal Day 6 (P6) corresponds to that of a human fetal brain at 28–32 weeks gestation and the neonatal rat is an ideal model in which to investigate effects of oxidative stress and neuroprotection of caffeine on the developing brain. Six-day-old Wistar rats were pre-treated with caffeine and exposed to 80% oxygen for 24 and 48 h. Caffeine reduced oxidative stress marker (heme oxygenase-1, lipid peroxidation, hydrogen peroxide, and glutamate-cysteine ligase catalytic subunit (GCLC)), promoted anti-oxidative response (superoxide dismutase, peroxiredoxin 1, and sulfiredoxin 1), down-regulated pro-inflammatory cytokines, modulated redox-sensitive transcription factor expression (Nrf2/Keap1, and NFκB), reduced pro-apoptotic effectors (poly (ADP-ribose) polymerase-1 (PARP-1), apoptosis inducing factor (AIF), and caspase-3), and diminished extracellular matrix degeneration (matrix metalloproteinases (MMP) 2, and inhibitor of metalloproteinase (TIMP) 1/2). Our study affirms that caffeine is a pleiotropic neuroprotective drug in the developing brain due to its anti-oxidant, anti-inflammatory, and anti-apoptotic properties.

Keywords: anti-oxidative response; caffeine; hyperoxia; oxidative stress; preterm infants; developing brain

1. Introduction

Advances in neonatal intensive care have led to a significant increase in the survival rate of premature infants, but extremely premature infants have a higher risk of dying or suffering permanent and serious damage [1,2]. Up to 50% of surviving extremely preterm infants show cognitive deficits or behavioral problems during the later stages of development [3]. The sequelae of prematurity are described to be triggered by oxidative stress and free radical-mediated cell and tissue damage, leading to the term "oxygen radical disease of the prematurity" [4–6].

There are several reasons for the high susceptibility of preterm infants to oxidative damage: (i) birth is associated with a dramatic change of intrauterine hypoxic milieu to a relatively hyperoxic extrauterine environment, and this relative hyperoxia can be enhanced by supplemental oxygen [7,8]; (ii) premature infants are less able to cope with the oxygen-rich environment of extrauterine life because their antioxidant defense system is poorly developed [9]; and (iii) preterm infants have increased susceptibility to infections [5].

Oxidative stress can be defined as an imbalance between the amount of reactive oxygen species (ROS) and the intracellular and extracellular antioxidant protection systems. The antioxidative defense system undergoes developmental changes during the neonatal period, resulting in a relevantly lower intracellular defense in preterm infants compared to term infants [10].

In addition to the understanding of the pathology of oxidative stress and the associated effects on the development of premature infants, additional strategies must be developed. Recent studies have proposed that caffeine presents antioxidant activity and therefore, protects human against disorders associated with oxidative stress [11,12]. The methylxanthine caffeine is used as a first-line pharmacotherapy against apnoea in preterm infants [13]. Caffeine has a higher therapeutic index and a longer half-life compared to other methylxanthines. In addition, to the reduction of the frequency of apnoea, caffeine has additional short- and long-term effects [14,15]. As an adenosine receptor antagonist caffeine improves neonatal outcome, shows neuroprotective effects in the developing brain [16,17], has anti-inflammatory effects [18,19], decreases rates of bronchopulmonary dysplasia (BPD) and death [14,20], and shortens the duration of mechanical ventilation [14,21]. Side effects of caffeine have been described [22,23] to include tachycardia, higher oxygen consumption, and transient decrease of the growth rate in very low birth weight infants [24].

Up to date, it is not yet clarified whether caffeine can also act as a free radical scavenger. Shi et al. reported that caffeine may act as an antioxidant scavenger, thus explaining the observed anticarcinogenic properties of caffeine and related methylxanthine compounds [25]. Furthermore, caffeine prevented lipid peroxidation, reduced oxidative DNA damage [26], modulated oxidative stress in rat liver [27], and showed immunmodulatory effects under oxidative stress in the neonatal rat brain [16] and immune cells [28]. Due to the anti-oxidant properties per se and/or by the anti-inflammatory and anti-apoptotic effects of caffeine [16,29,30], which seem to be adenosine receptor-mediated [30,31], caffeine would be a promising pleiotropic drug. Therefore, the aim of this in vivo study in a neonatal oxidative stress model was to investigate how caffeine affected the immature rodent brain against high oxygen exposure.

2. Results

2.1. Hyperoxia Induces Oxidative Stress Which Is Counteracted by Caffeine

Thiobarbituric acid reactive substances (TBARS) were increased in brain tissue of newborn rats exposed to 24 h of hyperoxia to 180% $\pm$ 27.4% ($p < 0.01$) compared to litter control mates kept in atmospheric air (Figure 1A). This was reduced by a single dose of caffeine (77% $\pm$ 11.8%; $p < 0.001$). In control animals, caffeine did not affect TBARS levels. Changes in TBARS after 48 h hyperoxia were not significant in comparison to matched normoxic litters. Interestingly, there was also a significant decrease in TBARS in normoxic newborn rats 48 h after the single administration of caffeine (51% $\pm$ 17.2%; $p < 0.05$).

To confirm the induction of oxidative stress by exposure to hyperoxia, we analyzed hydrogen peroxide concentrations through enzyme linked immunosorbent assay (ELISA) measurement in neonatal rodent brains after hyperoxia and normoxia with and without caffeine treatment (Figure 1B). Hyperoxia leads to a highly significant increase of hydrogen peroxide after 48 h exposure duration (489% $\pm$ 106.7%; $p < 0.001$) which was blocked by caffeine (147% $\pm$ 22.6%; $p < 0.001$). Treatment with caffeine in animals kept under normoxic conditions had no effect on hydrogen peroxide concentrations.

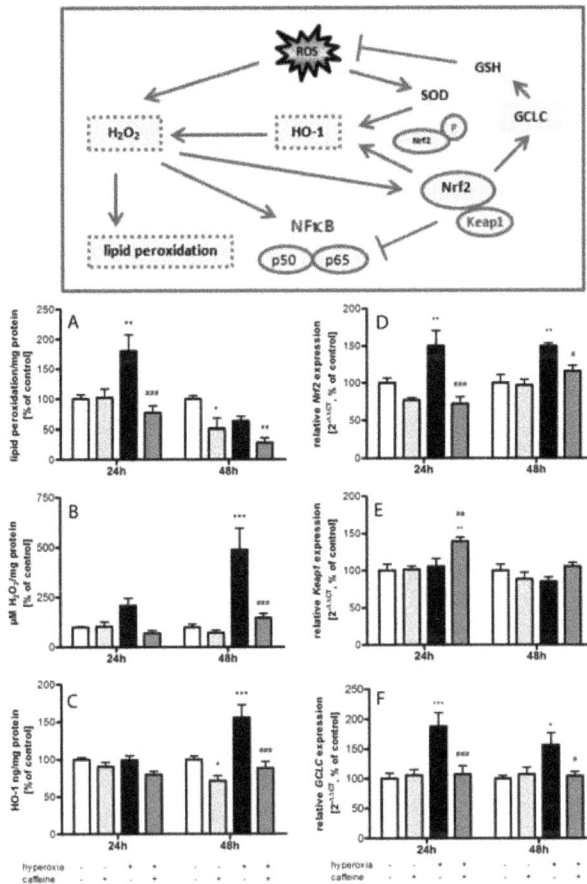

Figure 1. Caffeine reduces oxidative stress responses during exposure to hyperoxia. (**Box**) Reactive oxygen species imply hydrogen peroxide, which promotes lipid peroxidation. The antioxidant enzyme response is induced, which leads inter alia to an activation of heme oxigenase-1 (HO-1). The Nrf2-Keap1 system plays alongside the NFκB pathway an essential role in the implementation of the antioxidant gene regulation in response to oxidative stress. The catalytic subunit of glutamate-cysteine ligase (GCLC) is upregulated by Nrf2. Quantitation of brain homogenates by ELISA of: (**A**) TBARS/lipid peroxidation; (**B**) H_2O_2; and (**C**) HO-1; and mRNA expression by quantitative real-time PCR of: (**D**) *Nrf2*; (**E**) *Keap1*; and (**F**) *GCLC*. Groups are shown as normoxia (**white bars**), hyperoxia (**black bars**), with and without caffeine (**dark grey** and **light grey bars**, respectively) as mean ± SEM, n = 4–5 per group per time point. The 100% value is: (**A**) 1.268 and 1.959 μM/mg protein; (**B**) 0.887 and 2.946 μM/mg protein; (**C**) 3.604 and 2.891 ng/mg protein; (**D**) 1.008 and 1.029 C_T; (**E**) 1.014 and 1.029 C_T; and (**F**) 1.016 and 1.015 C_T for 24 and 48 h groups, respectively. Data were analyzed by two-way ANOVA with Bonferroni post hoc test, with * $p < 0.05$, ** $p < 0.01$, and *** $p < 0.001$ versus control (atmospheric air), and # $p < 0.05$, ## $p < 0.01$, and ### $p < 0.001$ versus hyperoxia (80% oxygen without caffeine).

Heme oxigenase-1 (HO-1) protein levels were markedly increased in brain tissue after 48 h of hyperoxic stimulation (156% ± 16.8%; $p < 0.001$) compared with brain tissue from rats under normoxia (Figure 1C). Caffeine reduced the protein level significantly after 48 h hyperoxia (88% ± 8.9%; $p < 0.001$). HO-1 is not increased after 24 h hyperoxia. However, caffeine treatment reduced HO-1 levels after 48 h in control litters kept in room air (71% ± 15.6%; $p < 0.05$).

NFE2-related factor 2 (*Nrf2*) (Figure 1D) was significantly induced both after 24 h (150% ± 20.3%; $p < 0.01$) and after 48 h (150% ± 3.9%; $p < 0.01$) of hyperoxia, which was blocked by caffeine (24 h to 72% ± 9.2%; $p < 0.001$, and 48 h to 116% ± 7.3%; $p < 0.05$). Contrary to *Nrf2*, gene expression of Kelch-like ECH-associated protein 1 (*Keap1*; Figure 1E) was not affected by hyperoxia alone, but was significantly elevated by caffeine after 24 h exposure to hyperoxia (140% ± 5.3%; $p < 0.01$). Caffeine did not influence the expression of *Keap1* at atmospheric air.

The glutamate-cysteine ligase (GCL) consists of two separate coded subunits, a catalytic (GCLC) and a modifier (GCLM), which catalyzes the rate-limiting phase of cellular antioxidant glutathione (GSH). *GCLC* (Figure 1F) mRNA expression was significantly induced both after 24 h (188% ± 23.0%; $p < 0.001$) and after 48 h (156% ± 20.4%; $p < 0.05$) of hyperoxia, which were significantly reduced by caffeine (24 h to 107% ± 14.4%; $p < 0.001$, and 48 h to 104% ± 7.8%; $p < 0.05$). Single treatment with caffeine in animals kept under normoxic conditions had no effect on *GCLC* expression.

2.2. Regulating Effect of Caffeine on the Imbalance of the Sulfiredoxin/Peroxiredoxin System after Hyperoxia

Acute exposure to high oxygen leads to exhaustion of superoxide dismutases 3 (*SOD3*). *SOD1* (Figure 2A) does reveal not statistically significant differences (at 24 h 66% ± 5.4% and at 48 h 72% ± 6.1%), *SOD2* (Figure 2B) to 63% ± 7.2% at 24 h, but *SOD3* (Figure 2C) was significantly reduced to 55% ± 9.6% ($p < 0.05$) at 24 h. In comparison to hyperoxia without caffeine, we found significant increases for *SOD1* to 111% ± 14.4% ($p < 0.05$) and 129% ± 18.5% ($p < 0.01$), for *SOD3* to 159% ± 13.1% ($p < 0.001$) and 171% ± 19.3% ($p < 0.001$) at 24 h and 48 h, respectively, and for *SOD2* to 217% ± 25.5% ($p < 0.001$) at 48 h. Under 48 h normoxic conditions, a single administration of caffeine leads to an increase of *SOD2* (188% ± 23.3%, $p < 0.001$) and *SOD3* (146% ± 11.2%, $p < 0.05$). Correspondingly, increasing peroxiredoxin (Prx) 1 and sulfiredoxin (Srx) 1 expression were observed at both time points. Prx1 protein expression (Figure 2D) was increased to 137% ± 2.8% ($p < 0.001$) at 24 h and to 145% ± 6.2% ($p < 0.001$) at 48 h, and Srx1 protein expression (Figure 2E) to 135% ± 4.3% ($p < 0.001$) and 150% ± 9.3% ($p < 0.001$), respectively. The administration of caffeine showed a significant decrease to 109% ± 6.2% ($p < 0.01$) and 75% ± 5.9% ($p < 0.001$) for Prx1, and to 99% ± 8.7% ($p < 0.001$) and 104% ± 9.3% ($p < 0.001$) for Srx1 to both times examined. Interestingly, there was a reduction in Prx1 at 48 h under normoxia to 56% ± 2.6% ($p < 0.001$) and an increase of protein expression after 24 h under normoxia to 126% ± 3.6% ($p < 0.01$).

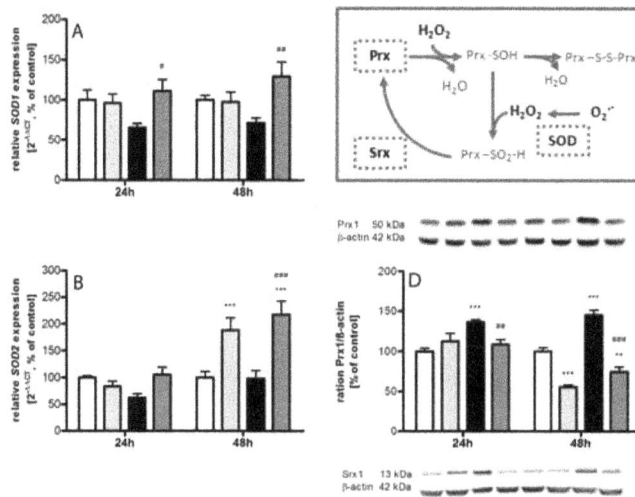

Figure 2. *Cont.*

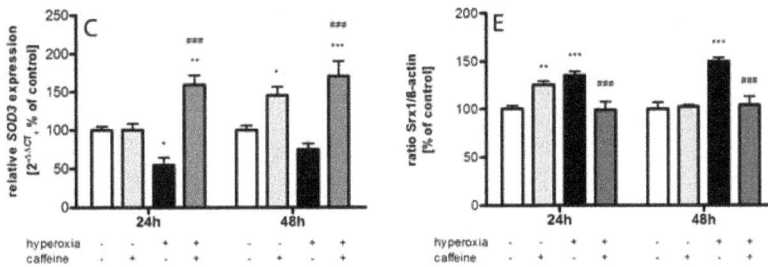

Figure 2. Caffeine modulates antioxidative enzymes. (**Box**) The high production of ROS under oxidative stress requires the existence of a set of ROS scavenger mechanisms. These are firstly, the group of superoxide dismutase (SOD), which are not only able to scavenge ROS but also repair cell damage and possibly serve as redox sensors, and secondly, the thiol-based antioxidants peroxiredoxin (Prx) as well as sulfiredoxin (Srx), which are major internal housekeeping antioxidant molecules that act as redox switches to modulate homeostasis. SOD isoform mRNA expression was analyzed in brain homogenates. Quantification of mRNA expression by quantitative real-time PCR of: (**A**) *SOD1*; (**B**) *SOD2*; and (**C**) *SOD3*; and protein expression measured by Western blot of: (**D**) Prx1; and (**E**) Srx1. Groups are shown as normoxia (**white bars**), hyperoxia (**black bars**), with and without caffeine (dark grey and light grey bars, respectively) as mean $\pm$ SEM, n = 4–5 per group per time point. The 100% value is: (**A**) 1.031 and 1.006 C_T; (**B**) 1.002 and 1.025 C_T; (**C**) 1.005 and 1.007 C_T; (**D**) 0.87 and 0.61 ratio intensity/mm^2; and (**E**) 0.39 and 0.67 ratio intensity/mm^2, for 24 h and 48 h groups, respectively. Data were analyzed by two-way ANOVA with Bonferroni post hoc test, with * $p < 0.05$, ** $p < 0.01$, and *** $p < 0.001$ versus control (atmospheric air), and # $p < 0.05$, ## $p < 0.01$, and ### $p < 0.001$ versus hyperoxia (80% oxygen without caffeine).

2.3. Effects of Caffeine on the Inflammatory Cytokine Expression

The cerebral neuro-inflammatory response in our neonatal oxidative stress model was analyzed by measuring changes in cytokine production and inducible nitric oxide synthase (iNOS) in the brain. Using qPCR for gene expression and ELISA for protein expression analysis, we determined levels of tumor necrosis factor α (TNFα), interleukin (IL)-1β, interferon γ (IFNγ), and IL-18 as pro-inflammatory cytokines, IL-12 as early pro- and late anti-inflammatory cytokine, and iNOS. Overall, significant differences between normoxia control group and the hyperoxia group were detected for all cytokines at 24 and/or 48 h (Figure 3).

Oxidative stress is usually thought to be responsible for tissue injury associated with a range of brain injury, inflammation, and degenerative processes. Moreover, inflammatory target proteins, including iNOS, are associated with oxidative stress induced by pro-inflammatory factors such as cytokines. Expression of *iNOS* (Figure 3A) was significantly increased by 24 h hyperoxia (179% $\pm$ 32.1%; $p < 0.01$), and significantly decreased at 48 h (49% $\pm$ 4.5%; $p < 0.05$), which was blocked by caffeine at 24 h (100% $\pm$ 8.0%; $p < 0.01$).

In response to brain injury by hyperoxia, TNFα production is rapidly increased. An increase of *TNFα* mRNA (127% $\pm$ 4.8%; $p < 0.05$) was found after 24 h of hyperoxia (Figure 3B). After 48 h of hyperoxia, the increase in protein expression of TNFα (Figure 3C) was largely pronounced (251% $\pm$ 50.8%; $p < 0.001$). The administration of caffeine drastically reduced *TNFα* mRNA expression under hyperoxic conditions at 24 h (31% $\pm$ 5.2%; $p < 0.001$) and at 48 h (51% $\pm$ 5.5%; $p < 0.01$), which was also observed under normoxic conditions (at 24 h 40% $\pm$ 7.6%; $p < 0.001$; at 48 h 57% $\pm$ 5.1%; $p < 0.001$). On translational protein levels, caffeine caused a marked reduction in the expression of TNFα protein at 48 h (from 251% $\pm$ 50.8% to 110% $\pm$ 21.7%; $p < 0.001$). However, TNFα protein was not affected by caffeine in newborn rats kept at atmospheric air (Figure 3C).

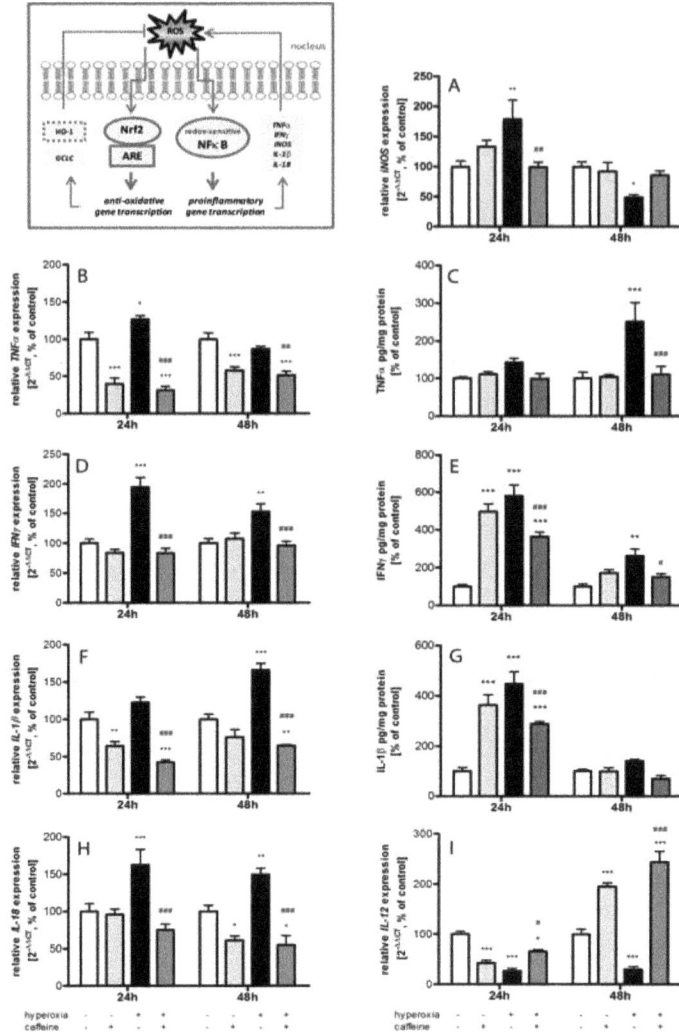

Figure 3. Changes in the expression of cytokines and iNOS in neonatal rat brains after hyperoxic injury with and without caffeine. (**Box**) Nrf2 activation promotes the expression of anti-oxidative gene transcription and reduces oxidative stress-induced inflammatory activation by blocking the redox-sensitive NFκB pathway (ARE; antioxidant response element). The protein concentration (pg/mg protein) and relative mRNA expression of cytokines were measured in brain homogenates from normoxia (**white bars**), caffeine with normoxia (**light grey bars**), hyperoxia (**black bars**), and hyperoxia with caffeine (dark grey bars) by ELISA assay and quantitative real-time PCR of: (**A**) *iNOS*; (**B,C**) TNFα; (**D,E**) IFNγ; (**F,G**) IL-1β; (**H**) *IL-18*; and (**I**) *IL-12*. Data are shown as mean ± SEM, $n = 4$–5 per group per time point. The 100% value is: (**A**) 1.016 and 1.015 C_T; (**B**) 1.021 and 1.018 C_T; (**C**) 21.22 and 29.19 pg/mg protein; (**D**) 1.012 and 1.013 C_T; (**E**) 1.701 and 2.121 pg/mg protein; (**F**) 1.021 and 1.010 C_T; (**G**) 16.58 and 8.37 pg/mg protein; (**H**) 1.026 and 1.016 C_T; and (**I**) 1.006 and 1.019 C_T for 24 and 48 h groups, respectively. Data were analyzed by two-way ANOVA with Bonferroni post hoc test, with * $p < 0.05$, ** $p < 0.01$, and *** $p < 0.001$ versus control (atmospheric air), and [#] $p < 0.05$, [##] $p < 0.01$, and [###] $p < 0.001$ versus hyperoxia (80% oxygen without caffeine).

In relation to the cytokines measured in these experiments, IFNγ showed the largest changes in mRNA and protein levels in the hyperoxic brain (Figure 3D,E). A significant increase was observed after 24 h (mRNA 194% ± 16.8%, $p < 0.001$; protein 580% ± 61.6%, $p < 0.001$) and 48 h (mRNA 153% ± 13.3%, $p < 0.01$; protein 262% ± 36.7%, $p < 0.01$) of hyperoxia exposure. A single dose of caffeine diminished this expression at both time points. At 24 and 48 h, there is a significant reduction in protein expression to 363% ± 24.1% ($p < 0.001$) and to 150% ± 18.5% ($p < 0.05$), respectively. Similarly, caffeine affects under hyperoxia *IFNγ* on the mRNA level with an attenuation to 83% ± 8.3% ($p < 0.001$) after 24 h and 96% ± 6.8% ($p < 0.001$) after 48 h. Remarkably, administration of caffeine in control animals showed a high increase of the IFNγ protein expression at 24 h (497% ± 40.8%; $p < 0.001$) compared to untreated control litters, whereas no changes were observed at the mRNA level.

As a further indication of pro-inflammatory responses, 24 h hyperoxia caused a large increase of IL-1β protein expression (448% ± 48.6%; $p < 0.001$) as compared to controls (Figure 3F,G). The application of caffeine in hyperoxic animal led to a reduction of IL-1β protein (at 24 h 289% ± 9.4%; $p < 0.001$). Caffeine administration alone under normoxia also showed a dramatic increase in the protein level after 24 h (363% ± 40.5%; $p < 0.001$). The mRNA expression significantly decreased during normoxia and caffeine administration after 24 h (64% ± 5.9%; $p < 0.01$). Hyperoxia exposure resulted in a non-significant increase (122% ± 7.4%) at 24 h, and a significant increase after 48 h hyperoxia exposure (166% ± 8.7%; $p < 0.001$). Here also, mRNA expression under hyperoxia is significantly reduced below the normoxic level by caffeine administration at 24 h (42% ± 3.2%; $p < 0.001$) and 48 h (65% ± 1.7%; $p < 0.001$).

Exposure to hyperoxia increased the concentration of *IL-18* mRNA at 24 h to 163% ± 20.6% ($p < 0.001$) and 48 h to 150% ± 8.6% ($p < 0.01$) compared with rat pups exposed to normoxia (Figure 3H). The application of caffeine significantly reduced cytokine level below hyperoxia at both time points to 72% ± 7.9% ($p < 0.001$) and to 55% ± 12.6% ($p < 0.001$), respectively. Caffeine under non-hyperoxic conditions resulted in a significant reduction in *IL-18* expression at 48 h (61% ± 5.9%; $p < 0.05$).

Interestingly, a drastic reduction in *IL-12* mRNA expression (Figure 3I; 28% ± 4.7%; $p < 0.001$, and 30% ± 5.7%; $p < 0.001$) is detected after 24 and 48 h of hyperoxia, respectively. Under hyperoxia exposure with prior application of caffeine, the expression increased significantly at 24 h at 66% ± 3.6% ($p < 0.05$) and at 48 h at 244% ± 21.5% ($p < 0.001$). In contrast, the expression of *IL-12* under normoxia with caffeine was reduced after 24 h (43% ± 5.3%; $p < 0.001$) and increased after 48 h (195% ± 7.5%; $p < 0.001$).

2.4. Hyperoxia Modulates Gene Expression of Transcription Factors and Caffeine Counteracts

As shown in Figure 4A the nuclear factor of kappa light polypeptide gene enhancer in B-cells (NFκB) protein expression was increased by caffeine under normoxia/hyperoxia at 24 h (237% ± 25.8%; $p < 0.001$, and 204% ± 36.7%; $p < 0.001$, respectively) and at 48 h (200% ± 6.0%; $p < 0.01$) in the cytosol compared to the control group. There were no changes under hyperoxic exposure alone, while the alteration in protein level in the nucleus indicated an up-regulation of NFκB under hyperoxic condition at both time points (224% ± 31.9%; $p < 0.001$, and 334% ± 36.3%; $p < 0.001$, respectively). In the nuclear protein fraction, a single dose of caffeine significantly reduced NFκB expression at 24 h to 114% ± 20.8% ($p < 0.001$), and at 48 h to 74% ± 4.1% ($p < 0.001$).

The mRNA expression ratios of *NFκB1* and *NFκB2* corresponded with the cytosolic protein data (Figure 4B,C). Here we showed that hyperoxia resulted in an mRNA reduction of both *NFκB* forms (*NFκB1* and 2) at both time points. Caffeine with hyperoxia exposure always led to an increase in mRNA expression (*NFκB1* at 24 h from 75% ± 6.5% to 154% ± 11.3% ($p < 0.001$), and at 48 h from 68% ± 6.1% to 115% ± 4.7% ($p < 0.01$); *NFκB2* at 24 h from 72% ± 4.0% to 114% ± 12.1% ($p < 0.01$), and at 48 h from 38% ± 3.5% to 79% ± 7.7% ($p < 0.01$)). *NFκB1* expression was bolstered by caffeine under normoxia after 24 h (142% ± 6.8%; $p < 0.01$).

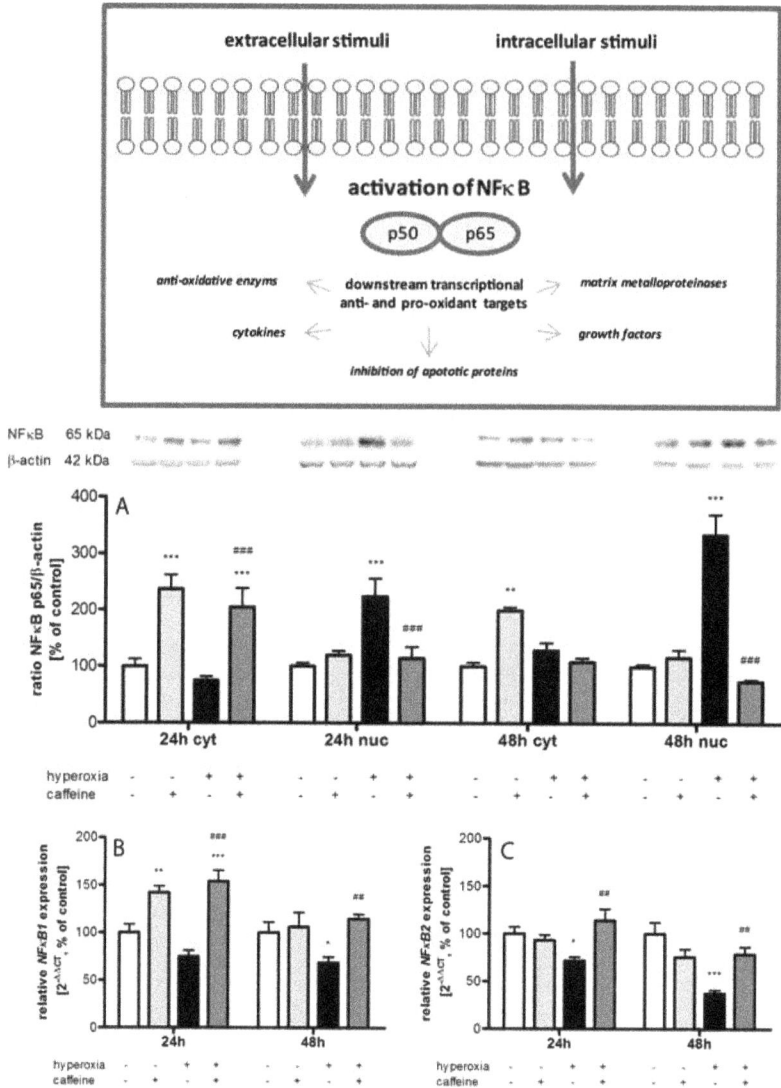

Figure 4. NFκB expression is affected by caffeine administration. (**Box**) Stimulation of the NFκB pathway is mediated by diverse signal transduction cascades. The change of protein expression of NFκB were measured in brain homogenates of normoxia (**white bars**), caffeine with normoxia (**light grey bars**), hyperoxia (**black bars**), and hyperoxia with caffeine (dark grey bars) by Western blot and quantitative real-time PCR with: (**A**) NFκB protein expression in cytosolic (cyt) and nuclear (nuc) fraction; and (**B,C**) *NFκB* mRNA expression after 24 h and 48 h of oxygen exposure. Data are shown as mean ± SEM, $n = 5$ per group per time point. The 100% value is: (**A**) (cyt) 0.035 and 0.024 ratio intensity/mm^2 and (nuc) 0.092 and 0.127 ratio intensity/mm^2; (**B**) 1.017 and 1.028 C_T; and (**C**) 1.010 and 1.032 C_T for 24 and 48 h groups, respectively. Data were analyzed by two-way ANOVA with Bonferroni post hoc test, with * $p < 0.05$, ** $p < 0.01$, and *** $p < 0.001$ versus control (atmospheric air), and ## $p < 0.01$ and ### $p < 0.001$ versus hyperoxia (80% oxygen without caffeine).

2.5. Caffeine Prevents Hyperoxia-Mediated Increase in Apoptotic Gene Expression

We measured the cytosolic and nuclear expression of cleaved poly (ADP-ribose) polymerase-1 (PARP-1) (Figure 5A). Hyperoxia increased nuclear PARP-1 at both times measured (24 h with 200% $\pm$ 24.3%; $p < 0.001$; 48 h with 188% $\pm$ 14.5%; $p < 0.001$) and no changes observed in cytosolic fraction. Caffeine reduced this increase to the normoxic level ($p < 0.001$). High oxygen also induced cytosolic apoptosis inducing factor (AIF) protein at 24 h (Figure 5B; 145% $\pm$ 4.9%; $p < 0.001$), increased caspase-3 cleavage (Figure 5C) at 24 h (143% $\pm$ 12.3%; $p < 0.001$) and at 48 h (133% $\pm$ 6.0%; $p < 0.01$), and enhanced *caspase-3* mRNA expression at 24 h (Figure 5D; 159% $\pm$ 9.0%; $p < 0.001$). For the targets investigated, the application of caffeine resulted in a significant reduction of protein and/or mRNA expression (AIF at 24 h from 145% $\pm$ 4.9% to 103% $\pm$ 4.6% ($p < 0.001$); cleaved caspase-3 at 24 h from 143% $\pm$ 12.3% to 91% $\pm$ 5.6% ($p < 0.001$), and at 48 h from 133% $\pm$ 6.0% to 109% $\pm$ 2.0% ($p < 0.05$); *caspase-3* mRNA expression at 24 h from 159% $\pm$ 9.0% to 108% $\pm$ 3.7% ($p < 0.001$), and at 48 h from to 74% $\pm$ 2.6% ($p < 0.05$)).

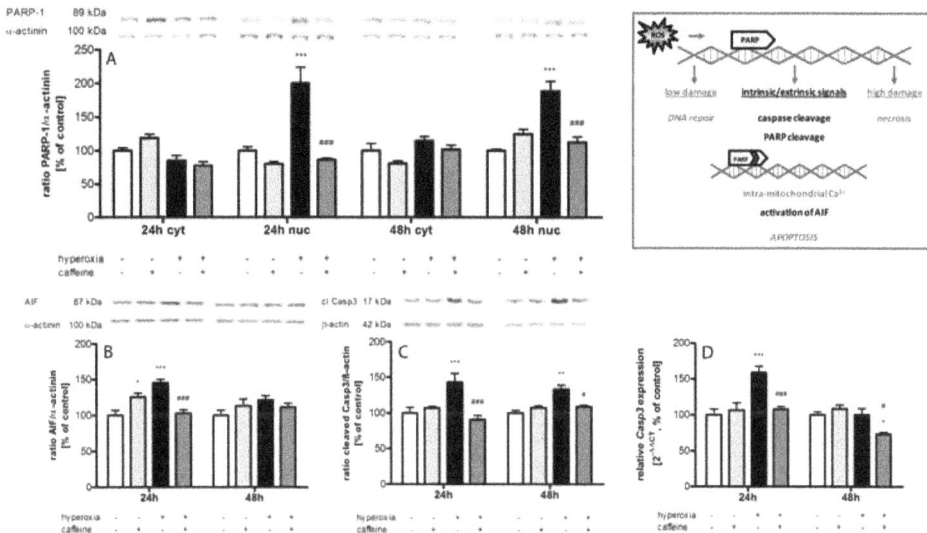

Figure 5. Caffeine inhibits hyperoxia-induced apoptotic key mediators. (**Box**) ROS-induced DNA damage activated PARP-1, an initially repair enzyme. PARP-1 is a substrate of caspase-3, a member of a highly specialized family of cysteinyl-aspartate proteases (caspases) involved in apoptosis. Upon cleavage and progression of cell death intramitochondrial calcium is released and AIF is induced. Analysis of the expression of apoptotic mediators were conducted in cytosolic (cyt) and nuclear (nuc) protein fraction and relative mRNA expression was measured in brain homogenates of normoxia (white bars), caffeine with normoxia (light grey bars), hyperoxia (black bars), and hyperoxia with caffeine (dark grey bars) by Western blot and quantitative real-time PCR with protein expression of: (**A**) cleaved PARP-1 (cyt and nuc); (**B**) AIF (cyt); and (**C**) cleaved caspase-3 (cyt); and mRNA expression of (**D**) *caspase-3*. Data are shown as mean $\pm$ SEM, $n = 5$ per group per time point. The 100% value is: (**A**) (cyt) 0.051 and 0.760 ratio intensity/mm^2 and (nuc) 0.064 and 0.043 ratio intensity/mm^2; (**B**) 0.216 and 0.309 ratio intensity/mm^2; (**C**) 0.691 and 0.823 ratio intensity/mm^2; and (**D**) 1.014 and 1.004 C$_T$ for 24 h and 48 h groups, respectively. Data were analyzed by two-way ANOVA with Bonferroni post hoc test, with * $p < 0.05$, ** $p < 0.01$, and *** $p < 0.001$ versus control (atmospheric air), and # $p < 0.05$, and ### $p < 0.001$ versus hyperoxia (80% oxygen without caffeine).

2.6. Caffeine Effects on Matrix Metalloproteinases and the tPa/Plasminogen System

Gel gelatin zymography was used to investigate the effect of oxidative stress and caffeine on the gelatinases expressions (Figure 6A). In the hyperoxic group a reduced level of pro-matrix metalloproteinases (MMP) 2 (72 kDa) was detected at 24 h (64% ± 3.1%; $p < 0.001$). After 48 h hyperoxia pro-MMP2 was increased (141% ± 2.2%; $p < 0.001$) and significantly decreased with caffeine application (79% ± 1.9%; $p < 0.001$). Caffeine alone under normoxic exposure showed a higher level on pro-MMP2 after 24 h (157% ± 5.6%; $p < 0.001$) and a lower level after 48 h (75% ± 1.2%; $p < 0.001$). At both time points acute hyperoxia exposure led to a drastic increase of active MMP2 (24 h with 281% ± 11.9%; $p < 0.001$; 48 h with 157% ± 3.7%; $p < 0.001$), which was again mitigated by caffeine (24 h to 211% ± 3.2%; $p < 0.001$; 48 h to 51% ± 1.0%; $p < 0.001$). Caffeine application at normoxia reduced the processing of active MMP2 (24 h to 76% ± 1.8%; $p < 0.001$; 48 h to 75% ± 1.2%; $p < 0.001$). Contrary to MMP2, no active MMP9 band was detected in any samples.

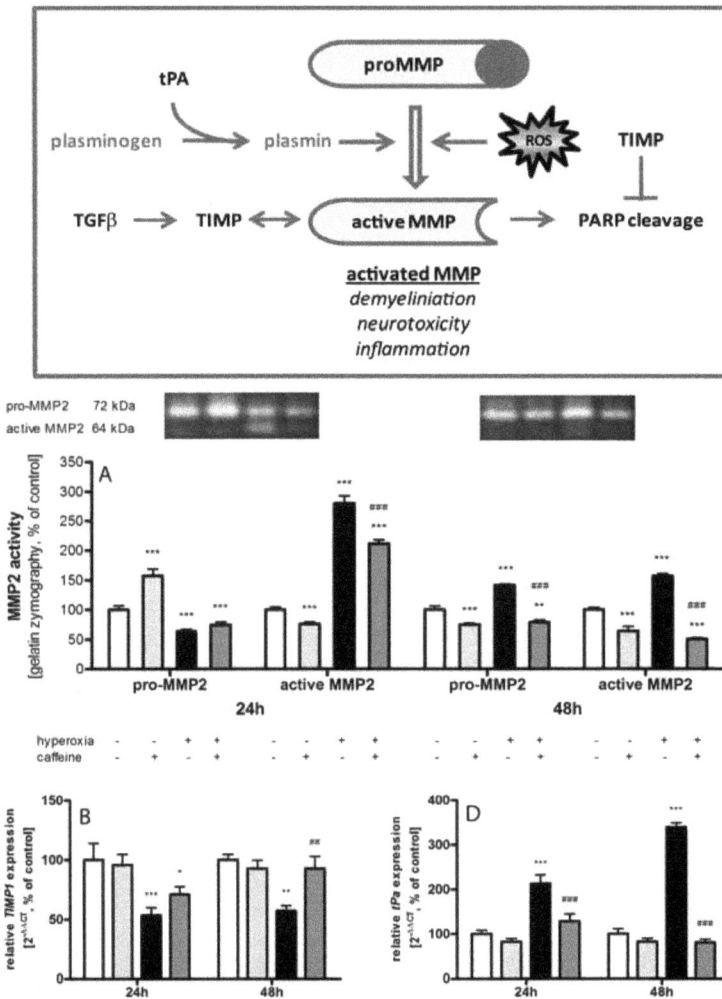

Figure 6. *Cont.*

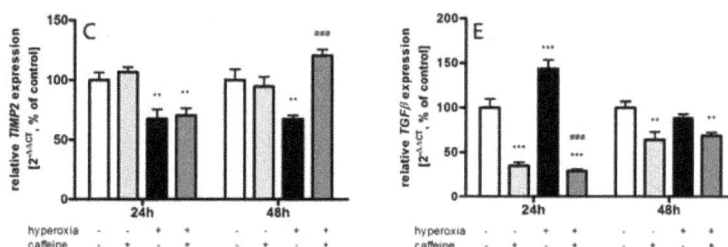

Figure 6. Hyperoxia-mediated imbalance in the MMP-TIMP system counteracted by caffeine. (**Box**) The matrix metalloproteinase (MMP) system can be activated via the plasminogen/plasmin system, and ROS. Active MMPs affect a variety of extracellular and immune regulatory proteins and are involved in modulating and degrading processes. Active MMPs can be inhibited by tissue inhibitors of metalloproteinases (TIMPs). TGF-β can enhance the expression of MMPs and TIMPs. MMP2 and MMP9 are able to cleave PARP-1. The MMP inhibitor TIMP2 is able to block PARP-1 degradation. The pro and active MMPs were analyzed with gelatin zymography and relative mRNA expression were measured in brain homogenates of normoxia (**white bars**), caffeine with normoxia (light grey bars), hyperoxia (**black bars**), and hyperoxia with caffeine (**dark grey bars**) by quantitative real-time PCR of: (**A**) pro/active MMP2; (**B**,**C**) *TIMP1/2* mRNA; (**D**) *tissue plasminogen activator (tPa)* mRNA; and (**E**) *TGF-β* mRNA. Data are shown as mean ± SEM, *n* = 4 (for gelatin zymography)-5 per group per time point. The 100% value is (**A**) (pro) 7.849×10^6 and 0.949×10^6 intensity/mm^2 and (active) 11.18×10^6 and 0.731×10^6 intensity/mm^2; (**B**) 1.043 and 1.004 C_T; (**C**) 1.008 and 1.017 C_T; (**D**) 1.013 and 1.026 C_T; and (**E**) 1.018 and 1.010 C_T for 24 and 48 h groups, respectively. Data were analyzed by two-way ANOVA with Bonferroni post hoc test, with * $p < 0.05$, ** $p < 0.01$, and *** $p < 0.001$ versus control (atmospheric air), and ## $p < 0.01$ and ### $p < 0.001$ versus hyperoxia (80% oxygen without caffeine).

The inhibitors of these metalloproteinases, especially tissue inhibitor of metalloproteinase (TIMP) 1/2 (Figure 6B,C), are down-regulated at 24 h (*TIMP1*: 54% ± 6.5%; $p < 0.001$; *TIMP2*: 68% ± 7.9%; $p < 0.01$) and 48 h (*TIMP1*: 57% ± 4.6%; $p < 0.01$; *TIMP2*: 67% ± 3.2%; $p < 0.01$). Caffeine increased the mRNA expression of both inhibitors only at 48 h (*TIMP1*: 93% ± 10.2%; $p < 0.01$; *TIMP2*: 121% ± 5.2%; $p < 0.001$). Caffeine alone had no significant influence. The tissue plasminogen activator (tPa) generates the active protease, plasmin, which is capable of degrading numerous substrates. Here (Figure 6D), hyperoxia led to a strong increase of *tPa* mRNA expression (24 h to 213% ± 19.9%; $p < 0.001$; 48 h to 339% ± 9.6%; $p < 0.001$) and a single dose of caffeine reduced *tPa* to normoxia level (24 h to 129% ± 16.6%; $p < 0.001$; 48 h to 80% ± 7.5%; $p < 0.001$).

The gene expression of *transforming growth factor-(TGF)-β* (Figure 6E) is highly increased after 24 h (143% ± 10.4%; $p < 0.001$) of hyperoxia. Caffeine strongly reduced *TGF-β* mRNA under normoxia (24 h to 35% ± 3.8%; $p < 0.001$; 48 h to 64% ± 8.8%; $p < 0.01$) as compared to control, as well as hyperoxic conditions compared to the hyperoxic group without caffeine (24 h to 29% ± 1.9%; $p < 0.001$; 48 h to 69% ± 3.6%; $p < 0.01$) at both time points.

3. Discussion

Neuronal injury and neurological maldevelopment in preterm infants is commonly ascribed to perinatal infection/inflammation and to oxidative stress hitting the immature brain at a vulnerable phase of development. In our neonatal rat model, oxidative stress is induced by early exposure to high oxygen levels for 24 and 48 h, in which neural cell injury has been shown to occur in a way that is comparable to the human neonatal situation [32].

In preterm infants, caffeine is used to stimulate breathing activity and prevent the onset of apnoea [24]. Based on some clinical studies, benefits of caffeine administration for the neurological outcome of preterm infants have been suggested [13,14,21,24]. To date, the mechanisms of neuroprotection afforded by caffeine are under investigated. Our study of the neonatal oxidative stress model elucidate that a single dose of caffeine in an acute hyperoxia model effectively inhibits pro-inflammatory cytokine production and pro-apoptotic effectors, modulates anti-oxidative enzymes, and affects components of the extracellular matrix.

3.1. Oxidative Stress Response

Oxidative stress is caused by excess of free radical formation and/or an overproduction of ROS, and is a major factor of injury in the developing brain leading to neurological sequelae of preterm birth [33,34]. In our study in six-day-old hyperoxia-exposed rats, oxidative stress was found to be caused by an increase in hydrogen peroxide, a reactive oxygen metabolic product and a major regulator of oxidative stress-related processes, and of the stress response protein HO-1, which confers protection against a variety of oxidant-induced cell and tissue injury [35], leading to an increase in lipid peroxidation [36].

The increased HO-1 expression suggests an activation of protective mechanisms in response to unwanted cellular oxidative conditions [35]. Anti-oxidative enzymes, such as glutathione peroxidase, catalase, and SODs are essential to preserving cells from exposure to oxidative damage [37]. SOD converts superoxide into hydrogen peroxide and oxygen, with hydrogen peroxide being less toxic. Several studies have demonstrated increases in oxidative stress markers after exposure to hyperoxia in animals and humans [4,38]. Changes in the gene expression and enzymatic activity of SODs have been characterized in the rodent brain as a consequence of aging, oxygen, or pro-oxidant drug exposure [39]. Our results revealed a drastic oxygen-induced reduction in SOD3 transcript in newborn rat brains. These findings are contrary to previous studies which showed an oxidative stress-induced up-regulation of the SOD transcript, but SOD activity was unaffected [37].

Nrf2 is a redox-sensitive transcription factor that mediates protection against oxidative stress via the transcriptional activation of several antioxidant enzymes through the antioxidant response element (ARE). Nrf2 is negatively regulated by Keap1 thereby providing inducible antioxidant defense. Under basal conditions Nrf2 is regulated by Keap1, but under oxidative stress, the Nrf2 pathways can also be regulated independently of Keap1 [40]. The experimentally determined *Nrf2* mRNA increase confirms the induction of antioxidant genes under hyperoxic conditions while the expression of Keap1 is unchanged.

Glutathione (GSH) plays an important role for antioxidant defense and for the regulation of intracellular redox homeostasis. The rate-limiting step of cellular antioxidant GSH is catalyzed by glutamate-cysteine ligase (GCL), which consists of a catalytic (GCLC) and a modifier (GCLM) subunit [41]. GCLC is regulated by Nrf2 via NFκB pathway [42]. Hyperoxia demonstrated a significant induction of GCLC mRNA expression in the developing brain. Together with the hyperoxia-induced increase of Nrf2 expression, these results underline the interplay of antioxidant responses [43].

Peroxiredoxin (Prx) and sulfiredoxin (Srx) are important proteins of the thioredoxin family and a key regulator of genes involved in oxidant defense, such as SOD3, Prx1, Srx1, and HO-1, and redox signaling. It is known that non-physiological oxygen concentrations change the balance of the Srx/Prx system in the developing brain [44]. Prx and Srx can act as an antioxidant and Srx solely reduces over-oxidized typical 2-Cys-Prx [45]. Hyperoxic changes elicit reactions of the antioxidant enzyme system, specifically the SODs, which play an essential role in antioxidant defense [37]. Our current data confirm previous studies [44] showing under oxidative stress a significant increase of peroxide reducing protein Prx1 and of antioxidant Srx1 protein expression.

An antioxidant activity of caffeine has been assumed in recent studies in which neuroprotection was demonstrated in patients suffering from neurodegenerative disorders [11,12]. In our study using acute hyperoxia, we demonstrated a highly significant anti-oxidative property of caffeine. A single dose of caffeine resulted in first line in a reduction of the oxidative response of lipid peroxidation, a marker of oxidative stress [46], in generation of hydrogen peroxide (H_2O_2), and in the relevant context the expression of HO-1 and *Nrf2*. In the second line, caffeine improved *SOD* RNA expression and suppressed Prx1 and Srx1 protein expressions. These results point to a broad interaction of caffeine with the antioxidant defense network. The reduced *SOD* expression under hyperoxic conditions could be explained by an exhaustion of the anti-oxidant system which was successfully counteracted by caffeine. Ahotupa et al. showed an initial decrease of SOD in hyperoxic rat brains, followed by a slight induction of its activity at a later time point [47]. Altogether, the findings suggest that the antioxidant defense capacity is diminished during exposure to hyperoxia and, moreover, represents a promising target for protective pharmacological strategies.

3.2. Inflammatory Response via NFκB Pathway

Dependent on the level of generated ROS, different redox-sensitive transcription factors are activated. Moderate oxidative stress is related to Nrf2 activation and intermediate amount of ROS initiates an inflammatory response through activation of NFκB pathway [48]. Our results in the developing brain underline a connection between induced Nrf2 expression and increased protein expression of HO-1, Prx1, and Srx1. Moreover, caffeine treatment seems to modify the redox-sensitive response on the Nrf2/Keap1 pathway, since it attenuated the hyperoxia-induced increase of NFκB expression. NFκB is activated by different endogenous stimuli, as well as by oxidative stress or cytokines [49,50]. In our hyperoxic rat pups, the pro-inflammatory genes TNFα, IFNγ, *iNOS*, IL-1β, and *IL-18* were induced. An interaction of oxidative stress and inflammatory responses has previously been described [51]. The pro-inflammatory cytokine IL-18 plays an essential role in IFNγ induction [52] and is associated with cell death after hyperoxia insult [38]. Excessive NO production causes tissues damage by formation of peroxynitrite. Consequently, nitrotyrosine is being formed and lipid peroxidation gets initiated [53]. Oxidative stress may enhance expression of the inducible isoform of NOS (iNOS) through the activation of NFκB [54]. Activation of NFκB may also increase the release of pro-inflammatory cytokines which again can induce iNOS, thus resulting in enhanced production of NO. Interestingly, Hoehn et al. [55] demonstrated that hyperoxia induces up-regulation of iNOS in the immature rat brain as a possible cause of cellular damage in the immature brain [56].

In our study, caffeine significantly reduced inflammatory responses. Surprisingly, *IL-12* mRNA expression was dramatically reduced under hyperoxia, which was blocked by caffeine. IL-12 is produced by microglia [57], and pro- and anti-inflammatory effects are reported [58,59]. It has to be discussed that caffeine not only exerted beneficial actions in hyperoxic rats but also triggered cellular changes and pro-inflammatory responses in normoxia control rats [60]. Caffeine reduced lipid peroxidation not only in response to hyperoxia but also under normoxic conditions. At the same time, the transcripts of *SOD2/3* and protein expression of Srx1 and pro-inflammatory cytokines (IL-1β and IFNγ) increased after caffeine administration at normoxia. Leon-Carmona and Galano disclosed in their functional studies that caffeine may act as radical scavenger [61]. Caffeine also reduced hydrogen levels in lung cells in vitro [62]. In mice, caffeine reduces lipid peroxidation [63]. Cytokine and anti-oxidative enzyme over expression in normoxic exposure could be protective in oxidative stress models [64,65]. Chavez-Valdez and co-workers discussed a range of caffeine plasma level appropriate for patient therapy, and higher plasma levels outside of these range coincided with an increase of inflammatory cytokines in a clinical study with preterm infants [18].

3.3. Apoptosis

High levels of ROS have been shown to induce apoptosis in the immature brain after hyperoxia [32]. In our study, exposure to high oxygen induced the nuclear acting of PARP-1, an enzyme which is activated by DNA strand breaks, and the caspase-independent AIF, and increased cleavage of caspase-3. In previous studies, exposure to postnatal hyperoxia also caused an increase in apoptotic cell death [16,44,51,66]. Anti-apoptotic properties of caffeine, also previously demonstrated by our group [16], were seen in this study by modified levels of PARP-1, AIF, and cleaved caspase-3 levels. In a hypoxic ischemic model of young rats, caffeine also revealed apoptosis-inhibiting effects [67]. Attenuation of PARP-1 inhibiting capacity has also been attributed to caffeine metabolites [68]. In our study, elevated PARP-1 and AIF were also found in controls after injection of 10 mg/kg caffeine. An increase of caspase-3 positive cells was also reported in sham animals of a hypoxia-ischemia study that received 20 mg/kg caffeine [67]. It seems possible that regulation of neuronal cell survival by caffeine occurs via direct activation of the NFκB pathway [69].

3.4. Fibrinolytic and Matrix Metalloproteinases System

Free radicals and ROS seem to activate matrix metalloproteinases (MMPs), with the biological function to degrade extracellular matrix of cerebral blood vessels and neurons [70], and active MMPs may disrupt the vital blood-brain barrier. Hyperoxia led to a drastic increase of MMP2 activation, while tissue inhibitors of MMPs (TIMP)-2, decreased. Caffeine might inhibit the processing of MMP2 through inhibition of TIMP2 expression, but molecular protection might also start further upstream through reduction of *tPa* transcript expression as found in our study, thus leading to less processing of active MMP. TGF-β is a multifunctional cytokine that regulates a broad diversity of physiological and pathological processes. Plasmin releases latent forms of growth factors such as TGF-β [71]. TGF-β effects on matrix degeneration are characterized by mixed reciprocally action of MMPs and inhibition of TIMPs to reduce the MMP/TIMP balance [72], so caffeine minimizes TGF-β expression. Sifringer et al. showed in traumatic and hyperoxic brain injury models a time-dependent increase of MMP activation in correlation to *TIMP* mRNA expression, and active MMP2 and MMP9 were reduced by protective treatment with erythropoietin [73].

3.5. Caffeine and Neuroprotection

For the clinical situation, it is important to precisely define the effects of caffeine on the otherwise healthy brain in order to avoid harm in preterm infants. The increase of apoptotic factors in normoxic newborn pups coincides with higher activity of MMP2 and also with increased NFκB protein expression in these animals. Since the injection of caffeine was performed 15 min before the beginning of hyperoxia exposure, it is possible that a certain degree of cellular stimulation is caused by caffeine that functions as a protective preconditioning in those animals challenged by hyperoxia later on. Preconditioning has been characterized as a protective procedure in brain injury models involving inflammatory stimulation [74], ischemia [75], and hypoxia [76] prior to the injurious events. Ischemic preconditioning is currently under evaluation for open heart surgery in various clinical centers and studies [77,78]. It is possible that caffeine represents a drug that in addition to its anti-oxidant and anti-inflammatory properties might also be useful for protective preconditioning of the brain. For prevention of brain injury in preterm infants, it has been suggested to investigate the use of anti-oxidant compounds such as melatonin, acetylcysteine, or allopurinol [79]. The strategy to use caffeine for brain protection in preterm infants would provide great advantages of a drug being licensed for this vulnerable patient population. The pharmacological properties and safety aspects of caffeine are well known to clinicians taking care of preterm infants, and ethical issues for clinical trials or routine care administration will be minor as compared to most other drug candidates.

4. Materials and Methods

4.1. Animals and Study Design

All procedures were approved by the state animal welfare authorities (LAGeSo G-0307/09) and followed institutional guidelines. Six-day-old *Wistar rats* from time-pregnant dams were obtained from Charité-Universitätsmedizin Berlin (Germany) and randomly assigned to cages and treatment. Animals were housed under controlled temperature and light conditions with food and water ad libitum. Experimental procedure and caffeine administration were carried out as described previously [16]. The rat pups were divided into two overarching experimental groups. From Postnatal Day 6 (P6), rat pups and their dams in all experimental groups were exposed to either 80% oxygen (hyperoxia, OxyCycler BioSpherix, Lacona, NY, USA) or atmospheric air (normoxia). In relation to the exposure to oxygen, the pups were divided into further subgroups, depending on the drug administration. Neonatal rats were administered vehicle (0.9% saline) or caffeine (10 mg/kg) as single application. Caffeine (Sigma-Aldrich, Steinheim, Germany) was dissolved in sterile distilled water. Each experimental group consisted of five animals with mixed gender. All injections of drug and vehicle were given intraperitoneally (i.p.) as a fixed proportion of body weight (100 µL/10 g). Caffeine or saline were administrated once 15 min before the start of atmospheric air or oxygen exposure.

4.2. Tissue Preparation

After 24 h (P7) or 48 h (P8) of exposure, rats were transcardially perfused with normal saline (pH 7.4) under anesthesia (i.p.) of ketamine (50 mg/kg), xylazine (10 mg/kg), and acepromazine (2 mg/kg), then decapitated, the olfactory bulb and cerebellum were removed, and brain hemispheres were snap-frozen in liquid nitrogen and stored at $-80\ ^{\circ}C$.

4.3. Protein Extraction

Protein was extracted as described in [16,44]. Briefly, snap-frozen brain tissue was homogenized in RIPA buffer solution for protein extraction. The homogenate was centrifuged at $3000\times g$ (4 °C) for 10 min, the microsomal fraction was subsequently centrifuged at $17,000\times g$ (4 °C) for 20 min, and stored at $-80\ ^{\circ}C$ until further analysis. After collecting the supernatant, protein concentrations were determined using the Pierce BCA kit (Pierce/Thermo Scientific, Rockford, IL, USA) with 30 min incubation at 37 °C prior to spectrophotometry at 562 nm.

4.4. Immunoblotting

Western blotting was performed as previously described [16,66] for following proteins: apoptosis inducing factor (AIF), cleaved caspase-3 (cleaved Casp3), nuclear factor of kappa light polypeptide gene enhancer in B-cells (NFκB), poly (ADP-ribose) polymerase-1 (PARP-1), peroxiredoxin 1 (Prx1), and sulfiredoxin 1 (Srx1). Briefly, protein extracts (20 µg per sample) were denatured in Laemmli sample loading buffer at 95 °C, size-fractionated by 12.5% (for AIF, NFκB, PARP-1, and Prx1) or 15% (for cleaved casp3 and Srx1) sodium dodecyl sulfate (SDS) polyacrylamide gel electrophoresis and electro transferred in transfer buffer to a nitrocellulose membrane (0.2 µm pore, Bio-Rad, Munich, Germany), or for NFκB to a PVDF membrane (0.45 µm pore, Merck Millipore, Darmstadt, Germany). Membranes were blocked for 1 h at room temperature (Srx1 and Prx1 with 1% (*v/v*) horse serum in Tris-buffered saline /0.1% (*v/v*) Tween 20 (TBST); AIF, cleaved Casp3, NFκB, and PARP-1 with 5% (*w/v*) bovine serum albumin (Serva, Heidelberg, Germany) in TBST). Equal loading and transfer of proteins was confirmed by staining the membranes with Ponceau S solution (Fluka, Buchs, Switzerland). The membranes were incubated overnight at 4 °C with the following antibodies: rabbit polyclonal anti-AIF (67 kDa; 1:1000; Cell Signaling, Cambridge, UK), rabbit monoclonal anti-cleaved caspase-3 (17 kDa; 1:1000; Cell Signaling), rabbit polyclonal anti-NFκB p65 (~60 kDa, 1:2500 for cytosolic and 1:750 for nuclear protein fraction; Merck Millipore), rabbit monoclonal anti-PARP-1 (89 kDa; 1:1000; Cell Signaling), goat polyclonal anti-peroxiredoxin 1 (Prx1; 50 kDa dimer; 1:400; Santa Cruz Biotechnology, Heidelberg, Germany), polyclonal goat anti-sulfiredoxin 1 (Srx1; 13 kDa; 1:200; Santa Cruz Biotechnology) diluted in 0.5% (*v/v*) horse serum or 1% (*w/v*) bovine serum albumin in TBST, corresponding to the blocking solution. As reference controls, were used monoclonal mouse anti-β-actin (42 kDa; 1:10,000; Sigma-Aldrich) and rabbit polyclonal anti-α-actinin (100 kDa; 1:1000; Cell Signaling) diluted in 1% (*w/v*) bovine serum albumin in TBST. Secondary incubations were performed with horseradish peroxidase-linked polyclonal donkey anti-goat (1:3000; Dianova, Hamburg, Germany), polyclonal rabbit anti-mouse (1:1000; DAKO, Glostrup, Denmark), or polyclonal donkey anti-rabbit (1:4000; Dianova) antibodies, diluted in 1% (*w/v*) bovine serum albumin in TBST, respectively. Positive signals were visualized using enhanced chemiluminescence (ECL; Amersham Biosciences, Freiburg, Germany) and quantified using a ChemiDoc™ XRS+ system and the software Image Lab™ (Bio-Rad). Each experiment was repeated three times.

4.5. Gelatin Zymography

Snap-frozen tissue was homogenized in working buffer (150 mM NaCl, 5 mM CaCl$_2$, 0.05% (*v/v*) Brij 35, 1% (*v/v*) Triton X-100, 0.02% (*w/v*) NaN$_3$, 50 mM Tris-HCl pH 7.4) on ice, followed by centrifugation for 5 min at 12,000× *g*. The protein concentration of the supernatant was measured with the Pierce BCA kit (Pierce/Thermo Scientific). One milligram of protein was used for enzyme enrichment and purification with Gelatin Sepharose™ 4B (GE Healthcare, Uppsala, Sweden) and eluted in 150 μL working buffer containing 10% (*v/v*) dimethylsulfoxide. Protein samples mixed with an equal volume of Novex® SDS-sample buffer (Life Technologies GmbH, Darmstadt, Germany), and incubated for 5 min at room temperature prior to loading to 10% Novex® gelatin-ready zymogram gel (Life Technologies GmbH). Gels were run at 125 V for 2 h. The gels were incubated in renaturation buffer (Novex®, Life Technologies GmbH) under gentle agitation for 30 min. They were then equilibrated for 30 min in developing buffer (Novex®, Life Technologies GmbH) and incubated in fresh developing buffer at 37 °C overnight, but at least for 18 h. Gels were washed three times with deionized water under gentle agitation for 5 min and stained with Coomassie Blue G 250 (SimplyBlue SafeStain, Life Technologies GmbH) for 30 min, and washed with deionized water for 30 min. Areas of protease activity appear as clear bands against a dark blue background where the protease has digested the gelatin substrate. Quantification of inverted matrix metalloproteinases (MMP) 2 band density was carried out using a ChemiDoc™ XRS+ system and the software Quantity One® (Bio-Rad).

4.6. Thiobarbituric Acid Reactive Substances (TBARS) Assay

Concentrations of markers of lipid peroxidation thiobarbituric acid reactive substances (TBARS) was determined using the TBARS assay kit (Cayman Chemical, Ann Arbor, MI, USA) according to manufacturer's instructions. Briefly, first a mixture of sample or standard and SDS solution was prepared. A color reagent was added to the mixture and boiled for one hour. The reaction was stopped on ice for 10 min and centrifuged at 1600× *g* (4 °C) for 10 min. The supernatant was transferred on a 96 well plate and absorbance was read at 530 nm in a microplate reader. TBARS concentration as a measure for lipid peroxidation was calculated from a malondialdehyde (MDA) standard curve and normalized to the amount of total protein.

4.7. Heme Oxygenase-1 (HO-1) Assay

HO-1 concentration was analyzed in samples of brain homogenate using Rat Heme Oxygenase-1 EIA Kit (Precoated, Takara Bio Europe/SAS, Saint-Germain-en-Laye, France) according to manufacturer's instructions. Briefly, 100 μL of sample or standard was first loaded and incubated for one hour at room temperature. Then the sample solution was removed and the wells washed three times with 400 μL of PBS containing 0.1% (*v/v*) Tween 20. Afterwards, 100 μL of antibody-POD conjugate solution was added and incubated at room temperature for one hour. The sample solution was removed and ensuing washed four times with 400 μL of PBS, aspirating thoroughly between washes. Substrate solution (100 μL) was added and incubated for 15 min at room temperature. The reaction was stopped by addition of 100 μL stop solution. The absorbance was read at 450 nm in a microplate reader. HO-1 concentration was determined by comparing sample to the standard curve and values were expressed as nanograms per milligram of protein.

4.8. Hydrogen Peroxide Assay

Hydrogen peroxide (H$_2$O$_2$) was measured in brain homogenates using OxiSelect Hydrogen peroxide assay kit (Cell Biolabs Inc., San Diego, CA, USA) according to manufacturer's instructions. Briefly, 25 μL of sample or standard was first added to the microtiter plate wells and then mixed thoroughly with 250 μL aqueous working reagent. Afterwards samples were incubated on a shaker for 30 min at room temperature and the absorbance was read at 540 nm. The peroxide content in unknown samples was determined by comparison with a predetermined H$_2$O$_2$ standard curve.

4.9. Enzyme-Linked Immunosorbent Assays (ELISAs)

Tumor necrosis factor α (TNFα), interleukin-1β (IL-1β), and interferon-γ (IFN-γ) concentrations were analyzed in samples of brain homogenate using rat TNFα/TNFSF1A, rat IL-1β/IL-1F2, and rat IFN-γ DuoSet ELISA (R&D Systems GmbH, Wiesbaden-Nordenstadt, Germany) according to manufacturer's instructions. Briefly, 96-well plates were coated with 100 μL of capture antibody, incubated overnight, washed with 400 μL wash buffer and incubated for 1 h for blocking with 300 μL of reagent diluent. After washing in wash buffer, 100 μL of samples or standards, diluted in reagent diluent, were added and incubation preceded for 2 h. Samples were then washed and incubated for 2 h in the presence of 100 μL of biotinylated detection antibody at room temperature. The supernatants were aspirated, the wells washed and 100 μL of horseradish peroxidase-conjugated streptavidin was added and incubation continued for 20 min. Samples were washed and 100 μL of substrate solution, a 1:1 mixture of H_2O_2 and tetramethylbenzidine (R&D Systems GmbH), was added. Samples were incubated in the dark for 20 min and the reaction was stopped using 50 μL 2 N H_2SO_4. Plates were read at 450 nm and TNFα, IL-1β, and IFN-γ concentrations were estimated from the standard curve and expressed as picograms per milligram protein.

4.10. RNA Extraction and Real-Time PCR

Gene expression analysis was performed as previously described [16,66]. Total RNA was isolated from snap-frozen tissue by acidic phenol/chloroform extraction (peqGOLD RNAPure™; PEQLAB Biotechnologie, Erlangen, Germany) and 2 μg of RNA was reverse transcribed. The PCR products of caspase-3 (*Casp3*), glutamate-cysteine ligase catalytic subunit (*GCLC*), interferon-γ (*IFN-γ*), interleukin-1β (*IL-1β*), interleukin-12b (*IL-12b*), interleukin-18 (*IL-18*), inducible nitric oxide synthase (*iNOS*), Kelch-like ECH-associated protein 1 (*Keap1*), nuclear factor of kappa light polypeptide gene enhancer in B-cells 1 and 2 (*NFκB1* and *NFκB2*), NFE2-related factor 2 (*Nrf2*), superoxide dismutase 1, 2 and 3 (*SOD1*, *SOD2*, and *SOD3*), tissue inhibitor of metalloproteinase 1 and 2 (*TIMP1* and *TIMP2*), tumor necrosis factor α (*TNFα*), tissue plasminogen activator (*tPa*), and hypoxanthine-guanine phosphoribosyltransferase (*HPRT*) were quantified in real time, using dye-labeled fluorogenic reporter oligonucleotide probes and primers (Table 1).

Table 1. Sequences of oligonucleotides.

cDNA	Oligonucleotide Sequence 5′–3′	Accession No.
	HPRT	
forward	GGAAAGAACGTCTTGATTGTTGAA	NM_012583.2
reverse	CCAACACTTCGAGAGGTCCTTTT	
probe	CTTTCCTTGGTCAAGCAGTACAGCCCC	
	Casp3	
forward	ACAGTGGAACTGACGATGATATGG	NM_012922.2
reverse	AATAGTAACCGGGTGCGGTAGA	
probe	ATGCCAGAAGATACCAGTGG	
	GCLC	
forward	GGAGGACAACATGAGGAAACG	NM_012815.2
reverse	GCTCTGGCAGTGTGAATCCA	
probe	GAGGCTACTTCTGTATTAGG	
	IFN-γ	
forward	GCAAAAGGACGGTAACACGAA	NM_138880.2
reverse	ATGGCCTGGTTGTCTTTCAAGA	
probe	TCTCTTTCTACCTCAGACTC	

Table 1. *Cont.*

cDNA	Oligonucleotide Sequence 5′–3′	Accession No.
	IL-1β	
forward	CTCCACCTCAATGGACAGAACA	NM_031512.2
reverse	CACAGGGATTTTGTCGTTGCT	
probe	CTCCATGAGCTTTGTACAAG	
	IL-12b	
forward	TGCTGCTCCACAAGAAGGAA	NM_022611.1
reverse	TTGGTGCTTCACACTTCAGGAA	
probe	ATGGAATTTGGTCCACCGAG	
	IL-18	
forward	CGGAGCATAAATGACCAAGTTCTC	NM_019165.1
reverse	TGGGATTCGTTGGCTGTTC	
probe	TTGACAAAAGAAACCCGCCTG	
	iNOS	
forward	AGCTGTAGCACTGCATCAGAAATG	NM_012611.3
reverse	CAGTAATGGCCGACCTGATGT	
probe	CAGACACATACTTTACGCCAC	
	Keap1	
forward	GATCGGCTGCACGGAACT	NM_057152.2
reverse	GCAGTGTGACAGGTTGAAGAACTC	
probe	CTCGGGAGTATATCTACATGC	
	NFκB1	
forward	GACCCAAGGACATGGTGGTT	NM_001276711.1
reverse	TCATCCGTGCTTCCAGTGTTT	
probe	CTGGGAATACTTCACGTGAC	
	NFκB2	
forward	GCCTAAACAGCGAGGCTTCA	NM_001008349.1
reverse	TCTTCCGGCCCTTCTCACT	
probe	TTTCGATATGGCTGTGAAGG	
	Nrf2	
forward	ACTCCCAGGTTGCCCACAT	NM_031789.2
reverse	GCGACTCATGGTCATCTACAAATG	
probe	CTTTGAAGACTGTATGCAGC	
	SOD1	
forward	CAGAAGGCAAGCGGTGAAC	NM_017050.1
reverse	CCCCATATTGATGGACATGGA	
probe	TACAGGATTAACTGAAGGCG	
	SOD2	
forward	GACCTACGTGAACAATCTGAACGT	NM_017051.2
reverse	AGGCTGAAGAGCAACCTGAGTT	
probe	ACCGAGGAGAAGTACCACGA	
	SOD3	
forward	GGAGAGTCCGGTGTCGACTTAG	NM_012880.1
reverse	CTCCATCCAGATCTCCAGGTCTT	
probe	CTGGTTGAGAAGATAGGCGA	
	TGF-β	
forward	CCTGCAGAGATTCAAGTCAACTGT	NM_021578.2
reverse	GTCAGCAGCCGGTTACCAA	
probe	CAACAATTCCTGGCGTT	
	TIMP1	
forward	CGGACCTGGTTATAAGGGCTAA	NM_053819.1
reverse	CGTCGAATCCTTTGAGCATCT	
probe	AGAAATCATCGAGACCACCT	

Table 1. *Cont.*

cDNA	Oligonucleotide Sequence 5′–3′	Accession No.
	TIMP2	
forward	GGCAACCCCATCAAGAGGAT	NM_021989.2
reverse	GGGCCGTGTAGATAAATTCGAT	
probe	AGATGTTCAAAGGACCTGAC	
	tPa	
forward	TCAGAAGAGGAGCTCGGTCCTA	NM_013151.2
reverse	TGGGACGTAGCCATGACTGAT	
probe	CAGAGATGAACAGACTCAGA	
	TNFα	
forward	CCCCCAATCTGTGTCCTTCTAAC	NM_012675.2
reverse	CGTCTCGTGTGTTTCTGAGCAT	
probe	TAGAAAGGGAATTGTGGCTC	

The FAM spectral data were collected from reactions carried out in separate tubes using the same stock of cDNA to avoid spectral overlap between FAM/TAMRA and limitations of reagents. PCR and detection were performed in triplicate and repeated two times for each sample in 11 μL reaction mix, which contained 5 μL of 2× KAPA PROBE FAST qPCR Master mix (PEQLAB Biotechnologie), 2.5 μL of 1.25 μM oligonucleotide mix, 0.5 μL (0.5 μM) of probe (BioTeZ, Berlin, Germany), and 3 to 17 ng of cDNA template with *HPRT* used as an internal reference. The PCR amplification was performed in 96-well optical reaction plates for 40 cycles with each cycle at 94 °C for 15 s and 60 °C for 1 min. The expression of target genes was analyzed with the StepOnePlus real-time PCR system (Life Technologies) according to the $2^{-\Delta\Delta Ct}$ method [80].

4.11. Statistical Analyses

All data are expressed as the mean ± standard error of the mean (SEM). Differences between the control group and experimental groups (oxygen exposure and/or caffeine application) were analyzed using a two-way analysis of variance (ANOVA) with the factors hyperoxia exposure, caffeine, and their interaction. To determine differences between individual groups, each two-way ANOVA was followed by a Bonferroni post-hoc test. A two-sided *p*-value of <0.05 was considered significant. Results of the analyses were represented normalized as 100% value of control (24 h and 48 h, respectively). For the real-time PCR data, the 100% value of control represented the C_T values, while C_T was defined as the threshold cycle number of PCRs at which amplified product was first detected. For protein data (immunoblotting, ELISA, gelatin zymography), the 100% value of control represented the protein concentration or molarity (ELISA), intensity ratio of pixel size (immunoblotting), or inverted intensity of pixel size (gelatin zymography). All graphics and statistical analyses were performed using the GraphPad Prism 6.0 software for Windows (GraphPad Software, La Jolla, CA, USA).

5. Conclusions

In summary, our study reveals that single administration of caffeine in oxygen-induced brain injury leads to improved neuroprotective response. Exposure to high oxygen in the developmental brain of six-day-old rat pups resulted in increased levels of lipid peroxidation, enhanced generation of hydrogen peroxidase and heme oxygenase-1, decreased anti-oxidative response, up-regulated pro-inflammatory cytokines expression, changed balanced redox-sensitive reply, and promoted extracellular matrix degeneration and apoptotic cascade. All these consequences of hyperoxia can be reversed by caffeine treatment.

Our study suggests that caffeine is a drug that exerts neuroprotective effects on the developing brain due to its anti-oxidant, anti-inflammatory, and anti-apoptotic properties. Future research should focus on the investigation of caffeine for preconditioning and further enhance the knowledge about safety of caffeine use in the immature brain.

Acknowledgments: We thank Giang Tong (German Heart Institute, Berlin, Germany) for proofreading the manuscript. Ulrike Weichelt is a fellow of the Promotionsstipendium Charité—Universitätsmedizin, Berlin, Germany and of the Hypatia Program, University of Applied Sciences, Berlin, Germany. This work was supported by Förderverein für frühgeborene Kinder an der Charité e.V., Berlin, Germany.

Author Contributions: Stefanie Endesfelder conceived and supervised the whole project, performed all animal experiments, co-performed most of the experiments, interpreted the results and performed statistical analysis, and wrote and prepared the manuscript. Ulrike Weichelt performed and analyzed immunoblotting and quantitative reverse transcription PCR experiments. Evelyn Strauß performed and analyzed ELISA experiments. Anja Schlör performed and analyzed immunoblotting experiments for Prx1 and Srx1. Marco Sifringer and Till Scheuer performed animal experiments, provided technical help, and contributed to the critical discussion of the manuscript. Christoph Bührer interpreted the results and revised the manuscript. Thomas Schmitz interpreted the results and prepared the manuscript.

Conflicts of Interest: The authors declare no conflict of interest.

References

1. Marlow, N.; Wolke, D.; Bracewell, M.A.; Samara, M.; Group, E.S. Neurologic and developmental disability at six years of age after extremely preterm birth. *N. Engl. J. Med.* **2005**, *352*, 9–19. [CrossRef] [PubMed]
2. Wilson-Costello, D.; Friedman, H.; Minich, N.; Fanaroff, A.A.; Hack, M. Improved survival rates with increased neurodevelopmental disability for extremely low birth weight infants in the 1990s. *Pediatrics* **2005**, *115*, 997–1003. [CrossRef] [PubMed]
3. Volpe, J.J. Perinatal brain injury: From pathogenesis to neuroprotection. *Ment. Retard. Dev. Disabil. Res. Rev.* **2001**, *7*, 56–64. [CrossRef]
4. Perrone, S.; Tataranno, M.L.; Stazzoni, G.; Buonocore, G. Biomarkers of oxidative stress in fetal and neonatal diseases. *J. Mater. Fetal Neonatal Med.* **2012**, *25*, 2575–2578. [CrossRef] [PubMed]
5. Saugstad, O.D. Update on oxygen radical disease in neonatology. *Curr. Opin. Obstet. Gynecol.* **2001**, *13*, 147–153. [CrossRef] [PubMed]
6. Stone, W.L.; Shah, D.; Hollinger, S.M. Retinopathy of prematurity: An oxidative stress neonatal disease. *Front. Biosci.* **2016**, *21*, 165–177. [CrossRef]
7. Buonocore, G.; Perrone, S.; Longini, M.; Vezzosi, P.; Marzocchi, B.; Paffetti, P.; Bracci, R. Oxidative stress in preterm neonates at birth and on the seventh day of life. *Pediatr. Res.* **2002**, *52*, 46–49. [CrossRef] [PubMed]
8. Vento, M.; Escobar, J.; Cernada, M.; Escrig, R.; Aguar, M. The use and misuse of oxygen during the neonatal period. *Clin. Perinatol.* **2012**, *39*, 165–176. [CrossRef] [PubMed]
9. O'Donovan, D.J.; Fernandes, C.J. Free radicals and diseases in premature infants. *Antioxid. Redox Signal.* **2004**, *6*, 169–176. [CrossRef] [PubMed]
10. Lee, Y.S.; Chou, Y.H. Antioxidant profiles in full term and preterm neonates. *Chang Gung Med. J.* **2005**, *28*, 846–851. [PubMed]
11. Prasanthi, J.R.; Dasari, B.; Marwarha, G.; Larson, T.; Chen, X.; Geiger, J.D.; Ghribi, O. Caffeine protects against oxidative stress and Alzheimer's disease-like pathology in rabbit hippocampus induced by cholesterol-enriched diet. *Free Radic. Biol. Med.* **2010**, *49*, 1212–1220. [CrossRef] [PubMed]
12. Ullah, F.; Ali, T.; Ullah, N.; Kim, M.O. Caffeine prevents d-galactose-induced cognitive deficits, oxidative stress, neuroinflammation and neurodegeneration in the adult rat brain. *Neurochem. Int.* **2015**, *90*, 114–124. [CrossRef] [PubMed]
13. Henderson-Smart, D.J.; de Paoli, A.G. Methylxanthine treatment for apnoea in preterm infants. *Cochrane Database Syst. Rev.* **2010**, CD000140. [CrossRef]
14. Schmidt, B.; Roberts, R.S.; Davis, P.; Doyle, L.W.; Barrington, K.J.; Ohlsson, A.; Solimano, A.; Tin, W. Caffeine therapy for apnea of prematurity. *N. Engl. J. Med.* **2006**, *354*, 2112–2121. [CrossRef] [PubMed]
15. Schmidt, B.; Roberts, R.S.; Davis, P.; Doyle, L.W.; Barrington, K.J.; Ohlsson, A.; Solimano, A.; Tin, W. Long-term effects of caffeine therapy for apnea of prematurity. *N. Engl. J. Med.* **2007**, *357*, 1893–1902. [CrossRef] [PubMed]
16. Endesfelder, S.; Zaak, I.; Weichelt, U.; Bührer, C.; Schmitz, T. Caffeine protects neuronal cells against injury caused by hyperoxia in the immature brain. *Free Radic. Biol. Med.* **2014**, *67*, 221–234. [CrossRef]

17. Rivkees, S.A.; Wendler, C.C. Adverse and protective influences of adenosine on the newborn and embryo: Implications for preterm white matter injury and embryo protection. *Pediatr. Res.* **2011**, *69*, 271–278. [CrossRef] [PubMed]

18. Chavez Valdez, R.; Ahlawat, R.; Wills-Karp, M.; Nathan, A.; Ezell, T.; Gauda, E.B. Correlation between serum caffeine levels and changes in cytokine profile in a cohort of preterm infants. *J. Pediatr.* **2011**, *158*, 57–64. [CrossRef] [PubMed]

19. Weichelt, U.; Cay, R.; Schmitz, T.; Strauss, E.; Sifringer, M.; Buhrer, C.; Endesfelder, S. Prevention of hyperoxia-mediated pulmonary inflammation in neonatal rats by caffeine. *Eur. Respir. J.* **2013**, *41*, 966–973. [CrossRef] [PubMed]

20. Taha, D.; Kirkby, S.; Nawab, U.; Dysart, K.C.; Genen, L.; Greenspan, J.S.; Aghai, Z.H. Early caffeine therapy for prevention of bronchopulmonary dysplasia in preterm infants. *J. Mater. Fetal Neonatal Med.* **2014**, *27*, 1698–1702. [CrossRef] [PubMed]

21. Patel, R.M.; Leong, T.; Carlton, D.P.; Vyas-Read, S. Early caffeine therapy and clinical outcomes in extremely preterm infants. *J. Perinatol.* **2013**, *33*, 134–140. [CrossRef] [PubMed]

22. Bauer, J.; Maier, K.; Linderkamp, O.; Hentschel, R. Effect of caffeine on oxygen consumption and metabolic rate in very low birth weight infants with idiopathic apnea. *Pediatrics* **2001**, *107*, 660–663. [CrossRef] [PubMed]

23. Steer, P.; Flenady, V.; Shearman, A.; Charles, B.; Gray, P.H.; Henderson-Smart, D.; Bury, G.; Fraser, S.; Hegarty, J.; Rogers, Y.; et al. High dose caffeine citrate for extubation of preterm infants: A randomised controlled trial. *Arch. Dis. Child. Fetal Neonatal Ed.* **2004**, *89*, F499–F503. [CrossRef] [PubMed]

24. Henderson-Smart, D.J.; Steer, P.A. Caffeine versus theophylline for apnea in preterm infants. *Cochrane Database Syst. Rev.* **2010**, CD000273. [CrossRef]

25. Shi, X.; Dalal, N.S.; Jain, A.C. Antioxidant behaviour of caffeine: Efficient scavenging of hydroxyl radicals. *Food Chem. Toxicol.* **1991**, *29*, 1–6. [CrossRef]

26. Devasagayam, T.P.; Kamat, J.P.; Mohan, H.; Kesavan, P.C. Caffeine as an antioxidant: Inhibition of lipid peroxidation induced by reactive oxygen species. *Biochim. Biophys. Acta* **1996**, *1282*, 63–70. [CrossRef]

27. Barcelos, R.P.; Souza, M.A.; Amaral, G.P.; Stefanello, S.T.; Bresciani, G.; Fighera, M.R.; Soares, F.A.; Barbosa, N.V. Caffeine supplementation modulates oxidative stress markers in the liver of trained rats. *Life Sci.* **2014**, *96*, 40–45. [CrossRef] [PubMed]

28. Tunc, T.; Aydemir, G.; Karaoglu, A.; Cekmez, F.; Kul, M.; Aydinoz, S.; Babacan, O.; Yaman, H.; Sarici, S.U. Toll-like receptor levels and caffeine responsiveness in rat pups during perinatal period. *Regul. Pept.* **2013**, *182*, 41–44. [CrossRef] [PubMed]

29. Chavez-Valdez, R.; Wills-Karp, M.; Ahlawat, R.; Cristofalo, E.A.; Nathan, A.; Gauda, E.B. Caffeine modulates TNF-α production by cord blood monocytes: The role of adenosine receptors. *Pediatr. Res.* **2009**, *65*, 203–208. [CrossRef] [PubMed]

30. Li, J.; Li, G.; Hu, J.L.; Fu, X.H.; Zeng, Y.J.; Zhou, Y.G.; Xiong, G.; Yang, N.; Dai, S.S.; He, F.T. Chronic or high dose acute caffeine treatment protects mice against oleic acid-induced acute lung injury via an adenosine A2A receptor-independent mechanism. *Eur. J. Pharmacol.* **2011**, *654*, 295–303. [CrossRef] [PubMed]

31. Tsutsui, S.; Schnermann, J.; Noorbakhsh, F.; Henry, S.; Yong, V.W.; Winston, B.W.; Warren, K.; Power, C. A1 adenosine receptor upregulation and activation attenuates neuroinflammation and demyelination in a model of multiple sclerosis. *J. Neurosci.* **2004**, *24*, 1521–1529. [CrossRef] [PubMed]

32. Ikonomidou, C.; Kaindl, A.M. Neuronal death and oxidative stress in the developing brain. *Antioxid. Redox Signal.* **2011**, *14*, 1535–1550. [CrossRef] [PubMed]

33. Chua, C.O.; Vinukonda, G.; Hu, F.; Labinskyy, N.; Zia, M.T.; Pinto, J.; Csiszar, A.; Ungvari, Z.; Ballabh, P. Effect of hyperoxic resuscitation on propensity of germinal matrix haemorrhage and cerebral injury. *Neuropathol. Appl. Neurobiol.* **2010**, *36*, 448–458. [CrossRef] [PubMed]

34. Waldbaum, S.; Patel, M. Mitochondrial dysfunction and oxidative stress: A contributing link to acquired epilepsy? *J. Bioenerg. Biomembr.* **2010**, *42*, 449–455. [CrossRef] [PubMed]

35. Gozzelino, R.; Jeney, V.; Soares, M.P. Mechanisms of cell protection by heme oxygenase-1. *Annu. Rev. Pharmacol. Toxicol.* **2010**, *50*, 323–354. [CrossRef] [PubMed]

36. Ho, E.; Karimi Galougahi, K.; Liu, C.C.; Bhindi, R.; Figtree, G.A. Biological markers of oxidative stress: Applications to cardiovascular research and practice. *Redox Biol.* **2013**, *1*, 483–491. [CrossRef] [PubMed]

37. Zaghloul, N.; Nasim, M.; Patel, H.; Codipilly, C.; Marambaud, P.; Dewey, S.; Schiffer, W.K.; Ahmed, M. Overexpression of extracellular superoxide dismutase has a protective role against hyperoxia-induced brain injury in neonatal mice. *FEBS J.* **2012**, *279*, 871–881. [CrossRef] [PubMed]

38. Sifringer, M.; Brait, D.; Weichelt, U.; Zimmerman, G.; Endesfelder, S.; Brehmer, F.; von Haefen, C.; Friedman, A.; Soreq, H.; Bendix, I.; et al. Erythropoietin attenuates hyperoxia-induced oxidative stress in the developing rat brain. *Brain Behav. Immun.* **2010**, *24*, 792–799. [CrossRef] [PubMed]

39. Morse, D.; Choi, A.M. Heme oxygenase-1: From bench to bedside. *Am. J. Respir. Crit. Care Med.* **2005**, *172*, 660–670. [CrossRef] [PubMed]

40. Bryan, H.K.; Olayanju, A.; Goldring, C.E.; Park, B.K. The Nrf2 cell defence pathway: Keap1-dependent and -independent mechanisms of regulation. *Biochem. Pharmacol.* **2013**, *85*, 705–717. [CrossRef] [PubMed]

41. Wu, G.; Fang, Y.Z.; Yang, S.; Lupton, J.R.; Turner, N.D. Glutathione metabolism and its implications for health. *J. Nutr.* **2004**, *134*, 489–492. [PubMed]

42. Yang, H.; Magilnick, N.; Lee, C.; Kalmaz, D.; Ou, X.; Chan, J.Y.; Lu, S.C. Nrf1 and Nrf2 regulate rat glutamate-cysteine ligase catalytic subunit transcription indirectly via NF-κB and AP-1. *Mol. Cell. Biol.* **2005**, *25*, 5933–5946. [CrossRef] [PubMed]

43. Ma, Q. Role of nrf2 in oxidative stress and toxicity. *Annu. Rev. Pharmacol. Toxicol.* **2013**, *53*, 401–426. [CrossRef] [PubMed]

44. Bendix, I.; Weichelt, U.; Strasser, K.; Serdar, M.; Endesfelder, S.; von Haefen, C.; Heumann, R.; Ehrkamp, A.; Felderhoff-Mueser, U.; Sifringer, M. Hyperoxia changes the balance of the thioredoxin/peroxiredoxin system in the neonatal rat brain. *Brain Res.* **2012**, *1484*, 68–75. [CrossRef] [PubMed]

45. Sandberg, M.; Patil, J.; D'Angelo, B.; Weber, S.G.; Mallard, C. NRF2-regulation in brain health and disease: Implication of cerebral inflammation. *Neuropharmacology* **2014**, *79*, 298–306. [CrossRef] [PubMed]

46. Esterbauer, H.; Cheeseman, K.H. Determination of aldehydic lipid peroxidation products: Malonaldehyde and 4-hydroxynonenal. *Methods Enzymol.* **1990**, *186*, 407–421. [PubMed]

47. Ahotupa, M.; Mantyla, E.; Peltola, V.; Puntala, A.; Toivonen, H. Pro-oxidant effects of normobaric hyperoxia in rat tissues. *Acta Physiol. Scand.* **1992**, *145*, 151–157. [CrossRef] [PubMed]

48. Halliwell, B.; Gutteridge, J.M.C. *Free Radicals in Biology and Medicine*, 5th ed.; Oxford University Press: Oxford, UK, 2015.

49. Trachootham, D.; Lu, W.; Ogasawara, M.A.; Nilsa, R.D.; Huang, P. Redox regulation of cell survival. *Antioxid. Redox Signal.* **2008**, *10*, 1343–1374. [CrossRef] [PubMed]

50. Yadav, S.; Gupta, S.P.; Srivastava, G.; Srivastava, P.K.; Singh, M.P. Role of secondary mediators in caffeine-mediated neuroprotection in maneb- and paraquat-induced Parkinson's disease phenotype in the mouse. *Neurochem. Res.* **2012**, *37*, 875–884. [CrossRef] [PubMed]

51. Sifringer, M.; von Haefen, C.; Krain, M.; Paeschke, N.; Bendix, I.; Buhrer, C.; Spies, C.D.; Endesfelder, S. Neuroprotective effect of dexmedetomidine on hyperoxia-induced toxicity in the neonatal rat brain. *Oxidative Med. Cell. Longev.* **2015**, *2015*, 530371. [CrossRef] [PubMed]

52. Yamada, M.; Kubo, H.; Kobayashi, S.; Ishizawa, K.; Sasaki, H. Interferon-gamma: A key contributor to hyperoxia-induced lung injury in mice. *Am. J. Physiol. Lung Cell Mol. Physiol.* **2004**, *287*, L1042–L1047. [CrossRef] [PubMed]

53. Beckman, J.S.; Koppenol, W.H. Nitric oxide, superoxide, and peroxynitrite: The good, the bad, and ugly. *Am. J. Physiol.* **1996**, *271*, C1424–C1437. [PubMed]

54. Griscavage, J.M.; Wilk, S.; Ignarro, L.J. Inhibitors of the proteasome pathway interfere with induction of nitric oxide synthase in macrophages by blocking activation of transcription factor NF-κB. *Proc. Natl. Acad. Sci. USA* **1996**, *93*, 3308–3312. [CrossRef] [PubMed]

55. Hoehn, T.; Felderhoff-Mueser, U.; Maschewski, K.; Stadelmann, C.; Sifringer, M.; Bittigau, P.; Koehne, P.; Hoppenz, M.; Obladen, M.; Buhrer, C. Hyperoxia causes inducible nitric oxide synthase-mediated cellular damage to the immature rat brain. *Pediatr. Res.* **2003**, *54*, 179–184. [CrossRef] [PubMed]

56. Ikeno, S.; Nagata, N.; Yoshida, S.; Takahashi, H.; Kigawa, J.; Terakawa, N. Immature brain injury via peroxynitrite production induced by inducible nitric oxide synthase after hypoxia-ischemia in rats. *J. Obstet. Gynaecol. Res.* **2000**, *26*, 227–234. [CrossRef] [PubMed]

57. Taoufik, Y.; de Goer de Herve, M.G.; Giron-Michel, J.; Durali, D.; Cazes, E.; Tardieu, M.; Azzarone, B.; Delfraissy, J.F. Human microglial cells express a functional IL-12 receptor and produce IL-12 following IL-12 stimulation. *Eur. J. Immunol.* **2001**, *31*, 3228–3239. [CrossRef]

58. Chang, H.D.; Radbruch, A. The pro- and anti-inflammatory potential of interleukin-12. *Ann. N. Y. Acad. Sci.* **2007**, *1109*, 40–46. [CrossRef] [PubMed]

59. Murphy, C.A.; Langrish, C.L.; Chen, Y.; Blumenschein, W.; McClanahan, T.; Kastelein, R.A.; Sedgwick, J.D.; Cua, D.J. Divergent pro- and anti-inflammatory roles for IL-23 and IL-12 in joint autoimmune inflammation. *J. Exp. Med.* **2003**, *198*, 1951–1957. [CrossRef] [PubMed]

60. Zeidan-Chulia, F.; Gelain, D.P.; Kolling, E.A.; Rybarczyk-Filho, J.L.; Ambrosi, P.; Terra, S.R.; Pires, A.S.; da Rocha, J.B.; Behr, G.A.; Moreira, J.C. Major components of energy drinks (caffeine, taurine, and guarana) exert cytotoxic effects on human neuronal SH-SY5Y cells by decreasing reactive oxygen species production. *Oxidative Med. Cell. Longev.* **2013**, *2013*, 791795. [CrossRef] [PubMed]

61. Leon-Carmona, J.R.; Galano, A. Is caffeine a good scavenger of oxygenated free radicals? *J. Phys. Chem. B* **2011**, *115*, 4538–4546. [CrossRef] [PubMed]

62. Tiwari, K.K.; Chu, C.; Couroucli, X.; Moorthy, B.; Lingappan, K. Differential concentration-specific effects of caffeine on cell viability, oxidative stress, and cell cycle in pulmonary oxygen toxicity in vitro. *Biochem. Biophys. Res. Commun.* **2014**, *450*, 1345–1350. [CrossRef] [PubMed]

63. Pohanka, M. Caffeine alters oxidative homeostasis in the body of BALB/c mice. *Bratislavske Lekarske Listy* **2014**, *115*, 699–703. [CrossRef] [PubMed]

64. Motterlini, R.; Foresti, R.; Bassi, R.; Green, C.J. Curcumin, an antioxidant and anti-inflammatory agent, induces heme oxygenase-1 and protects endothelial cells against oxidative stress. *Free Radic. Biol. Med.* **2000**, *28*, 1303–1312. [CrossRef]

65. Ran, Q.; Liang, H.; Gu, M.; Qi, W.; Walter, C.A.; Roberts, L.J., 2nd; Herman, B.; Richardson, A.; van Remmen, H. Transgenic mice overexpressing glutathione peroxidase 4 are protected against oxidative stress-induced apoptosis. *J. Biol. Chem.* **2004**, *279*, 55137–55146. [CrossRef] [PubMed]

66. Schmitz, T.; Krabbe, G.; Weikert, G.; Scheuer, T.; Matheus, F.; Wang, Y.; Mueller, S.; Kettenmann, H.; Matyash, V.; Buhrer, C.; et al. Minocycline protects the immature white matter against hyperoxia. *Exp. Neurol.* **2014**, *254*, 153–165. [CrossRef] [PubMed]

67. Kilicdag, H.; Daglioglu, Y.K.; Erdogan, S.; Zorludemir, S. Effects of caffeine on neuronal apoptosis in neonatal hypoxic-ischemic brain injury. *J. Mater. Fetal Neonatal Med.* **2014**, *27*, 1470–1475. [CrossRef] [PubMed]

68. Geraets, L.; Moonen, H.J.; Wouters, E.F.; Bast, A.; Hageman, G.J. Caffeine metabolites are inhibitors of the nuclear enzyme poly(ADP-ribose)polymerase-1 at physiological concentrations. *Biochem. Pharmacol.* **2006**, *72*, 902–910. [CrossRef] [PubMed]

69. Barkett, M.; Gilmore, T.D. Control of apoptosis by Rel/NF-κB transcription factors. *Oncogene* **1999**, *18*, 6910–6924. [CrossRef] [PubMed]

70. Gu, Y.; Dee, C.M.; Shen, J. Interaction of free radicals, matrix metalloproteinases and caveolin-1 impacts blood-brain barrier permeability. *Front. Biosci.* **2011**, *3*, 1216–1231. [CrossRef]

71. Coutts, A.; Chen, G.; Stephens, N.; Hirst, S.; Douglas, D.; Eichholtz, T.; Khalil, N. Release of biologically active TGF-β from airway smooth muscle cells induces autocrine synthesis of collagen. *Am. J. Physiol. Lung Cell Mol. Physiol.* **2001**, *280*, L999–L1008. [PubMed]

72. Millis, A.J.; Hoyle, M.; McCue, H.M.; Martini, H. Differential expression of metalloproteinase and tissue inhibitor of metalloproteinase genes in aged human fibroblasts. *Exp. Cell Res.* **1992**, *201*, 373–379. [CrossRef]

73. Sifringer, M.; Genz, K.; Brait, D.; Brehmer, F.; Lober, R.; Weichelt, U.; Kaindl, A.M.; Gerstner, B.; Felderhoff-Mueser, U. Erythropoietin attenuates hyperoxia-induced cell death by modulation of inflammatory mediators and matrix metalloproteinases. *Dev. Neurosci.* **2009**, *31*, 394–402. [CrossRef] [PubMed]

74. Hayakawa, K.; Okazaki, R.; Morioka, K.; Nakamura, K.; Tanaka, S.; Ogata, T. Lipopolysaccharide preconditioning facilitates M2 activation of resident microglia after spinal cord injury. *J. Neurosci. Res.* **2014**, *92*, 1647–1658. [CrossRef] [PubMed]

75. Saad, M.A.; Abdelsalam, R.M.; Kenawy, S.A.; Attia, A.S. Ischemic preconditioning and postconditioning alleviates hippocampal tissue damage through abrogation of apoptosis modulated by oxidative stress and inflammation during transient global cerebral ischemia-reperfusion in rats. *Chem. Biol. Interact.* **2015**, *232*, 21–29. [CrossRef] [PubMed]

76. Parmar, J.; Jones, N.M. Hypoxic preconditioning can reduce injury-induced inflammatory processes in the neonatal rat brain. *Int. J. Dev. Neurosci.* **2015**, *43*, 35–42. [CrossRef] [PubMed]

77. Heusch, G. Cardioprotection: Chances and challenges of its translation to the clinic. *Lancet* **2013**, *381*, 166–175. [CrossRef]
78. Hausenloy, D.J.; Yellon, D.M. Ischaemic conditioning and reperfusion injury. *Nat. Rev. Cardiol.* **2016**, *13*, 193–209. [CrossRef] [PubMed]
79. Robertson, N.J.; Tan, S.; Groenendaal, F.; van Bel, F.; Juul, S.E.; Bennet, L.; Derrick, M.; Back, S.A.; Valdez, R.C.; Northington, F.; et al. Which neuroprotective agents are ready for bench to bedside translation in the newborn infant? *J. Pediatr.* **2012**, *160*, 544–552. [CrossRef] [PubMed]
80. Livak, K.J.; Schmittgen, T.D. Analysis of relative gene expression data using real-time quantitative PCR and the 2(-Delta Delta C(T)) Method. *Methods* **2001**, *25*, 402–408. [CrossRef] [PubMed]

International Journal of
Molecular Sciences

MDPI

Article

Synthetic Secoisolariciresinol Diglucoside (LGM2605) Protects Human Lung in an Ex Vivo Model of Proton Radiation Damage

Anastasia Velalopoulou [1,†], Shampa Chatterjee [2,†], Ralph A. Pietrofesa [1], Cynthia Koziol-White [1], Reynold A. Panettieri [1], Liyong Lin [3], Stephen Tuttle [3], Abigail Berman [3], Constantinos Koumenis [3] and Melpo Christofidou-Solomidou [1,*]

1   Division of Pulmonary, Allergy, and Critical Care, Department of Medicine, University of Pennsylvania Perelman School of Medicine, 3450 Hamilton Walk, Stemmler Hall, Office Suite 227, Philadelphia, PA 19104, USA; avelalopoulou@gmail.com (A.V.); ralphp@pennmedicine.upenn.edu (R.A.P.); cjk167@rbhs.rutgers.edu (C.K.-W.); rp856@rbhs.rutgers.edu (R.A.P.)
2   Department of Physiology, University of Pennsylvania Perelman School of Medicine, Philadelphia, PA 19104, USA; shampac@pennmedicine.upenn.edu
3   Department of Radiation Oncology, University of Pennsylvania Perelman School of Medicine, Philadelphia, PA 19104, USA; Liyong.Lin@uphs.upenn.edu (L.L.); Steve.Tuttle@uphs.upenn.edu (S.T.); abigail.berman@uphs.upenn.edu (A.B.); Costas.Koumenis@uphs.upenn.edu (C.K.)
*   Correspondence: melpo@pennmedicine.upenn.edu; Tel.: +1-215-573-9917; Fax: +1-215-746-0376
†   These authors contributed equally to this work.

Received: 6 November 2017; Accepted: 16 November 2017; Published: 25 November 2017

Abstract: Radiation therapy for the treatment of thoracic malignancies has improved significantly by directing of the proton beam in higher doses on the targeted tumor while normal tissues around the tumor receive much lower doses. Nevertheless, exposure of normal tissues to protons is known to pose a substantial risk in long-term survivors, as confirmed by our work in space-relevant exposures of murine lungs to proton radiation. Thus, radioprotective strategies are being sought. We established that LGM2605 is a potent protector from radiation-induced lung toxicity and aimed in the current study to extend the initial findings of space-relevant, proton radiation-associated late lung damage in mice by looking at acute changes in human lung. We used an ex vivo model of organ culture where tissue slices of donor living human lung were kept in culture and exposed to proton radiation. We exposed donor human lung precision-cut lung sections (huPCLS), pretreated with LGM2605, to 4 Gy proton radiation and evaluated them 30 min and 24 h later for gene expression changes relevant to inflammation, oxidative stress, and cell cycle arrest, and determined radiation-induced senescence, inflammation, and oxidative tissue damage. We identified an LGM2605-mediated reduction of proton radiation-induced cellular senescence and associated cell cycle changes, an associated proinflammatory phenotype, and associated oxidative tissue damage. This is a first report on the effects of proton radiation and of the radioprotective properties of LGM2605 on human lung.

Keywords: antioxidant; cell cycle; human lung sections; inflammation; LGM2605; organ culture; oxidative stress; phase II enzymes; proton radiation; reactive oxygen species; senescence

1. Introduction

Lung cancer remains the leading cause of cancer-related death in the USA. Radiation therapy plays a prominent role in the treatment of lung cancer patients, and the standard of care for locally-advanced non-small cell lung cancer (NSCLC) is chemoradiation [1,2]. Innovative advances in radiation therapy have resulted in novel modalities such as proton beam therapy, whereby higher doses of radiation are maintained on the tumor target, and tissue regions beyond the target volume are exposed to relatively low doses [3].

Despite these advances, there is a substantial risk of sub-acute and late side effects from damage to normal tissues, such as the pulmonary conditions of radiation pneumonitis and lung fibrosis, which can cause significant morbidity and mortality. More importantly, proton therapy, despite its focus on target areas of the tumor, has not shown "a survival benefit" for cancer patients due to a significant detrimental effect of radiation on critical organs at risk (OAR), which negates its benefit [4,5]. As of yet no effective intervention or cure for radiation-induced late effects has been developed.

Over the past decade, we have evaluated radiation effects on the lung and our work has identified dose-dependent distinct lung, pathological, and physiological changes. Specifically, our work in identifying space-radiation-relevant normal tissue damage in lungs exposed to proton radiation revealed significant long term damage. These alterations resemble emphysema, a phenotype of chronic obstructive pulmonary disease (COPD), accompanied by reduced oxygenation induced by proton radiation at 800 days (26 months) following a single exposure to radiation (1 Gy, 2 Gy, and 3 Gy). In this chronic stage of lung injury, we discovered marked imbalances in lung sphingolipid signaling pathways induced by both radiation types, with severely reduced anti-apoptotic and pro-proliferative levels of sphingosine-1 phosphate (S1P), associated with increases in tissue senescence markers. We proposed that the loss of S1P, regulated enzymatically via sphingosine kinases 1 and 2 (SphK1/2) and sphingosine lyase (SphL), is directly linked to maladaptive lung repair and premature/accelerated cellular senescence, and it plays a critical role in radiation-induced lung disease [6].

Therefore, this pathway provides the opportunity to develop useful preventive interventions. A major candidate for protection against radiation induced injury as described above is secoisolariciresinol diglucoside (SDG), a biphenolic that scavenges free radicals and triggers a potent endogenous antioxidant response (EAR), preventing radiation-induced acute lung injury [7–10]. We developed a novel synthetic SDG called LGM2605 [11] that retains the antioxidant properties of the natural compound, but with the advantage of a consistent pharmacodynamic profile [12].

We hypothesize that proton radiation induces oxidative stress-mediated lung injury with impaired repair due to maladaptive accelerated senescence, culminating in lung tissue loss and dysfunction; we postulate that lung injury can be prevented by treatment with LGM2605. To test this, we employed human lung tissue in the form of human, precision-cut lung slices (huPCLS), a novel ex vivo human lung organ culture model using tissue from deceased donors [13,14] and investigated the effects of 4 Gy proton radiation exposure of these tissues with and without LGM2605 (50–100 μM) pre-treatment. The radiation dose and LGM2605 concentrations were chosen based on proton therapy protocols (where about 30–60% of the lung receives approximately 4 Gy) and the radioprotective efficacy of LGM2605 (as reported in our earlier studies) [15], respectively.

This first report of the effects of proton radiation in a model of the human lung establishes that proton radiation-induced inflammation, oxidative stress, and senescence can be significantly diminished by LGM2605. This work lays the groundwork for future studies on radioprotection, as well as mitigation, in response to proton beam exposure in cancer therapy, as well as in proton exposure from solar particle events (SPE) in deep space exploration, whereby crew members are exposed to proton radiation and other harmful galactic cosmic radiation (GCR) [16].

2. Results

We utilized an ex vivo model of organ culture whereby human donor lung slices that are representative of the in vivo tissue were kept alive in culture, as confirmed by the beating cilia of airway epithelial cells lining the bronchioles [17], to study the effect of LGM2605 treatment on proton radiation-induced lung damage, inflammation, and oxidative stress. LGM2605 treatment (50 μM) was initiated 4 h prior to exposure to 4 Gy proton radiation and lung sections were collected at 30 min and 24 h post radiation exposure (Scheme 1).

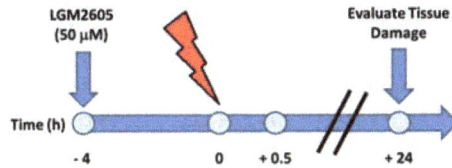

Scheme 1. Schematic presentation of the experimental protocol. Human, precision-cut lung sections were exposed to 50 μM synthetic secoisolariciresinol diglucoside (SDG) (LGM2605) 4 h prior to exposure to 4 Gy proton radiation. Lung sections were harvested at 30 min and 24 h post radiation exposure.

2.1. LGM2605 Prevents the Expression of Proinflammatory Cytokine Gene Levels and Reduces the Induction COX-2 by Proton Radiation in huPCLS

The acute inflammatory response to proton radiation exposure of lung tissue (huPCLS) was further characterized by determining the mRNA levels of proinflammatory cytokines (*IL-1β*, *IL-6*, *TNFα*, and *IL-1α*) and cyclooxygenase-2 (*COX-2*), an enzyme responsible for the formation of pro-inflammatory moieties [18]. The transcript levels of inflammatory markers, *IL-1β*, *IL-6*, *TNFα*, and *IL-1α*, and *COX-2* in huPCLS were monitored at 30 min and 24 h post 4 Gy proton radiation with or without LGM2605 pretreatment for 4 h and are shown in Figure 1A–E. LGM2605 pretreatment significantly dampened the proton radiation-induced inflammation in huPCLS (36.7% reduction for *IL-6* and 25.1% reduction for *IL-1β*) and significantly ($p < 0.05$) decreased the induction of *COX-2* by 25.0% when evaluated 30 min post radiation exposure.

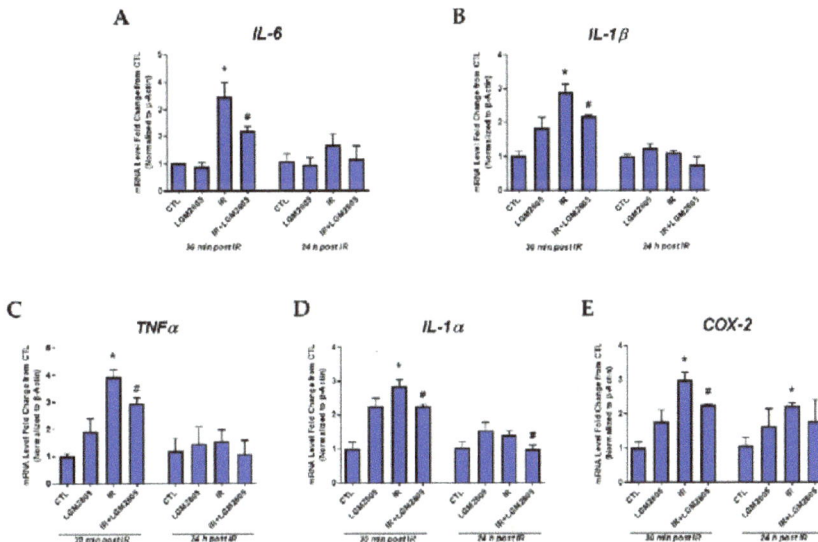

Figure 1. LGM2605 reduces inflammation in proton-irradiated human donor, lung precision-cut lung sections (huPCLS). Human, precision-cut lung sections ($n = 3$–4) from one donor were evaluated at 30 min and 24 h post 4 Gy proton radiation with or without LGM2605 pretreatment for 4 h. Data from one representative donor lung are presented as the average fold change from non-irradiated control ± SEM. Transcript levels of *IL-6* (**A**), *IL-1β* (**B**), *TNFα* (**C**), *IL-1α* (**D**), and *COX-2* (**E**) are normalized to *β-Actin*. * Indicates a statistically significant ($p < 0.05$) difference from the respective non-irradiated control. # Indicates a statistically significant ($p < 0.05$) difference from IR (ionizing radiation-only exposure) (4 Gy proton radiation). IL, interleukin; TNF, tumor necrosis factor; COX, cyclooxygenase; CTL, non-irradiated control.

2.2. LGM2605 Boosts Antioxidant Gene Levels by Proton Radiation in huPCLS

Figure 2 depicts the expression of antioxidant genes *HMOX1* and *NQO1* in huPCLS after 30 min and 24 h post 4 Gy proton radiation with or without LGM2605 pretreatment for 4 h. At 30 min post radiation, mRNA levels of the antioxidant gene *NQO1* among proton-exposed huPCLS show a significant ($p < 0.05$) increase over non-irradiated samples (1.47 ± 0.13-fold increase), which was further significantly boosted by the action of LGM2605 (2.16 ± 0.10-fold increase). The effects of both the radiation exposure and the test article were less profound by 24 h post exposure.

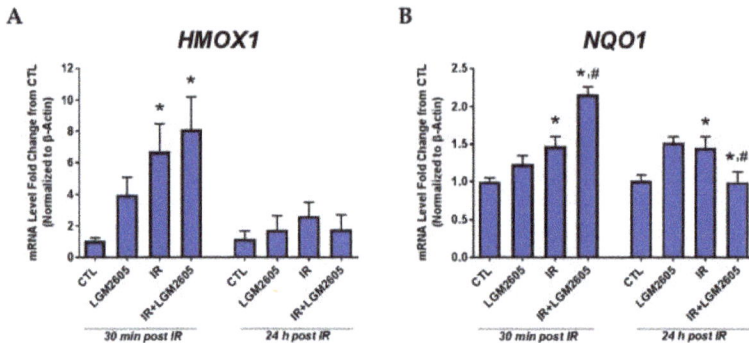

Figure 2. LGM2605 increases expression levels of antioxidant and cell protective genes in proton-irradiated huPCLS. qPCR analysis of *HMOX1* (**A**) and *NQO1* (**B**) was performed on human, precision-cut lung sections ($n = 3$–4) from one donor after 30 min and 24 h post 4 Gy proton radiation with or without LGM2605 pretreatment for 4 h. Data from one representative donor lung are presented as the average fold change from non-irradiated control $\pm$ standard error of the mean (SEM). Transcript levels of tested genes are normalized to *β-Actin*. * Indicates a statistically significant ($p < 0.05$) difference from the respective non-irradiated control. # Indicates a statistically significant ($p < 0.05$) difference from IR (4 Gy proton radiation).

2.3. LGM2605 Decreases Proton Radiation-Induced Senescence and Biomarkers of Cellular Senescence in huPCLS

Next, we evaluated cellular senescence in huPCLS by staining for senescence-associated β-galactosidase. As shown in Figure 3, exposure to 4 Gy proton radiation significantly ($p < 0.05$) increased senescence-associated β-galactosidase (SA-β-gal) staining (blue staining in Figure 3A), which was significantly reduced by LGM2605 pretreatment to levels comparable to non-irradiated huPCLS. After observing such a robust improvement in proton radiation-induced cellular senescence with LGM2605 pretreatment, we next evaluated molecular biomarkers of cellular senescence both at the message (gene) and the induced protein level. The oncosuppressor gene *TP53* orchestrates the transcriptional response that culminates in senescence and cell death [19]; the protein TP53 that is induced gets stabilized (via post translational modifications) in response to damage/radiation. TP53 can activate cell cycle-arresting genes like cyclin-dependent kinase (*CDKN1A*) [20]. The CDK family of proteins drives the cell cycle by phosphorylating various substrates involved in cell growth and differentiation. CDK4/6 and CDK2 inactivate the retinoblastoma tumor suppressor protein (pRb), which is a "gatekeeper" of the G1-S phase transition. Another member of the CDK family is the *CDKN2A* gene that encodes different transcripts involved mostly in cell cycle regulation and cellular senescence such as the tumor suppressor proteins p16 and p19. Figures 4 and 5 show proton radiation-induced increases in the mRNA and protein levels of TP53, as well as CDKN2A (p16), and that this increase, which is an index of increased senescence, is significantly ($p < 0.05$) abrogated by LGM2605 pretreatment. Additionally, phosphorylation of pRb was boosted by as much as 80% by LGM2605 as compared to untreated irradiated lung slices, in which expression was decreased by 20% from respective non-irradiated controls. This reversal of the hypophosphorylated state of pRb is associated with inhibition of cellular senescence [21]. *CDK2, CDK4,* and *CDK6*, which trigger the G1/S phase and thus drive cell proliferation, are significantly ($p < 0.05$) elevated among huPCLS pretreated with LGM2605 and exposed to proton radiation compared to huPCLS, which is exposed to proton radiation alone.

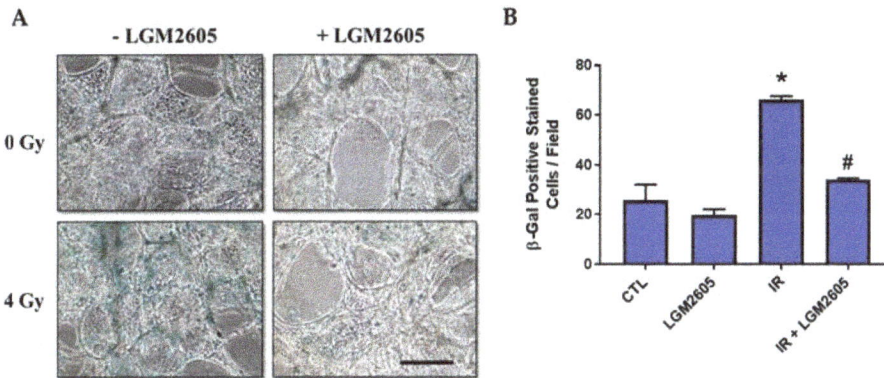

Figure 3. LGM2605 reduces senescence-associated β-galactosidase (SA-β-gal) staining in huPCLS following exposure to 4 Gy proton radiation. huPCLS were exposed to 4 Gy proton radiation and evaluated 24 h later for the detection of senescence-associated β-galactosidase (**A**); 3 random fields per slide were selected and positive-stained cells were counted through the entire thickness of the section by selecting 8 different focus levels. The average number of SA-β-gal positive-stained cells per field are presented as the mean ± SEM (**B**). Scale bar = 150 μm. * Indicates a statistically significant ($p < 0.05$) difference from the respective non-irradiated control. # Indicates a statistically significant ($p < 0.05$) difference from IR (4 Gy proton radiation).

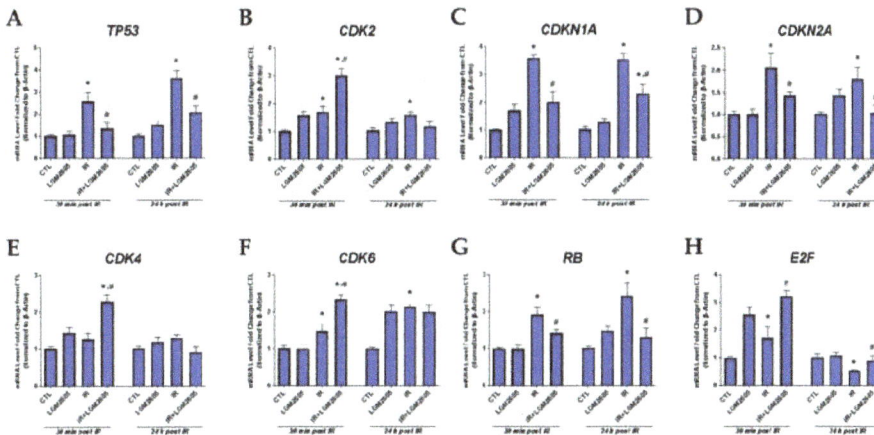

Figure 4. Senescence genes induced by proton radiation are decreased with LGM2605 treatment. qPCR analysis in human, precision-cut lung sections ($n = 3$–4) from one donor after 30 min and 24 h post 4 Gy proton radiation with or without LGM2605 pretreatment for 4 h (**A–H**). Data from one representative donor lung are presented as the average fold change from non-irradiated control ± SEM. Transcript levels of tested genes are normalized to *β-Actin*. * Indicates a statistically significant ($p < 0.05$) difference from the respective non-irradiated control. # Indicates a statistically significant ($p < 0.05$) difference from IR (4 Gy proton radiation).

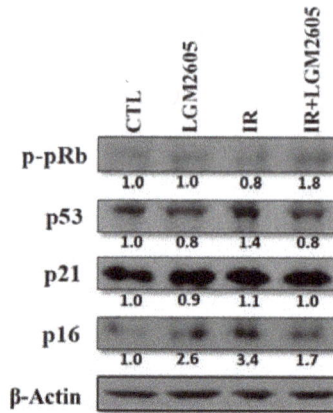

Figure 5. Senescence proteins induced by 4 Gy proton radiation in huPCLS. Western Blot (WB) analysis 30 min post 4 Gy proton radiation exposure of whole tissue lysates from human, precision-cut lung sections (*n* = 3–4) from one donor treated with or without LGM23605 pretreatment. Results shown are from one representative donor lung. Numbers below each WB band designates fold change over control (CTL) after values have been normalized to β-Actin.

2.4. LGM2605 Reduces Proton Radiation-Induced Oxidative Stress in huPCLS

As shown in Figure 6, increase in green fluorescence that arises from the oxidation of the fluorogenic probe that represents cellular oxidative stress implies proton radiation-induced oxidative stress in huPCLS. This cell-permeant dye is non-fluorescent while in a reduced state, and exhibits bright fluorescence in the presence of reactive oxygen species (ROS) or oxidized lipids. LGM2605 at both 50 and 100 µM reduced CellROX Green fluorescence, implying that oxidative stress induced by proton radiation can be significantly reduced by this agent. Figure 7 indicates a 3D image of the irradiated huPCLS, depicting that oxidative damage from proton exposure is full-thickness (i.e., though the entire 300–350 µM lung section) and protection by LGM2605, although added to the medium, i.e., externally, extends through the entire thickness of the exposed tissue.

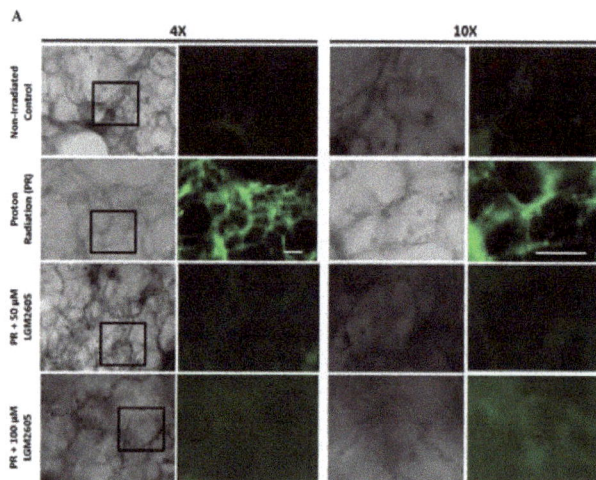

Figure 6. *Cont.*

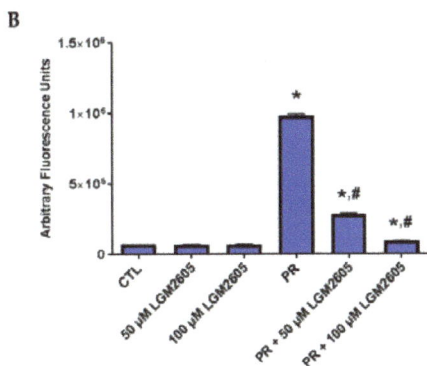

Figure 6. LGM2605 reduces proton radiation-induced oxidative stress in huPCLS. Radiation-induced oxidative stress in huPCLS was monitored by CellROX Green staining and fluorescence. huPCLS were exposed to proton radiation and treated with LGM2605, after which they were labeled with CellROX Green dye (5 μM) for 20 min. This was followed by removal of the dye and resuspension of huPCLS in fresh media and fluorescence imaging of the sections at λ = 488 nm. Images were acquired by 4× and 10× lens. The boxed regions of images acquired at 4× were visualized in 10× to show more structural details. Scale bar = 100 μm (**A**); All images were acquired at the same settings using MetaMorph acquisition software (Version 7.7, Molecular Devices, Downington, PA, USA). The images acquired by 4× lens were used to quantify using MetaMorph and ImageJ software (Fiji Version, National Institutes of Health, Bethesda, MD, USA). The data was acquired from images acquired in 6–7 fields (**B**). Results from one representative donor lung are presented. * Indicates a statistically significant ($p < 0.05$) difference from the respective non-irradiated control. # Indicates a statistically significant ($p < 0.05$) difference from IR (4 Gy proton radiation).

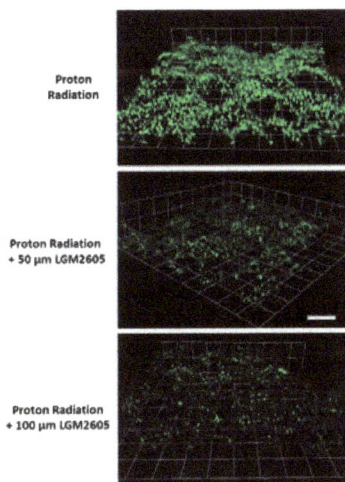

Figure 7. LGM2605 reduces proton radiation-induced oxidative stress in huPCLS. As the huPCLS were ~300 to 350 μm thick, the oxidative stress along the depth of the tissue and its diminution by LGM2605 were also assessed by imaging along the z-axis at 10 μm interval throughout the tissue. The z stack images acquired on a multiphoton fluorescent microscope were converted into 3D display stacks (using Volocity® Visualization Program, Version 6.3, PerkinElmer, Waltham, MA, USA). Movies of 3D stack were made by ImageJ analysis software. Scale bar = 160 μm. Results from one representative donor lung are presented.

2.5. LGM2605 Reduces a Proinflammatory Phenotype in Proton-Irradiated huPCLS

Having observed an increase in the mRNA levels of proinflammatory cytokines, such as *IL-1β* and *IL-6*, following 4 Gy proton radiation exposure, we proceeded to further investigate the proinflammatory phenotype induced by proton radiation by evaluating the expression of intercellular adhesion molecule-1 (ICAM-1) in huPCLS following radiation. As shown in Figure 8, ICAM-1 expression is increased in huPCLS at 24 h post radiation exposure, noted by the increase in green fluorescence (Figure 8A). LGM2605 pretreatment led to a dose-dependent decrease of proton radiation-induced ICAM-1 expression (50 μM LGM2605 pretreatment led to an 87% reduction in proton radiation-induced ICAM-1 expression and 100 μM LGM2605 pretreatment led to a 96% reduction in proton radiation-induced ICAM-1 expression) (Figure 8B).

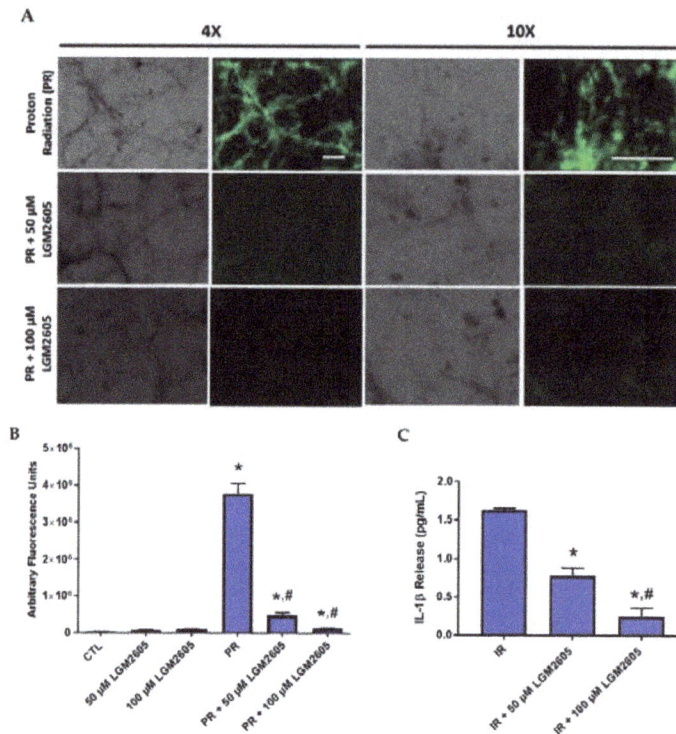

Figure 8. LGM2605 prevents proton-radiation induced inflammation in huPCLS. Radiation-induced inflammation in huPCLS was monitored by intercellular adhesion molecule-1 (ICAM-1) staining and fluorescence. Human, precision-cut lung sections were permeabilized and immunostained with anti-ICAM antibody 24 h post 4 Gy proton radiation exposure. Fluorescence imaging of the sections at $\lambda = 488$ nm using 4× and 10× lens was carried out. Scale bar = 100 μm (**A**); All images were acquired at the same settings using MetaMorph acquisition software. The images acquired by 4× lens were used to quantify using MetaMorph and ImageJ software. The data was acquired from images acquired in 6–7 fields (**B**); IL-1β release, as determined by enzyme-linked immunosorbent assay (ELISA), was determined at 24 h post radiation (**C**). Results from one representative donor lung are presented. * Indicates a statistically significant ($p < 0.05$) difference from the respective non-irradiated control. # Indicates a statistically significant ($p < 0.05$) difference from IR (4 Gy proton radiation).

The observed decrease in ICAM-1-induced expression following proton radiation exposure prompted us to look at the release of a key inflammatory cytokine, IL-1β, by proton-irradiated huPCLS.

As shown in Figure 8C, LGM2605 pretreatment significantly ($p < 0.05$) reduced IL-1β release by 52% (50 μM LGM2605) and 85% (100 μM LGM2605).

2.6. LGM2605 Does Not Impair Tumor Cell Eradication by Proton Radiation

Exposure of cancer cells to ionizing radiation induces cellular senescence, thus inhibiting tumor growth and progression [22]. Specifically, modulating radiation-induced senescence of non-small cell lung cancer cells has been investigated as a mechanism of overcoming radioresistance [23,24]. As a confirmation that LGM2605 pretreatment does not impair the tumoricidal effect of proton radiation exposure, A549 human lung adenocarcinoma cells were exposed to 4 Gy proton radiation and evaluated for radiation-induced cellular senescence by staining for senescence-associated β-galactosidase.

As shown in Figure 9, treatment with LGM2605-alone induced a dose-dependent increase in cellular senescence in A549 cells without radiation proton radiation exposure. Regardless of the dose of LGM2605, proton radiation induced a significant ($p < 0.05$) increase in the percentage of SA-β-gal positive-stained cells (31.8 ± 6.0% with 0 μM LGM2605, 34.8 ± 6.0% with 50 μM LGM2605, and 35.2 ± 3.8% with 100 μM LGM2605). No significant differences between A549 irradiated cells treated with different doses of LGM2605 were observed. These findings are supported by our previous work, where we showed how the radioprotective properties of the lignan SDG did not prevent eradiation of lung tumors from radiation exposure [25].

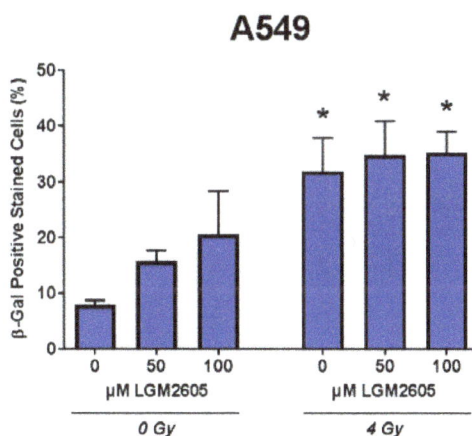

Figure 9. LGM2605 does not impair proton radiation-induced cellular senescence in A549 human lung adenocarcinoma cells. 4 h prior to 4 Gy proton radiation exposure, A549 cells were treated with 0 μM, 50 μM, or 100 μM LGM2605. 24 h following radiation exposure, the cells were stained for senescence-associated β-galactosidase. The blue-stained senescent cells were viewed by light microscopy and positive-stained cells were counted from 5 randomly selected fields. The average number of β-galactosidase positive-stained cells per field are presented as the mean ± SEM. * Indicates a statistically significant ($p < 0.05$) difference from the respective non-irradiated control.

3. Discussion

Unwanted exposure to radiation of normal tissues, and especially of radiosensitive organs such as the lung, can occur during radiotherapy to treat malignancies, in radiological accidents, and in terroristic acts involving radioactive material, or during space travel where crewmembers may be exposed to galactic cosmic radiation (GCR) and protons from solar particle events (SPEs). Radiation exposure is an identified potential health risk to crewmembers during space travel; however, its effects on the pulmonary system are largely unexplored, constituting a noticeable knowledge gap. Specifically, unrecognized GCR-induced lung damage could lead to chronic breathing problems in crewmembers, such as cough, dyspnea, and decreased exercise capacity [26]. Using a mouse model of GCR and proton radiation exposure, our group identified dose-dependent, distinct, lung pathological and physiological changes resembling chronic obstructive pulmonary disease (COPD) emphysema, accompanied by reduced oxygenation, induced by all GCR and proton radiation types tested, and evaluated more than two years following a single exposure to radiation [6]. The current study aimed to extend those findings using a novel human lung model whereby ultrathin slices (<200 μm) from human donor lungs are kept alive in an ex vivo organ culture system for up to 7 days from procurement. This allowed us to test acute changes to lung tissues following exposure to radiation and to elucidate mechanisms of tissue damage inked to delayed effects such as the ones observed earlier [6]. Importantly, this study evaluated the properties of a novel synthetic agent found to be radioprotective by ROS scavenging, nuclear factor (erythroid-derived 2)-like 2 (Nrf2) pathway activation, and endogenous antioxidant defense boosting, as discovered more recently by scavenging of radiation-induced active chlorine species. We show here for the first time that LGM2605 is an effective radioprotecting agent in human lung tissues exposed to proton radiation. Additionally, we identified the antioxidant and anti-senescence action of this agent, a mechanism that could translate to downstream tissue protection from late effects of radiation (Figure 10). This work provides the basis for such work to be undertaken in future studies.

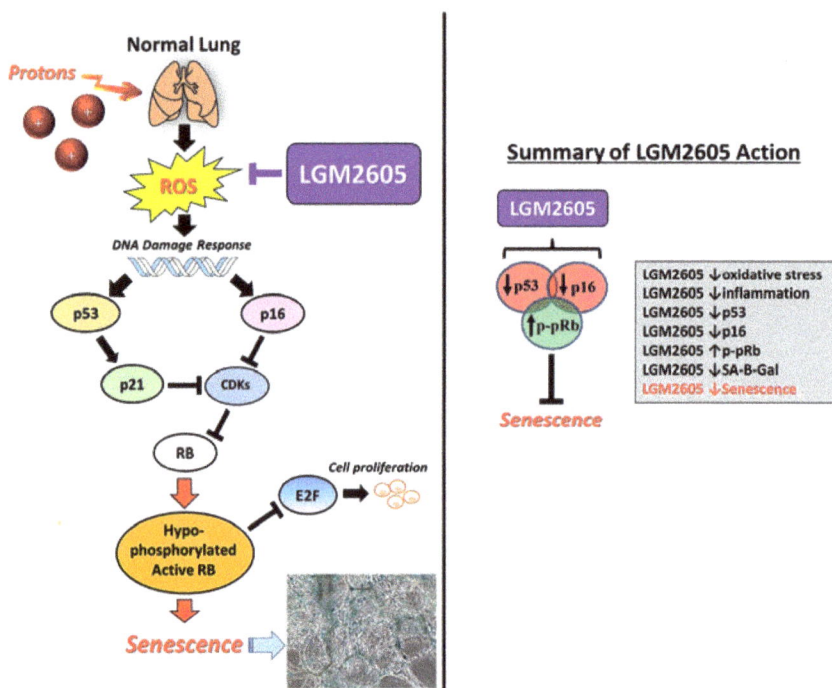

Figure 10. Schematic representation of LGM2605's blockade of senescence markers induced by proton radiation of normal lung tissue/cell. Arrow-headed lines indicate activation and bar-headed lines indicate inhibition. ROS, Reactive oxygen species; CDK, cyclin-dependent kinase; RB, retinoblastoma; E2F, E2 factor.

It has long been known that exposure of normal tissues surrounding the tumor regions to radiation is an unwanted side effect of radiotherapy, which is a major treatment option in lung cancer management [27]. To prevent these complications in the form of fibrosis and the spread of secondary or metastatic cancers, novel innovative radiotherapies such as proton therapy have become an area of growing research. Proton therapy involves less scatter of protons as compared to X-ray/γ radiation. The energy and acceleration of protons can be controlled precisely if maximum (tumoricidal) energy is delivered on the cancer tissue. Despite better control on radiation, proton therapy does not necessarily translate into a survival benefit for the cancer patient due to a significant detrimental effect of radiation on critical organs such as the lung [28,29]. Additionally, radiation-induced complications have very long latency times. For instance, the development of cardiovascular complications after radiotherapy for breast cancer irradiation generally takes between 5–20 years. Therefore, interventions pre- and post-radiation that can regulate proton radiation-induced signaling in normal tissue are being sought. Findings from the current study could potentially benefit cancer patients undergoing radiotherapy as pre-treatment with LGM2605 may prove radioprotective from unwanted side effects.

Indeed, the lignan SDG was shown in rodent models of lung damage to be a potent protector from radiation-induced lung toxicity when given prior to radiation exposure [30], and even post-radiation exposure [31]. SDG, now synthesized as LGM2605 to ensure constant pharmacodynamics, can be employed as a protective agent in proton therapy. The huPCLS preparation used in this study emulates and preserves essential characteristics of human lung; thus, we reasoned that it would be a valuable tool used to detect proton radiation-induced damage to the human lung. In this study, we monitored inflammation, oxidative stress (and antioxidant status), and senescent markers post protein beam exposure. Radiation-induced senescence in non-malignant, normal cells such as non-proliferating endothelial cells has been known and is mediated in part by ROS generation and p53 activation [32,33]. Of note, exposure of normal tissues and cells to moderate levels of radiation (<10 Gy), as encountered in normal tissues outside the targeted tumor tissue in radiotherapy patients or during space travel, induces senescence via multiple pathways [33]. Berman and coworkers [34] have reported an in-silico comparative analysis of passive scattering proton therapy (PSPT) and intensity modulated proton therapy (IMPT) with intensity-modulated photon beam radiotherapy (IMRT) radiation treatment. The study reported that 30–60% of the lung receives ~400 cGy and was thus selected as a relevant dose to study lung responses to proton exposures.

4. Materials and Methods

4.1. Human, Precision-Cut Lung Slices (huPCLS)

huPCLS were prepared as previously described [13,14]. Briefly, whole human lungs from non-asthma donors were obtained from the National Disease Research Interchange (Philadelphia, PA, USA) and were dissected and inflated using 2% ($w \cdot v^{-1}$) low melting point agarose. Once the agarose set, the lobe was sectioned, and cores of 8 mm diameter were made. The cores that contained a small airway by visual inspection were sliced at a thickness of 350 µm (Precisionary Instruments VF300 Vibratome, Greenville, NC, USA) and collected in wells containing supplemented Ham's F-12 medium. The cores generated were randomized as to the location in the lungs they were derived from, so the slices generated came from throughout the lungs and not one specific area. Slices containing contiguous segments of the same airway served as controls and were incubated at 37 °C in a humidified air:CO_2 (ratio of 95:5) incubator. Sections were rinsed with fresh media 2–3 times on days 1 and 2 to remove agarose and endogenous substances released that may confound the production of inflammatory mediators [13].

4.2. Irradiation Procedure

huPCLS were irradiated with a spread out proton Bragg peak (SOBP) in the Robert's Proton Therapy facility at the University of Pennsylvania with a 20 cm × 20 cm beam area encompassing a 3 × 3 square array of 60 mm dishes with huPCLS. Delivery modality was a scanned pencil beam, with dose delivered in sequentially shallower energy layers to generate an SOBP. Beam energy and SOBP width are chosen to produce a dose-averaged LET at the cell irradiation depth of 3–5 keV/µm for a total dose 4 Gy. The irradiation depth was determined with proton range-verified water equivalent thickness. The proton beam output is calibrated according to the International Atomic Energy Agency code of practice for absorbed dose to water [35] using ionization chambers with traceable calibration certificates from the National Institute of Standards and Technology (NIST).

4.3. LGM2605 Treatment

LGM2605 (50 and 100 μM) was added to the huPCLS 4 h prior to proton radiation. The schema for LGM2605 treatment, as well as for the measuring of the various indices of inflammation and oxidative damage, are shown in Scheme 1.

4.4. Western Blots

Total protein content from huPCLS was isolated using Trizol by following the manufacturer's suggested protocol for sequential precipitation of RNA, DNA, and proteins (Invitrogen, Carlsbad, CA, USA). Briefly, after homogenization of the lung sections and incubation with chloroform, samples were allowed to form three different phases from which the lower organic phase was isolated. Protein content was precipitated using isopropanol, washed to remove impurities, and resuspended and quantified using BCA Protein Assay (Thermo Scientific, Waltham, MA, USA). Samples were loaded on 8–12% NuPAGE gel (Invitrogen, Carlsbad, CA, USA) and proteins were transferred on PolyScreen PV transfer membrane (PerkinElmer Life Sciences, Boston, MA, USA). After membrane blocking with 5% BSA, protein levels of human senescent proteins p21 Waf1/Cip1 (product number 2948, Cell Signaling Technology, Danvers, MA, USA), p16 INK4A (product number 4824, Cell Signaling Technology), p53 (product number 9282, Cell Signaling Technology), and Phospho-Rb (product number 9307, Cell Signaling Technology) were detected using rabbit and mouse monoclonal antibodies, following manufacturer recommended dilutions. Peroxidase-conjugated donkey anti-rabbit IgG (code 711-035-152, Jackson ImmunoResearch Laboratories, Inc., West Grove, PA, USA) and goat anti-mouse IgG (code 115-035-003, Jackson ImmunoResearch Laboratories, Inc.) were used as secondary antibodies. Bands were visualized using Western Lighting Chemiluminescence Reagent Plus (PerkinElmer Life Sciences) and quantified with ImageJ software (Fiji Version, National Institutes of Health, Bethesda, MD, USA). Tissue preparation from the limited lung slices available from each donor lung resulted in restricting protein amounts, which only allow a single western blot with no possibility of repeat.

4.5. Oxidative Stress

Monitoring of huPCLS for oxidative stress was carried out by fluorescence using CellROX Green (Thermofisher Scientific, Eugene, OR, USA). After proton exposure, huPCLS were labeled with 5 μM CellROX Green dye and incubated for about 20 min, after which these were washed with RPMI buffer. HuPCLS were placed on clear glass coverslips and imaged using a Nikon fluorescence microscope (Nikon Diaphot TMD, Melville, NY, USA). For each condition, the data was quantified over 3–4 fields using MetaMorph acquisition software (Version 7.7, Molecular Devices, Downington, PA, USA). Images were acquired at excitation of 488 nm. All images were acquired at the same settings and the CellROX Green fluorescence quantified by integrating the intensity of fluorescence across the entire fields.

4.6. Images in Z-Stack

As the huPCLS were ~300–350 μm thick, the oxidative stress along the depth of the tissue and its diminution by LGM2605 were also assessed by imaging along the z-axis at 10 μm micron interval throughout the tissue. Images were acquired at excitation of 488 nm to record the CellROX Green fluorescence. The imaging data was presented as a 3D image, as well as in the form of a video file at fifteen frames, single z plane acquisition. All z stack images were acquired using a Leica STED 3X Super-Resolution Microscope equipped with a LAS X software version 3.1 (Wetzlar, Germany) for image acquisition and Huygens Professional software (Version 4.5, Scientific Volume Imaging,

Hilversum, The Netherlands) for deconvolution. Images were converted into single stacks using ImageJ analysis software (NIH). This software was also used to make movies, whereby the images were rotated along the xy axes. Finally, 3D display stacks were made by fitting in the ImageJ stacks into Volocity® Visualization Program (Version 6.3, PerkinElmer, Waltham, MA, USA). Scale bar = 160 μm.

4.7. RNA Isolation and Gene Expression Analysis

Total RNA was isolated from huPCLS and qPCR analysis was performed as previously described [15,25,36] using individual TaqMan® Probe-Based Gene Expression Assays (Applied Biosystems, Life Technologies, Carlsbad, CA, USA) selected for proinflammatory cytokines (*IL-1β, IL-6, TNFα,* and *IL-1α*), inflammation (*COX-2*), relevant cytoprotective and antioxidant enzymes (heme oxygenase-1 (*HMOX1*) and NADPH:quinone oxidoreductase-1 (*NQO1*), and cell cycle markers (*TP53, CDK2, CDKN1A, CDKN2A, CDK4, CDK6, RB,* and *E2F*). Transcript levels of tested genes were normalized to *β-Actin*.

4.8. Senescence-Associated β-Galactosidase Staining

Proton radiation-induced cellular senescence of A549 human lung adenocarcinoma cells was evaluated by senescence-associated β-galactosidase (SA-β-gal) staining. Briefly, 30,000 cells were seeded in 60 mm dishes and pretreated with either 0, 50, or 100 μM LGM2605 4 h prior to 4 Gy proton radiation. At 24 h post radiation, cells were fixed and stained using the Senescence β-Galactosidase Staining Kit (Cell Signaling Technology) according to the manufacturer's instructions. The blue-stained senescent cells were viewed by light microscopy and positive-stained cells were counted from 5 randomly selected fields. SA-β-gal-stained huPCLS were similarly evaluated by light microscopy. Eight focus levels in 3 randomly-selected fields were evaluated per slide under 400× magnification.

4.9. Intercellular Adhesion Molecule (ICAM) Quantification

Human, precision-cut lung sections were permeabilized and immunostained with anti-ICAM antibody (1:250) 24 h post irradiation with 4 Gy protons, huPCLS. Secondary antibody (goat anti mouse) with FITC label was used. Fluorescence imaging of the sections at λ = 488 nm using 4× and 10× lens was carried out. All images were acquired at the same settings using MetaMorph acquisition software. The images acquired by 4× lens were used to quantify using MetaMorph and ImageJ software. The data was acquired from images acquired in 6–7 fields.

4.10. IL-1β Cytokine Release

Levels of proinflammatory cytokine, IL-1β, were determined in tissue culture medium at 24 h post 4 Gy proton radiation exposure using an enzyme-linked immunosorbent assay (ELISA). Samples were run undiluted in triplicate, and the assay was performed according to manufacturer's instructions (BD biosciences, San Jose, CA, USA). Levels of IL-1β are reported as picograms per milliliter (pg/mL) of culture medium.

4.11. Statistical Analysis

All results shown are from multiple lung sections sampled from different sites, yet from the same core biopsy (n = 3–4 sections), and are reported as the mean $\pm$ the standard error of the mean (SEM). Of note, due to large inter-donor variability, 5 different donor lungs were evaluated for their sensitivity to radiation exposure using induction (>2-fold change over non-irradiated control) of senescence-relevant genes using qPCR at both 30 min and 24 h as a measure of selection, and just two IR-responder lungs were selected to be evaluated further with WB. The selected donor lungs were then evaluated for the protective effects of the test agent (LGM2605). Data are normally distributed and statistically significant differences were determined by one-way analysis of variance (ANOVA), followed by Tukey's multiple comparisons tests using GraphPad Prism version 6.00 for

Windows, GraphPad Software, La Jolla, CA, USA, www.graphpad.com. All pairwise comparisons were performed within each respective time point post radiation exposure, and statistically significant differences are reported as: * indicates a statistically significant ($p < 0.05$) difference from the respective non-irradiated control, and # indicates a statistically significant ($p < 0.05$) difference from IR (4 Gy proton radiation).

Gene expression data are reported as the fold change from CTL (no LGM2605 and no exposure to proton radiation) for each respective time point. Statistically significant differences in mRNA levels were determined using nonparametric tests (Mann–Whitney tests) due to the non-normality of the fold change data. Statistically significant differences were determined at $\alpha = 0.05$.

5. Conclusions

In summary, our findings provide evidence that LGM2605 treatment significantly reduces proton radiation-induced oxidative lung damage and cellular senescence, while ameliorating an overall proinflammatory phenotype in a human lung organ culture model system. Additionally, LGM2605 pretreatment of proton-irradiated human lung slices significantly upregulates antioxidant genes and downregulates proinflammatory cytokine gene levels. LGM2605 protects huPCLS from a senescent-like phenotype induced by proton radiation. The senescent-like phenotype regulated at the gene and protein level by p53, members of the CDK family, p21, and p16 is significantly reduced by LGM2605 pre-treatment. Additionally, radiation-induced cellular senescence has been associated with radiation-induced late effects. We, therefore, speculate that abrogation of senescence acutely, as shown by the current study, may have downstream protective effects. LGM2605 may be a candidate countermeasure agent for space-relevant radiation exposures, as well as in adverse effects of radiotherapy for cancer eradication.

Acknowledgments: This work was funded in part by: NIH-R01 CA133470 (Melpo Christofidou-Solomidou (MCS)), NIH-1R21CA178654-01 (MCS), NIH-1R21 AT008291-01 (MCS), NIH-R03 CA180548 (MCS), 1P42ES023720-01 (MCS), and by pilot project support from 1P30 ES013508-02 awarded to MCS (its contents are solely the responsibility of the authors and do not necessarily represent the official views of the NIEHS, NIH).

Author Contributions: Anastasia Velalopoulou performed the experiments and data analysis and assisted in writing the manuscript; Shampa Chatterjee performed all morphometry studies and assisted in data analysis and in writing the manuscript; Ralph A. Pietrofesa conducted data analysis, and statistical analyses and interpretation, and assisted with manuscript preparation; Cynthia Koziol-White and Reynold A. Panettieri provided huPCLS and assisted with data analysis and writing of the manuscript; Liyong Lin, Stephen Tuttle, Abigail Berman, and Constantinos Koumenis assisted with the proton irradiation exposures and assisted with manuscript preparation; Melpo Christofidou-Solomidou designed the study and the individual experiments, analyzed and interpreted data, wrote the manuscript, and supervised lab personnel. All co-authors reviewed the manuscript before submission and approved the final version.

Conflicts of Interest: Melpo Christofidou-Solomidou reports grants from the NIH during the conduct of the study. In addition, Melpo Christofidou-Solomidou has patents No. PCT/US14/41636 and No. PCT/US15/22501 pending and has a founders equity position in LignaMed, LLC. All other coauthors declare no conflict of interest.

Abbreviations

CDK	cyclin-dependent kinase
COPD	chronic obstructive pulmonary disease
COX-2	cyclooxygenase-2
EAR	endogenous antioxidant response
ELISA	enzyme-linked immunosorbent assay
GCR	galactic cosmic radiation
HMOX1	heme oxygenase-1
huPCLS	human, precision-cut lung slices
ICAM	intercellular adhesion molecule
IL-1α	interleukin-1α
IL-1β	interleukin-1β

IL-6	interleukin-6
IMPT	intensity modulated proton therapy
IMRT	intensity modulated photon beam radiotherapy
LGM2605	synthetic SDG
Nrf2	nuclear factor (erythroid-derived 2)-like 2
NQO1	NADPH: quinone oxidoreductase-1
OAR	organs at risk
PBS	phosphate-buffered saline
qPCR	quantitative polymerase chain reaction
RNS	reactive nitrogen species
ROS	reactive oxygen species
SA-β-gal	senescence-associated β-galactosidase
S1P	sphingosine-1 phosphate
SDG	secoisolariciresinol diglucoside
SphK1	sphingosine kinase 1
SphK2	sphingosine kinase 2
SpHL	sphingosine lyase
TNFα	tumor necrosis factor α

References

1. Feliciano, J.; Feigenberg, S.; Mehta, M. Chemoradiation for definitive, preoperative, or postoperative therapy of locally advanced non-small cell lung cancer. *Cancer J.* **2013**, *19*, 222–230. [CrossRef] [PubMed]
2. Rosenzweig, K.E.; Gomez, J.E. Concurrent chemotherapy and radiation therapy for inoperable locally advanced non-small-cell lung cancer. *J. Clin. Oncol.* **2017**, *35*, 6–10. [CrossRef] [PubMed]
3. Cotter, S.E.; McBride, S.M.; Yock, T.I. Proton radiotherapy for solid tumors of childhood. *Technol. Cancer Res. Treat.* **2012**, *11*, 267–278. [CrossRef] [PubMed]
4. Brodin, N.P.; Munck Af Rosenschold, P.; Aznar, M.C.; Kiil-Berthelsen, A.; Vogelius, I.R.; Nilsson, P.; Lannering, B.; Bjork-Eriksson, T. Radiobiological risk estimates of adverse events and secondary cancer for proton and photon radiation therapy of pediatric medulloblastoma. *Acta Oncol.* **2011**, *50*, 806–816. [CrossRef] [PubMed]
5. Fontenot, J.D.; Lee, A.K.; Newhauser, W.D. Risk of secondary malignant neoplasms from proton therapy and intensity-modulated X-ray therapy for early-stage prostate cancer. *Int. J. Radiat. Oncol. Biol. Phys.* **2009**, *74*, 616–622. [CrossRef] [PubMed]
6. Christofidou-Solomidou, M.; Pietrofesa, R.A.; Arguiri, E.; Schweitzer, K.S.; Berdyshev, E.V.; McCarthy, M.; Corbitt, A.; Alwood, J.S.; Yu, Y.; Globus, R.K.; et al. Space radiation-associated lung injury in a murine model. *Am. J. Physiol. Lung Cell. Mol. Physiol.* **2015**, *308*, L416–L428. [CrossRef] [PubMed]
7. Pietrofesa, R.; Turowski, J.; Tyagi, S.; Dukes, F.; Arguiri, E.; Busch, T.M.; Gallagher-Colombo, S.M.; Solomides, C.C.; Cengel, K.A.; Christofidou-Solomidou, M. Radiation mitigating properties of the lignan component in flaxseed. *BMC Cancer* **2013**, *13*, 179. [CrossRef] [PubMed]
8. Pietrofesa, R.A.; Turowski, J.B.; Arguiri, E.; Milovanova, T.N.; Solomides, C.C.; Thom, S.R.; Christofidou-Solomidou, M. Oxidative lung damage resulting from repeated exposure to radiation and hyperoxia associated with space exploration. *J. Pulm. Respir. Med.* **2013**, *3*, 1000158.
9. Pietrofesa, R.A.; Velalopoulou, A.; Arguiri, E.; Menges, C.W.; Testa, J.R.; Hwang, W.T.; Albelda, S.M.; Christofidou-Solomidou, M. Flaxseed lignans enriched in secoisolariciresinol diglucoside prevent acute asbestos-induced peritoneal inflammation in mice. *Carcinogenesis* **2015**, *37*, 177–187. [CrossRef] [PubMed]
10. Pietrofesa, R.A.; Velalopoulou, A.; Lehman, S.L.; Arguiri, E.; Solomides, P.; Koch, C.J.; Mishra, O.P.; Koumenis, C.; Goodwin, T.J.; Christofidou-Solomidou, M. Novel double-hit model of radiation and hyperoxia-induced oxidative cell damage relevant to space travel. *Int. J. Mol. Sci.* **2016**, *17*, 953. [CrossRef] [PubMed]
11. Mishra, O.P.; Simmons, N.; Tyagi, S.; Pietrofesa, R.; Shuvaev, V.V.; Valiulin, R.A.; Heretsch, P.; Nicolaou, K.C.; Christofidou-Solomidou, M. Synthesis and antioxidant evaluation of (*S*,*S*)- and (*R*,*R*)-secoisolariciresinol diglucosides (SDGs). *Bioorg. Med. Chem. Lett.* **2013**, *23*, 5325–5328. [CrossRef] [PubMed]

12. Mishra, O.P.; Pietrofesa, R.; Christofidou-Solomidou, M. Novel synthetic (*S*,*S*) and (*R*,*R*)-secoisolariciresinol diglucosides (SDGs) protect naked plasmid and genomic DNA from γ radiation damage. *Radiat. Res.* **2014**, *182*, 102–110. [CrossRef] [PubMed]

13. Cooper, P.R.; Lamb, R.; Day, N.D.; Branigan, P.J.; Kajekar, R.; San Mateo, L.; Hornby, P.J.; Panettieri, R.A., Jr. TLR3 activation stimulates cytokine secretion without altering agonist-induced human small airway contraction or relaxation. *Am. J. Physiol. Lung Cell. Mol. Physiol.* **2009**, *297*, L530–L537. [CrossRef] [PubMed]

14. Koziol-White, C.J.; Jia, Y.; Baltus, G.A.; Cooper, P.R.; Zaller, D.M.; Crackower, M.A.; Sirkowski, E.E.; Smock, S.; Northrup, A.B.; Himes, B.E.; et al. Inhibition of spleen tyrosine kinase attenuates IgE-mediated airway contraction and mediator release in human precision cut lung slices. *Br. J. Pharmacol.* **2016**, *173*, 3080–3087. [CrossRef] [PubMed]

15. Velalopoulou, A.; Tyagi, S.; Pietrofesa, R.A.; Arguiri, E.; Christofidou-Solomidou, M. The flaxseed-derived lignan phenolic secoisolariciresinol diglucoside (SDG) protects non-malignant lung cells from radiation damage. *Int. J. Mol. Sci.* **2015**, *17*, 7. [CrossRef] [PubMed]

16. Chancellor, J.C.; Scott, G.B.; Sutton, J.P. Space radiation: The number one risk to astronaut health beyond low earth orbit. *Life* **2014**, *4*, 491–510. [CrossRef] [PubMed]

17. Delmotte, P.; Sanderson, M.J. Ciliary beat frequency is maintained at a maximal rate in the small airways of mouse lung slices. *Am. J. Respir. Cell Mol. Biol.* **2006**, *35*, 110–117. [CrossRef] [PubMed]

18. Lee, J.M.; Yanagawa, J.; Peebles, K.A.; Sharma, S.; Mao, J.T.; Dubinett, S.M. Inflammation in lung carcinogenesis: New targets for lung cancer chemoprevention and treatment. *Crit. Rev. Oncol. Hematol.* **2008**, *66*, 208–217. [CrossRef] [PubMed]

19. Green, D.R.; Kroemer, G. Cytoplasmic functions of the tumour suppressor p53. *Nature* **2009**, *458*, 1127–1130. [CrossRef] [PubMed]

20. Levine, A.J.; Oren, M. The first 30 years of p53: Growing ever more complex. *Nat. Rev. Cancer* **2009**, *9*, 749–758. [CrossRef] [PubMed]

21. Chandler, H.; Peters, G. Stressing the cell cycle in senescence and aging. *Curr. Opin. Cell Biol.* **2013**, *25*, 765–771. [CrossRef] [PubMed]

22. Tsuboi, K.; Moritake, T.; Tsuchida, Y.; Tokuuye, K.; Matsumura, A.; Ando, K. Cell cycle checkpoint and apoptosis induction in glioblastoma cells and fibroblasts irradiated with carbon beam. *J. Radiat. Res.* **2007**, *48*, 317–325. [CrossRef] [PubMed]

23. Luo, H.; Wang, L.; Schulte, B.A.; Yang, A.; Tang, S.; Wang, G.Y. Resveratrol enhances ionizing radiation-induced premature senescence in lung cancer cells. *Int. J. Oncol.* **2013**, *43*, 1999–2006. [CrossRef] [PubMed]

24. He, X.; Yang, A.; McDonald, D.G.; Riemer, E.C.; Vanek, K.N.; Schulte, B.A.; Wang, G.Y. MiR-34a modulates ionizing radiation-induced senescence in lung cancer cells. *Oncotarget* **2017**, *8*, 69797–69807. [CrossRef] [PubMed]

25. Christofidou-Solomidou, M.; Tyagi, S.; Pietrofesa, R.; Dukes, F.; Arguiri, E.; Turowski, J.; Grieshaber, P.A.; Solomides, C.C.; Cengel, K.A. Radioprotective role in lung of the flaxseed lignan complex enriched in the phenolic secoisolariciresinol diglucoside (SDG). *Radiat. Res.* **2012**, *178*, 568–580. [CrossRef] [PubMed]

26. Saha, J.; Wilson, P.; Thieberger, P.; Lowenstein, D.; Wang, M.; Cucinotta, F.A. Biological characterization of low-energy ions with high-energy deposition on human cells. *Radiat. Res.* **2014**, *182*, 282–291. [CrossRef] [PubMed]

27. Benveniste, M.F.; Welsh, J.; Godoy, M.C.; Betancourt, S.L.; Mawlawi, O.R.; Munden, R.F. New era of radiotherapy: An update in radiation-induced lung disease. *Clin. Radiol.* **2013**, *68*, e275–e290. [CrossRef] [PubMed]

28. Remick, J.S.; Schonewolf, C.; Gabriel, P.; Doucette, A.; Levin, W.P.; Kucharczuk, J.C.; Singhal, S.; Pechet, T.T.V.; Rengan, R.; Simone, C.B., II; et al. First clinical report of proton beam therapy for postoperative radiotherapy for non-small-cell lung cancer. *Clin. Lung Cancer* **2017**, *18*, 364–371. [CrossRef] [PubMed]

29. Rwigema, J.M.; Verma, V.; Lin, L.; Berman, A.T.; Levin, W.P.; Evans, T.L.; Aggarwal, C.; Rengan, R.; Langer, C.; Cohen, R.B.; et al. Prospective study of proton-beam radiation therapy for limited-stage small cell lung cancer. *Cancer* **2017**, *123*, 4244–4251. [CrossRef] [PubMed]

30. Lee, J.C.; Krochak, R.; Blouin, A.; Kanterakis, S.; Chatterjee, S.; Arguiri, E.; Vachani, A.; Solomides, C.C.; Cengel, K.A.; Christofidou-Solomidou, M. Dietary flaxseed prevents radiation-induced oxidative lung damage, inflammation and fibrosis in a mouse model of thoracic radiation injury. *Cancer Biol. Ther.* **2009**, *8*, 47–53. [CrossRef] [PubMed]

31. Christofidou-Solomidou, M.; Tyagi, S.; Tan, K.S.; Hagan, S.; Pietrofesa, R.; Dukes, F.; Arguiri, E.; Heitjan, D.F.; Solomides, C.C.; Cengel, K.A. Dietary flaxseed administered post thoracic radiation treatment improves survival and mitigates radiation-induced pneumonopathy in mice. *BMC Cancer* **2011**, *11*, 269. [CrossRef] [PubMed]

32. Lafargue, A.; Degorre, C.; Corre, I.; Alves-Guerra, M.C.; Gaugler, M.H.; Vallette, F.; Pecqueur, C.; Paris, F. Ionizing radiation induces long-term senescence in endothelial cells through mitochondrial respiratory complex II dysfunction and superoxide generation. *Free Radic. Biol. Med.* **2017**, *108*, 750–759. [CrossRef] [PubMed]

33. Wang, Y.; Boerma, M.; Zhou, D. Ionizing radiation-induced endothelial cell senescence and cardiovascular diseases. *Radiat. Res.* **2016**, *186*, 153–161. [CrossRef] [PubMed]

34. Berman, A.T.; Teo, B.K.; Dolney, D.; Swisher-McClure, S.; Shahnazi, K.; Both, S.; Rengan, R. An in-silico comparison of proton beam and IMRT for postoperative radiotherapy in completely resected stage IIIA non-small cell lung cancer. *Radiat. Oncol.* **2013**, *8*, 144. [CrossRef] [PubMed]

35. Andreo, P.; Burns, D.T.; Hohlfeld, K.; Huq, M.S.; Kanai, T.; Laitano, F.; Smyth, V.; Vynckier, S. *Absorbed Dose Determination in External Beam Radiotherapy: An international Code of Practice for Dosimetry Based on Standards of Absorbed Dose to Water*; Technical Report; IAEA: Vienna, Austria, 2000.

36. Pietrofesa, R.A.; Velalopoulou, A.; Albelda, S.M.; Christofidou-Solomidou, M. Asbestos induces oxidative stress and activation of Nrf2 signaling in murine macrophages: Chemopreventive role of the synthetic lignan secoisolariciresinol diglucoside (LGM2605). *Int. J. Mol. Sci.* **2016**, *17*, 322. [CrossRef] [PubMed]

MDPI

St. Alban-Anlage 66

4052 Basel, Switzerland

Tel. +41 61 683 77 34

Fax +41 61 302 89 18

http://www.mdpi.com

International Journal of Molecular Sciences Editorial Office

E-mail: ijms@mdpi.com

http://www.mdpi.com/journal/ijms

www.ingramcontent.com/pod-product-compliance
Lightning Source LLC
Chambersburg PA
CBHW041214220326
41597CB00033BA/5897